PROGRAMMING AND CUSTOMIZING THE MULTICORE PROPELLER™ MICROCONTROLLER

D1268352

PROGRAMMING AND CUSTOMIZING THE MULTICORE PROPELLER™ MICROCONTROLLER

THE OFFICIAL GUIDE

PARALLAX INC.

Shane Avery   Chip Gracey   Vern Graner
Martin Hebel   Joshua Hintze   André LaMothe
Andy Lindsay   Jeff Martin   Hanno Sander

New York   Chicago   San Francisco   Lisbon   London   Madrid
Mexico City   Milan   New Delhi   San Juan   Seoul
Singapore   Sydney   Toronto

The McGraw·Hill Companies

Cataloging-in-Publication Data is on file with the Library of Congress.

McGraw-Hill books are available at special quantity discounts to use as premiums and sales promotions, or for use in corporate training programs. To contact a representative please e-mail us at bulksales@mcgraw-hill.com.

Programming and Customizing the Multicore Propeller™ Microcontroller

2 3 4 5 6 7 8 9 0   DOC/DOC   1 9 8 7 6 5 4 3 2 1 0

ISBN 978-0-07-166450-9
MHID 0-07-166450-5

This book is printed on acid-free paper.

Sponsoring Editor
Judy Bass

Acquisitions Coordinator
Michael Mulcahy

Editorial Supervisor
David E. Fogarty

Project Manager
Somya Rustagi,
Glyph International

Copy Editor
Lisa McCoy

Proofreader
Shivani Arora,
Glyph International

Production Supervisor
Pamela A. Pelton

Composition
Glyph International

Art Director, Cover
Jeff Weeks

For Ellie, my lovely wife.

S. A.

To my wife Kym, my children Nic and Sami,
and my Mom and Dad for all the support.

V. L. G.

For BJ and Kris, thanks for being so supportive.

M. H.

For my wife Crystal and son Hunter.

J. M. H.

To inventors everywhere.

A. L.

For Stacy, Kaitlin, and Kaylani
who were wonderfully supportive
despite being deprived of my attention.

J. M.

For Mami and Papi, thanks for everything.

H. S.

CONTENTS

ABOUT THE AUTHORS

Shane Avery graduated from Cal Poly, San Luis Obispo, with a bachelor's degree in computer engineering and from Cal State Northridge with masters in electrical engineering. His graduate work focused on system-on-chip design in FPGAs and ASICs. At Ziatech/Intel he debugged single-board CompactPCI computers for the telephony industry. He then worked with a toy company, Logic-Plus, developing the Verilog code for the FPGA inside a toy video camera that would insert static images onto the background. Currently, he works for the United States Navy, designing and developing embedded hardware for new military weapons. This year he began his first company focusing on embedded electronics, called Avery Digital.

Chip Gracey, President of Parallax Inc., has a lifelong history of invention and creativity. His early programming projects included the famous ISEPIC software duplication device for the Commodore 64 and various microprocessor development tools, yet he is most well-known for creating the BASIC Stamp®. Chip Gracey is the Propeller chip's chief architect and designer. His formal educational background is nearly empty, with all of his experience being the result of self-motivation and personal interest. Chip continues to develop custom silicon at Parallax.

Vern Graner has been in the computer industry since being recruited by Commodore Business Machines in 1987. He has had a lifelong relationship with electronics, computers, and entertainment encompassing multiple fields including animatronics, performance art, computer control, network systems design, and software integration. As president of The Robot Group Inc., Vern's work has been featured at Maker Faire, First Night Austin, SxSW, Linucon, Dorkbot, Armadillocon, and even in *SPIN* magazine. In 2007, he was awarded The Robot Group's DaVinci Award for his contributions to the arts and technology community. Though some of his writings have been featured in *SERVO* magazine, he is currently best known for his regular contributions to *Nuts and Volts* magazine as the author of the monthly column "Personal Robotics". He currently is employed as a senior software engineer in Austin, Texas.

Martin Hebel holds a master of science in education (MS) and bachelor of science in electronics technologies (BS), obtained from Southern Illinois University Carbondale (SIUC) following 12 years of service as a nuclear technician on submarines. He is an associate professor in Information Systems and Applied Technologies at SIUC, instructing in the Electronic Systems Technologies program where he teaches microcontroller programming, industrial process control, and networking. His research in wireless sensor networks, using controllers including the Propeller chip, has led to agricultural research with SIUC's agricultural sciences, University of Florida, and funding through a USDA grant. He has also collaborated with researchers at the University of Sassari, Italy in biological wireless monitoring using parallel processing, which was presented at a NATO conference in Vichy, France.

Joshua Hintze graduated from Brigham Young University with bachelor's and master's degrees in electrical engineering. His main research focus was unmanned aerial vehicles where he helped create the world's smallest fully functional autopilot (patent received). Before graduating, Josh took a research position at NASA Ames Research Center in Moffett Field, California. At NASA, Josh designed algorithms for landing autonomous helicopters by scanning potential landing locations with stereo cameras and machine vision algorithms. Josh is cofounder of Procerus Technologies that builds and ships autopilots all over the world, many of which end up in military applications. He has written numerous published articles and was the author of "Inside The XGS PIC 16-Bit," published by Nurve Networks.

André LaMothe holds degrees in mathematics, computer science, and electrical engineering. He is a computer scientist, 3D game developer, and international best-selling author. He is the creator of Waite Group's "Black Art Series" as well as the series editor of Course PTR's "Game Development Series." Best known for his works in computer graphics and game development, he is currently the CEO of Nurve Networks LLC, which develops and manufactures embedded systems for educational and entertainment channels. Additionally, he holds a teaching position currently at Game Institute.

Andy Lindsay is an applications engineer and a key member of Parallax's Education Department. To date, Andy has written eight Parallax educational textbooks, including *What's a Microcontroller?*, *Robotics with the Boe-Bot*, and *Propeller Education Kit Labs: Fundamentals*. These books and their accompanying kits have gained widespread acceptance by schools in the United States and abroad, and some have been translated into many languages. Andy earned a Bachelor of Science in electrical and electronic engineering from California State University, Sacramento. He has worked for Parallax Inc. for over ten years, where he continues to write, teach, and develop educational products.

Jeff Martin is Parallax's senior software engineer and has been with the company for 13 years and counting. He attended California State University Sacramento, studied many areas of computer science, and earned a Bachelor of Science in system's software. In 1995, Jeff visited Parallax to purchase a BASIC Stamp 1. He rapidly learned and mastered the original BASIC Stamp and was shortly after offered a position to support it and other Parallax products. He is currently in the R & D department and is responsible for Parallax software IDEs, key printed manuals, and hardware maintenance for core product lines. Jeff collaborates closely with Chip Gracey and the rest of the R & D group on the Propeller product line, from hardware to software and documentation.

Hanno Sander has been working with computers since he programmed a lunar lander game for the z80 when he was six. Since then he graduated from Stanford University with a degree in computer science and then started his corporate career as an Internet entrepreneur. He moved to New Zealand in 2005 to spend time with his growing family and developed sophisticated, yet affordable robots, starting with the DanceBot. His technical interests include computer vision, embedded systems, industrial control, control theory, parallel computing, and fuzzy logic.

About Parallax Inc.

Parallax Inc., a privately held company, designs and manufactures microcontrollers, embedded system development tools, small single-board computers, and robots that are used by electronic engineers, educational institutions, and hobbyists.

FOREWORD

In the early and mid-1970s, semiconductor companies offered only a few rudimentary microprocessors that most people have never heard of or have long forgotten. Now, though, engineers, scientists, entrepreneurs, students, and hobbyists can choose from a wide spectrum of processors, some of which include two, four, or more processor "cores" that let a chip perform several tasks simultaneously. Then why do we need a new type of eight-core processor developed by Parallax, a small company in Rocklin, California? The reasons are many.

The Propeller chip takes a different approach and offers developers eight processors with identical architectures. That means any 32-bit processor, or cog, can run code that could run on any other cog equally well. You can write code for one cog and simply copy it to run exactly the same way on another cog. This type of copy-and-paste operation works well if you have, say, several identical servos, sensors, or displays that run on one Propeller chip.

And unlike many multicore devices, a Propeller chip needs no operating system, so you don't have to learn Linux, Windows CE, VxWorks, or another operating system to jump in and write useful code. Code-development tools are free, and you don't need add-ons that cut into your budget. The many projects in this book will help you better understand how to take advantage of the Propeller chip's capabilities.

The Propeller chip also simplifies operations and coordination of tasks because it offers both shared and mutually exclusive resources. The former includes the chip's 32 I/O pins and its system counter, which gives all cogs simultaneous access to information used to track or time events. Any cog can control any I/O pin, which means you can assign I/O pins as needed and easily change assignments late in a project schedule.

A central "Hub" controls access to the mutually exclusive resources such that each cog can access them exclusively, one at a time. Think of the Hub as a spinner—or propeller!—that rotates and gives each cog access to key resources for a set time. Hub operations use the Propeller Assembly language instructions rather than the aptly named higher-level Spin language. The Propeller's main memory is one of the mutually exclusive resources. You would not want two programs to try to access memory simultaneously or to modify a value in use by another cog.

The Propeller chip and Parallax offer users another, less tangible, asset: a devoted cadre of users and developers. Parallax has an active Propeller Chip forum, with more than 430 pages of posts that go back to early 2006. Parallax forum membership has reached more than 17,000 registered members. Run a Google search for "Parallax Propeller," and you'll find individual projects, discussions, products, and code. If you run into a problem getting your hardware or software to work, someone on the Internet usually has a suggestion or comment—often within a few minutes. Parallax also offers many add-on devices, such as accelerometers, ultrasonic range sensors, GPS receivers, and an image sensor, that help bring projects and designs to fruition rapidly.

Over the years I have worked with many types of processors, from early eight-bit devices to new ARM-based chips. But none of the chip suppliers has offered such a wide variety of practical educational information as Parallax offers for its processors. Anyone interested in the Propeller will find many articles, application notes, lab experiments, and manuals on the company's Web site to ensure they get off to a good start and maintain their interest and momentum as they learn more. I like the Parallax Propeller chip and have enjoyed working with it, although my coding skills are still somewhat basic. You'll like the Propeller, too, even if you only have a basic curiosity about how computer chips can easily control things in the real world. As you learn how to measure things such as voltage, temperature, sound, and so on, you'll get hooked. The Propeller chip is not only powerful and capable—it's easy and fun to work with.

JON TITUS
Friend of Parallax Inc. and microcomputer inventor
Herriman, Utah

INTRODUCTION

Parallax Inc. brought together nine experienced authors to write 12 chapters on many aspects and applications of multicore programming with the Propeller chip. This book begins with an introduction to the Propeller chip's architecture and Spin programming language, debugging techniques, and sensor interfacing. Then, the remainder of the book introduces eight diverse and powerful applications, ending with a speech synthesis demonstration written by the Propeller chip's inventor, Chip Gracey. We hope you find this book to be informative and inspirational. For more Propeller-related resources, visit www.parallax.com/propeller, and to join in the conversation with the Propeller community, visit forums.parallax.com.

ADDITIONAL RESOURCES FOR THIS BOOK

The software, documentation, example code, and other resources cited in the following chapters are available for free download from PCMProp directory at ftp://ftp.propeller-chip.com.

About the Example Code   Code listings for projects in this text come from diverse sources. The Propeller chip's native languages are object-based, and many prewritten objects are included with the Propeller Tool programming software or are posted to the public Propeller Object Exchange at obex.parallax.com.

Copyright for Example Code   All Spin and Propeller Assembly code listings included in this book, including those sourced from the Propeller Object Exchange, are covered by the MIT Copyright license, which appears below.

SPECIAL CONTRIBUTORS

Parallax Inc. would like to thank their team members: Chip Gracey for inventing the amazing Propeller chip; Ken Gracey for envisioning this book; Joel "Bump" Jacobs for creating the original cartoons in Chapters 1 and 2; Rich Allred for the majority of the illustrations in Chapters 1 through 5; Jeff Martin, Andy Lindsay, and Chip Gracey for authoring their chapters; and Stephanie Lindsay for coordinating production with McGraw-Hill. Parallax also thanks André LaMothe for authoring his chapter and for heading up the team of authors from the Propeller community: Martin Hebel, Hanno Sander, Shane Avery, Joshua Hintze, and Vern Graner. Special thanks go to Jon Titus for so generously providing the Foreword, and to Judy Bass at McGraw-Hill for finding this project worthwhile and making it happen.

THE PROPELLER CHIP MULTICORE
MICROCONTROLLER

Jeff Martin

Introduction

In the 1990s, the term "multicore" had more to do with soldering equipment than it did with computer processors. Though the concept was young and relatively nameless, this was the time many silicon engineers began focusing their efforts on multiprocessor technology. By the middle of the following decade, "multicore" had become the industry buzzword and the first consumer-accessible products arrived on the market. The short years to follow would mark a time of extreme evolution, innovation, and adoption of multicore technology that is bound to continue at a fast pace.

So what exactly is multicore and why is it so important? These are just two of the many questions we'll answer throughout this book, with insightful examples and exciting projects you can build yourself. We'll reveal how this technology is changing the way problems are solved and systems are designed. Most importantly, we'll show just how accessible multicore technology is to you.

Caution: This book is a collaboration of many enthusiastic authors who are eager to demonstrate incredible possibilities well within your reach. Reading this material may leave you feeling inspired, exhilarated, and empowered to invent new products and explore new ideas; prepare yourself!

In this chapter, we'll do the following:

- Learn what multicore means and why it's important
- Introduce the multicore Propeller™ microcontroller
- Explore Propeller hardware we'll use throughout this book

Multicore Defined

A multicore processor is a system composed of two or more independent CPUs, usually in the same die, that achieves multiprocessing in a single physical package (see Fig. 1-1). Put simply, a multicore chip can do many things simultaneously!

MULTIPROCESSING VERSUS MULTITASKING

Are you thinking "multitasking"? For computers, *multitasking* is a method of sharing a single CPU for multiple, possibly unrelated tasks. A single-core device that is multitasking fast enough gives the illusion of things happening at once.

A multicore device, however, achieves *true multiprocessing:* performing multiple tasks simultaneously. In comparison, a multicore device can run at a slower speed, consume less power, and achieve better results than a fast-running single-core device can.

In this book we will usually refer to "tasks," "functions," and "processes" in the context of multiprocessing, rather than multitasking.

It may seem quite daunting to get multiple processors all working together in a single, coherent application. In fact, once upon a time the task was notably treacherous since it was unclear exactly how to apply multicore technology. Many complex systems were devised that either stripped developers of power or burdened them with unruly multithread obstacles.

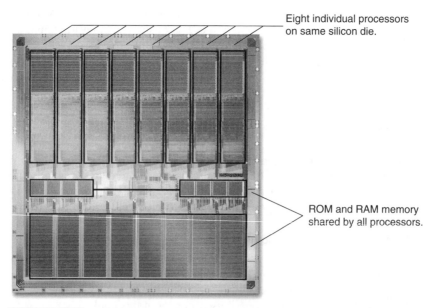

Eight individual processors on same silicon die.

ROM and RAM memory shared by all processors.

Figure 1-1   **Close-up of multicore Propeller chip silicon.**

Lucky for us, technique and technology have evolved to give us devices like the Propeller microcontroller. It is now quite natural to create multicore applications. After reading this book, you may find yourself wondering how you ever got along without it!

Why Multicore?

Why is multicore so important? After all, thousands of applications have been built using single-core devices.

While that's true, despite the successes, there have always been two obstacles impeding progress: *asynchronous events* and *task fidelity*.

- Asynchronous events are things that occur when you least expect them. They are inherently hard to handle in a timely manner with a single-core device.
- Task fidelity is the level of attention given to an activity. The lower the fidelity, the lower the precision with which the task is carried out.

These are two opposing entities, each vying for the precious time of a single processor. As asynchronous events increase, the fidelity of existing tasks suffers. As the demand for task fidelity increases, fewer asynchronous events are handled.

With a single-core device, balancing these two competing demands often means requiring processor interrupts and specialized hardware. Processor interrupts allow asynchronous events to be addressed while specialized hardware remains focused on high-fidelity tasks.

But is that the best solution? It means the "brains" of an application must rely on other hardware for high-speed tasks and relegate itself to lower-priority tasks while waiting for the interrupt of asynchronous events. It means systems become more expensive and complex to build, often with multiple chips to support the demands. It also means designers have the difficult challenge of finding the right "special" hardware for the job, learning that hardware, and dealing with any limitations it imposes, all in addition to programming the brains of the application in the first place!

Perhaps the best solution is most apparent in our everyday lives. How many times in your life have you wished there were two of you? Or three or more? Ever needed to "finish that report," "make that call," and "do those chores" while being pressed for quality time with your spouse, friends, kids, or your hobbies? (See Fig. 1-2.)

Wouldn't it be great, even for a short time, if you could do multiple things at once completely without distraction or loss of speed? Maybe we cannot, but a multicore device can!

Multicore Propeller Microcontroller

The Propeller microcontroller realizes this dream in its ability to clone its "mind" into two, three, or even eight individual processors, each working simultaneously with no distractions. Moreover, it can do this on a temporary or permanent basis with each

Figure 1-2   Clones can multiprocess!

processor sleeping until needed, consuming almost no power, yet waking in less than 10 millionths of a second to handle events.

CLEAR YOUR MIND OF INTERRUPTIONS

If you know all about interrupts on single-core devices, forget it now! Interrupts can be troublesome for real-time applications and are nonexistent in the multicore Propeller. Why? With a device like the Propeller, you don't need them. Just focus a processor on a task that needs such handling; it can sleep until the needed moment arrives and won't negatively affect the rest of the application's efforts.

The multicore Propeller is a system of homogenous processors and general-purpose I/O pins. Learn to use one processor and you know how to use them all. There's no specialized hardware to learn for demanding tasks; just assign another processor to the job. This incredibly useful hardware has inspired many who may otherwise have not considered multicore technology for an embedded system application.

Tip: Since its inception, multicore technology has continued to evolve to give us many kinds of devices, tools, and schemes. A quick review of "multicore" on Wikipedia (www.wikipedia.org) reveals the many ways the term is applied to a variety of unique hardware designs. We will focus on a solid foundation built with simple rules and proven results. These concepts can help you regardless of the multicore platform you use.

CONCEPT

Demanding jobs require a highly skilled team of workers, a fine-tuned force that performs in harmony aiming for a single goal. The multicore Propeller wraps this team of processors into one tiny package. These processors, called cogs, are at your service waiting to be called upon as the need arises. Both powerful and flexible, they lie dormant until needed, sleep and wake on a moment's notice, or run continuously.

The flat memory architecture, homogenous processor design, and built-in languages make for a simple architecture that is easy to learn.

Use in Practice   Here's how you'd use the multicore Propeller microcontroller in an application.

■ Build a Propeller Application out of objects (see Fig. 1-3).

> **Tip:** *Objects* are sets of code and data that are self-contained and have a specific purpose. Many objects already exist for various tasks; you can choose from those and can also create new ones.

■ Compile the Propeller Application and download it to the Propeller's RAM or EEPROM (see Fig. 1-4).
■ After download, the Propeller starts a cog to execute the application from Main RAM, as in Fig. 1-5.
■ The application may run entirely using only one cog or may choose to launch additional cogs to process a subset of code in parallel, as in Fig. 1-6. Of course, this performs as explicitly designed into each specialized object by you and other developers.

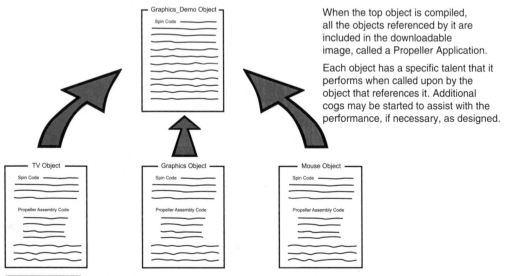

Figure 1-3   **A Propeller application's object hierarchy.**

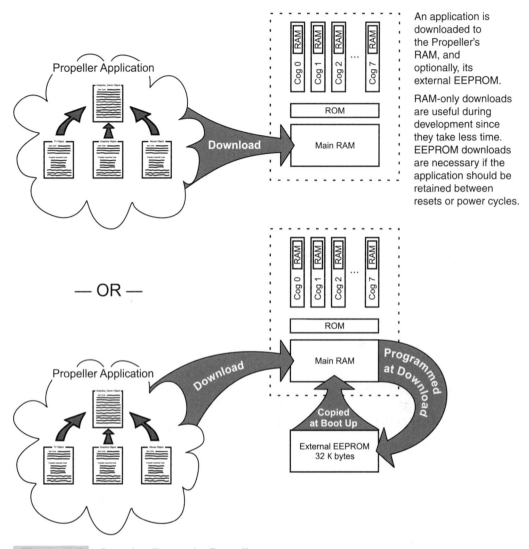

Figure 1-4   **Downloading to the Propeller.**

Tip: If the application was downloaded to EEPROM, it will also start in the same way whenever a power-up or reset event occurs.

Propeller applications may use multiple cogs all the time or just sometimes, as the need requires. Cogs may also be activated and then put to sleep (consuming very little power) so they may wake up instantly when needed.

Tip: The next chapter will take you step-by-step through hardware connections and an example development process.

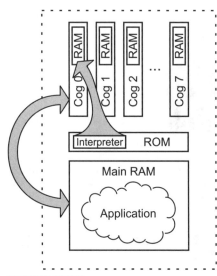

Upon application launch, a single cog runs the ROM-based Spin Interpreter and executes the application from Main RAM. From there, the application can launch additional cogs on Spin or Assembly code, as needed.

Figure 1-5   Application launch.

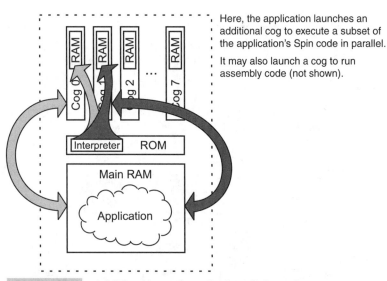

Here, the application launches an additional cog to execute a subset of the application's Spin code in parallel.

It may also launch a cog to run assembly code (not shown).

Figure 1-6   Additional cog launched on Spin code.

PROPELLER HARDWARE

We'll briefly discuss the Propeller chip's hardware and how it works in the rest of this chapter, and then you'll be introduced to multicore programming and debugging in the following two chapters. What you learn here will be put to good use in the exciting projects filling the remainder of this book.

Packages   The Propeller chip itself is available in the three package types shown in Fig. 1-7.

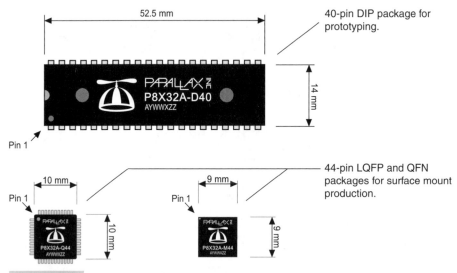

Figure 1-7   Propeller chip's packages.

Pin Descriptions and Specifications   Each package features the same set of pins, with the exception that the surface mount packages (LQFP and QFN) have four extra power pins (see Fig. 1-8 and Table 1-1). All functional attributes are identical regardless of package (see Table 1-2).

Architecture   Physically, the Propeller is organized as shown in Fig. 1-1, but functionally, it is like that shown in Fig. 1-9.

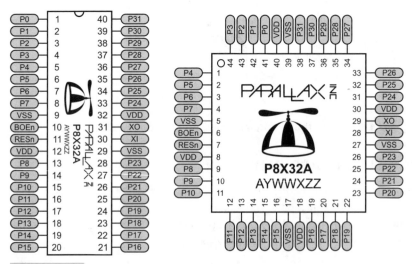

Figure 1-8   Propeller chip's pin designators.

TABLE 1-1   PIN DESCRIPTIONS

PIN NAME	DIRECTION	DESCRIPTION
P0 – P31	I/O	General-purpose I/O Port A. Can source/sink 40 mA each at 3.3 VDC. Logic threshold is ≈ ½ VDD; 1.65 VDC @ 3.3 VDC
		Pins P28 – P31 are special purpose upon power-up/reset but are general purpose I/O afterwards. P28/P29 are I2C SCL/SDA to external EEPROM. P30/P31 are serial Tx/Rx to host.
VDD	---	3.3-volts DC (2.7 – 3.3 VDC)
VSS	---	Ground.
BOEn	I	Brown Out Enable (active low). Must connect to VDD or VSS. If low, RESn outputs VDD (through 5 KΩ) for monitoring; drive low to reset. If high, RESn is CMOS input with Schmitt Trigger.
RESn	I/O	Reset (active low). Low causes reset: all cogs disabled and I/O floating. Propeller restarts 50 ms after RESn transitions high.
XI	I	Crystal Input. Connect to crystal/oscillator pack output (XO left disconnected) or to one leg of crystal/resonator with X0 connected to the other, depending on CLK register settings. No external resistors or capacitors are required.
XO	O	Crystal Output. Feedback for an external crystal. May leave disconnected depending on CLK register settings. No external resistors or capacitors are required.

TABLE 1-2   SPECIFICATIONS

ATTRIBUTE	DESCRIPTION
Model	P8X32A
Power Requirements	3.3 volts DC (max current draw must be ≤ 300 mA)
External Clock Speed	DC to 80 MHz (4 MHz to 8 MHz with Clock PLL running)
System Clock Speed	DC to 80 MHz
Internal RC Oscillator	12 MHz or 20 kHz (may range from 8 MHz – 20 MHz or 13 kHz – 33 kHz, respectively)
Main RAM/ROM	64 KB: 32 KB RAM + 32 KB ROM
Cog RAM	512 × 32 bits each
RAM/ROM Organization	Long (32-bit), Word (16-bit), or Byte (8-bit) addressable
I/O pins	32 CMOS signals with VDD/2 input threshold
Current Source/Sink per I/O	40 mA
Current Draw @ 3.3 VDC, 70 °F	500 µA per MIPS (MIPS = Freq (MHz) / 4 * Active Cogs)

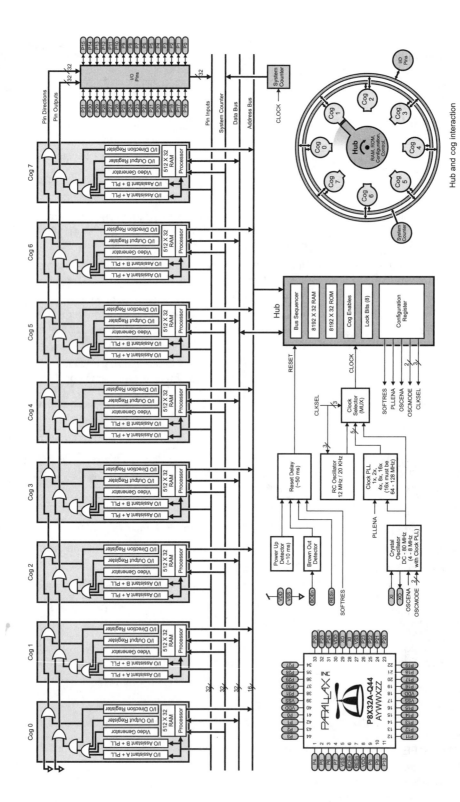

Figure 1-9 Propeller block diagram.

The cogs (processors) are all alike and work together as a team, sharing access to all system hardware, main memory, System Counter, configuration registers, I/O pins, etc. Let's look closely at some notable components shown in Fig. 1-9.

Cogs (processors)   The Propeller contains eight processors, called cogs, numbered 0 to 7. Each cog contains the same components and can run tasks independent of the others. All use the same clock source so they each maintain the same time reference and all active cogs execute instructions simultaneously.

> **Tip:** Propeller processors are called cogs because they are simple and uniform, like the cogs on gears that mesh with others of their kind to induce change. Their simplicity assures reliability and their collective delivers powerful results.

Cogs start and stop at runtime to perform independent or cooperative tasks simultaneously. As the developer, you have full control over how and when each cog is employed; there is no compiler-driven or operating system–based splitting of tasks between multiple cogs. This explicit parallelism empowers you to deliver deterministic timing, power consumption, and response to the embedded application.

Each cog has its own RAM, called Cog RAM, containing 512 registers of 32 bits each. Cog RAM is used for both code and data, except for the last 16 special-purpose registers (see Table 1-3) that provide an interface to the System Counter, I/O pins, and local cog peripherals.

TABLE 1-3   COG RAM SPECIAL-PURPOSE REGISTERS

ADDRESS	NAME	TYPE	DESCRIPTION
$1F0	PAR	Read-Only	Boot Parameter
$1F1	CNT	Read-Only	System Counter
$1F2	INA	Read-Only	Input States for P31–P0
$1F3	INB	Read-Only	<reserved>
$1F4	OUTA	Read/Write	Output States for P3–P0
$1F5	OUTB	Read/Write	<reserved>
$1F6	DIRA	Read/Write	Direction States for P31–P0
$1F7	DIRB	Read/Write	<reserved>
$1F8	CTRA	Read/Write	Counter A Control
$1F9	CTRB	Read/Write	Counter B Control
$1FA	FRQA	Read/Write	Counter A Frequency
$1FB	FRQB	Read/Write	Counter B Frequency
$1FC	PHSA	Read/Write	Counter A Phase
$1FD	PHSB	Read/Write	Counter B Phase
$1FE	VCFG	Read/Write	Video Configuration
$1FF	VSCL	Read/Write	Video Scale

When a cog is started, registers 0 ($000) through 495 ($1EF) are loaded sequentially from Main RAM/ROM, its special-purpose registers are cleared to zero, and it begins executing instructions starting at Cog RAM register 0. It continues to execute code until it is stopped or rebooted by either itself or another cog, or a reset occurs.

Hub   The Hub maintains system integrity by ensuring that mutually exclusive resources are accessed by only one cog at a time. Mutually exclusive resources include things like Main RAM/ROM and configuration registers.

The Hub gives each cog access to such resources once every 16 clock cycles in a round-robin fashion, from Cog 0 through Cog 7 and back to Cog 0 again. If a cog tries to access a mutually exclusive resource out of order, it will simply wait until its next hub access window arrives. Since most processing occurs internally in each of the cogs, this potential for delay is not too frequent.

> **Information:** The Hub is our friend. It prevents shared memory from being clobbered by multiple cogs attempting simultaneous access, which would lead to catastrophic failure. In Chap. 3, you will see examples of how the Propeller's programming languages allow the developer to coordinate read/write timing among multiple cogs. Search for "Hub" in the Propeller Manual or Propeller Tool Help (www.parallax.com) to find out more about the Hub.

Memory   There are three distinct blocks of memory inside the Propeller chip.

■ Main RAM (32 K bytes; 8 K longs)
■ Main ROM (32 K bytes; 8 K longs)
■ Cog RAM (512 longs × 8 cogs)

Both Main RAM and Main ROM are shared (mutually exclusively) by all cogs, each able to access any part of those two blocks in turn. Main RAM is where the Propeller Application resides (code and data); Main ROM contains support data and functions (see Fig. 1-10). Every location is accessible as a byte (8 bits), word (2 bytes), or long (2 words).

Cog RAM is located inside a cog itself (see Fig. 1-11). Cog RAM is for exclusive use by the cog that contains it. Every register in Cog RAM is accessible only as a long (32 bits, 2 words, 4 bytes).

I/O Pins   One of the beauties of the Propeller lies within its I/O pin architecture. While the Propeller chip's 32 I/O pins are shared among all cogs, they are not a mutually exclusive resource. Any cog can access any I/O pins at any time—no need to wait for a hub access window! The cogs achieve this by gating their individual I/O signals through a set of AND and OR gates, as seen at the top of each cog in Fig. 1-9.

The cog collective affects the I/O pins as described by these simple rules:

■ A pin is an output only if an active cog sets it to an output.
■ A pin outputs high only if the aforementioned cog sets it high.

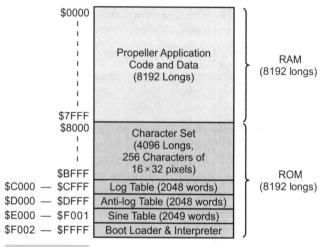

Figure 1-10   Propeller Main RAM/ROM.

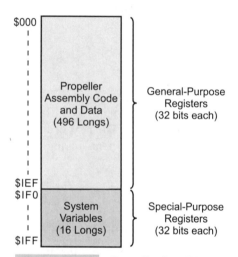

Figure 1-11   Propeller Cog RAM.

When executing a well-behaved application, the team of cogs has flexible control over the I/O pins without causing conflicts between them. A pin is an input unless a cog makes it an output and an output pin is low unless a cog sets it high.

Tip: An active cog is one that is executing instructions or sleeping. An inactive cog is one that is completely shut down. Only active cogs influence the direction and state of I/O pins.

System Counter   The System Counter is the Propeller chip's time-base. It's a global, read-only, 32-bit counter that increments once every System Clock cycle. Cogs read the System Counter via their CNT register to perform timing calculations and accurate delays. The System Counter is not a mutually exclusive resource; every cog can read it simultaneously.

Information: We use the System Counter in nearly every exercise in the next chapter.

Counter Modules and Video Generators   These are some of the Propeller's secret weapons. Each cog contains two counter modules and one video generator. They are simple state-machines that perform operations in parallel with the cogs that contain them.

Using its video generator, a cog can display graphics and text on a TV or computer monitor display. Using its counter modules, possibly in concert with software objects, a cog can perform functions that might require dedicated hardware in other systems, such as measuring pulses, frequency, duty cycle, or signal decay, or performing delta-sigma A/D and duty-modulated D/A conversion.

Information: We put these powerful state-machines to good use in later chapters.

Summary

We learned what multicore is about and why it's important. We also explored our multicore hardware in preparation for the journey ahead. The next chapter will apply this hardware a step at a time while teaching simple problem-solving and multicore programming techniques.

Exercises

To further your learning experience, we recommend trying the following exercises on your own:

1 Carefully consider opportunities for multicore devices. How could a self-propelled robot be enhanced using multicore? What if it had multiple legs and arms?
2 Think about ways humans exhibit multicore traits. Yes, we have only one "mind," but what keeps our heart beating and our lungs pumping while we are busy thinking about this? What about "learned" reflexes? Keep this in mind when applying multicore hardware to future applications.

2

INTRODUCTION TO PROPELLER PROGRAMMING

Jeff Martin

Introduction

The most reliable systems are built using simple, proven concepts and elemental rules as building blocks. In truth, those basic principles are valuable for solving many everyday problems, leading to solid and dependable results. Together, we'll apply those principles in step-by-step fashion as we learn to program the multicore Propeller in the following exercises.

You'll be running a Propeller application in no time, and writing your own in mere minutes! You'll learn that you can achieve quick results, perform time-sensitive tasks, and use objects and multiple processors to build amazing projects in little time! In addition, you'll know where to find an ever-growing collection of documentation, examples, and free, open-source objects that an entire community of developers is eager to share with you!

In this chapter, we'll do the following:

- Cover the available forms of the Propeller
- Install the development software and connect our Propeller
- Explore the Spin language with simple, single-core examples
- Run our examples on a Propeller in RAM and EEPROM
- Create a simple multicore example
- Make a building block object
- Adjust timing and stack size
- Find out where to learn more

Resources: Demo code and other resources for this chapter are available for free download from ftp.propeller-chip.com/PCMProp/Chapter_02.

What's the Secret?

How can you build incredible, multicore systems without getting lost in the details? Surprisingly, there's no real secret. In fact, teachers have been drilling the solution into our heads for years.

> **Solution:** Break big problems into smaller, simpler ones and solve them individually. Then put the individual solutions together to tackle the original problem (see Fig. 2-1).

That's it! For applications dealing with many things at once, often a separate, focused process (core) can address each task individually, and the collective of separate processes can achieve amazing things with little work on the developer's part. It's easier than you may think. If you start every application with this in mind, you'll be most successful.

Let's test this out with an example. Suppose you have an assembly line that produces thousands of products per minute. The machinery to do it is expensive, so both quantity and quality must be high to keep the business profitable. You need a system that inspects the product at various critical stages, discarding the bad, keeping the good, and adjusting the speed of assembly along the way to maximize throughput. The workers and managers can't be left out of the loop; they need reports of some kind and the ability to adjust settings as the need arises. And, most importantly, each of these things should behave consistently without any bottlenecks introduced by the activities of another.

Figure 2-1 **Big problem solved a piece at a time.**

This may sound horribly complex, but breaking it down into separate problems eases the pain:

- First, concentrate on a system that only inspects the product and gives a pass or fail response to each one.
- Then, devise a process for discarding units deemed bad while keeping the good.
- Build a component whose sole task is to adjust assembly line speed based on a ratio of good versus bad product produced.
- Create a system to display production status.
- Finally, build a feature that takes human input to change operating parameters.

The task is much easier now. Solve each of these five smaller problems, one at a time, with little or no regard for the others. Each can be a specialized process that focuses most of its energy on the given task.

A multicore device like the Propeller can then perform all of these specialized functions in separate processors, each faithfully fulfilling its "simple" duty despite the complexities and timing requirements of the other functions running concurrently. The final system, completely controlled by a single Propeller chip, may use a camera for visual product inspection, solenoids to kick bad units off the assembly line, actuators to adjust the speed, one or more Video Graphics Array (VGA) displays to show system status, and one or more keyboards for user input. All equipment is standard, inexpensive, and readily available.

Ready to Dive In?

As you follow this chapter's exercises, keep in mind that every function we have the Propeller perform for us is just an "example process." We will use simple example processes, like blinking light-emitting diodes (LEDs), to demonstrate an application while focusing on the concepts of Propeller programming.

In place of each example, we could use audio, video, analog-to-digital, or any number of other possible processes, but they would obscure the point. The point is that the concepts in this short chapter serve as building blocks for many types of "processes," from simple to sophisticated, from single-core to multicore.

The rest of this book will show you fantastic examples of those capabilities, taking full advantage of what the multicore Propeller has to offer. The foundation we build in this chapter will provide you with a strong understanding of the things to come.

Let's Get Connected!

Let's get started by connecting the Propeller and testing it out! You can get a Propeller chip in many different forms to suit your needs (see Fig. 2-2).

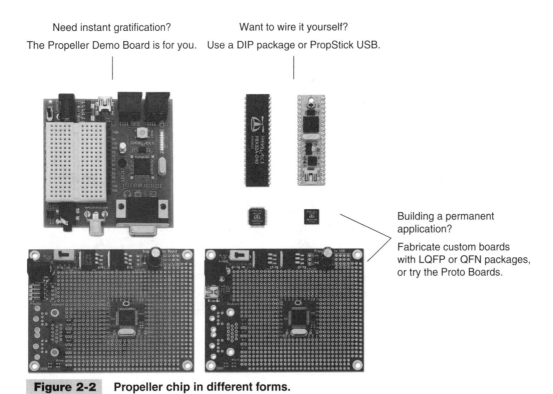

Need instant gratification?
The Propeller Demo Board is for you.

Want to wire it yourself?
Use a DIP package or PropStick USB.

Building a permanent application?

Fabricate custom boards with LQFP or QFN packages, or try the Proto Boards.

Figure 2-2   Propeller chip in different forms.

Tip: Check out www.parallax.com/propeller for current product offerings.

For demonstration purposes and ease-of-use, we'll focus on the Propeller Demo Board in this chapter's exercises. Don't worry—the other products can also perform the same, simple examples we show here with the addition of some simple circuitry.

SOFTWARE: PROPELLER TOOL—INSTALL THIS FIRST

Now that we've selected a Propeller product, to develop applications for it, we first need to install the Propeller Tool software and connect the Propeller to a computer and power supply.

✓ Download and install the Propeller Tool software.
  o The Propeller Tool software is available free from Parallax Inc. Go to www. parallax.com/Propeller and select the Downloads link.
  o Install with the default options. The software will automatically load the Windows Universal Serial Bus (USB) drivers needed for the next steps.
✓ Start the Propeller Tool software.
  o When installation is complete, start the software by double-clicking the Propeller Tool icon on your desktop, or follow the Start → Programs → Parallax Inc. menu item. A window should appear similar to Fig. 2-3.

Object View—shows application structure.

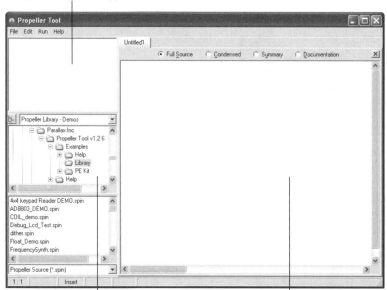

Folder and File views—provides   Edit pane—where you enter your code.
access to objects on the computer.

Figure 2-3   **Propeller Tool software (Windows).**

Tip: In addition to being a code editor, the Propeller Tool is a launch point for a wealth of Propeller information. The Help menu includes not only Propeller Help, but the manual, datasheet, schematics, and educational labs as well.

Tip: Linux and Mac software (Fig. 2-4) is also available but may not include the USB drivers for your system. See instructions with the software to install USB drivers before connecting hardware.

HARDWARE: PROPELLER DEMO BOARD—CONNECT THIS SECOND

✓ Connect a standard USB cable (A to mini B type cable).
   o Insert the cable's "A" connector into an available USB port on your computer and the "mini B" connector to the Propeller Demo Board's USB jack as shown in Fig. 2-5. The computer will indicate that it found new hardware and should automatically configure itself since we installed the USB drivers in the previous step.
✓ Connect a power supply (6–9 VDC wall-pack with center-positive, 2.1-mm plug).
   o Insert the power supply's plug into the jack next to the power switch.
✓ Turn on the power.
   o Slide the power switch to the <u>ON</u> position and verify that the power light, located near the switch, illuminates.

Edit pane—where you enter your code.

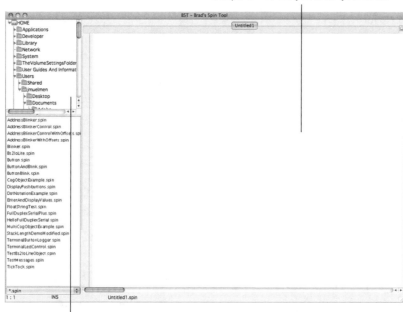

Folder and File views—provides
access to objects on the computer.

Figure 2-4   Brad's Spin Tool (BST) on a Macintosh.

Figure 2-5   Propeller Demo Board connected and powered.

PROP PLUG USB INTERFACE TOOL FOR CUSTOM PROPELLER BOARDS

Many Propeller boards feature a USB interface. If you are not using the Propeller Demo Board, see the product's documentation for connection details.

For discrete Propeller chips, Fig. 2-6 shows the connection using the Propeller Plug (available from www.parallax.com). Refer to the chip's pin names if translating from the DIP to the LQFP or QFN packages.

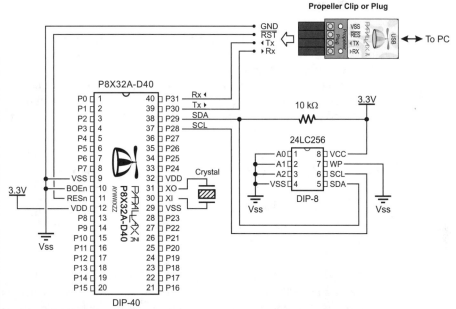

Figure 2-6   Propeller DIP to Prop Plug connections.

If you don't have a Propeller Plug or USB port, look for "Hardware Connections" in Propeller Tool Help for an example connection to an RS-232 serial port.

Now test the connection.

✓ Perform the Propeller Identification process.
  ○ Press the F7 key or select the Run → Identify Hardware. . . menu item. The Propeller Tool will scan for the Propeller and display results similar to Fig. 2-7.

Figure 2-7   Identification dialog showing version
and port of Propeller chip.

Tip: If the software was unable to find the Propeller chip, check all cable connections and the power switch/light, then try the identification process again. You may also verify the USB connection and driver using the Serial Port Search List; select the Edit → Preferences menu item, choose the Operation tab, click the Edit Ports button, then connect/disconnect the USB cable while watching the Serial Port Search List window. When working properly, your Propeller connection will appear as a "USB Serial Port."

Your First Propeller Application

Now that we can talk to the Propeller, let's write a short program to test it. We'll start our exercises slow and easy and accelerate into more advanced topics as we build up our knowledge.

✓ Type the following code into the blank edit pane of the Propeller Tool software.
  o Make sure the PUB line begins at the leftmost edge of the edit pane. Note that the case of letters (uppercase/lowercase) does not matter but indention often does; we indented the lines under PUB LED_On by two spaces.

```
PUB LED_On

  dira[16] := 1
  outa[16] := 1
  repeat
```

Tip: This source is from: PCMProp/Chapter_02/Source/LED_On.spin.

When done, your screen should look something like Fig. 2-8.

✓ Now compile the code by pressing the F9 key, or by selecting the Run → Compile Current → Update Status menu item.
  o If everything is correct, "Compilation Successful" should appear briefly on the status bar. If you entered something incorrectly, an error message will appear indicating the problem; recheck your work and compile again.

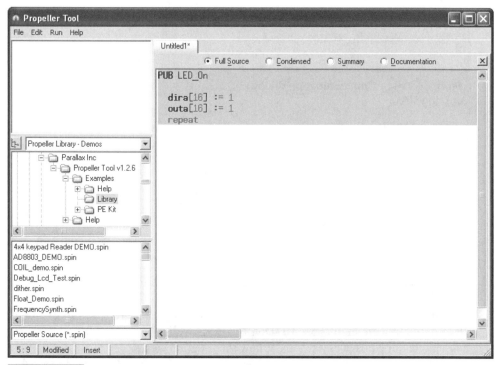

Figure 2-8   Propeller Tool with LED_On application entered.

LED SCHEMATIC

If you are not using the Propeller Demo Board, add the circuit shown in Fig. 2-9 to your setup for the following exercises. Pxx labels refer to Propeller input/output (I/O) pins, not physical pin numbers.

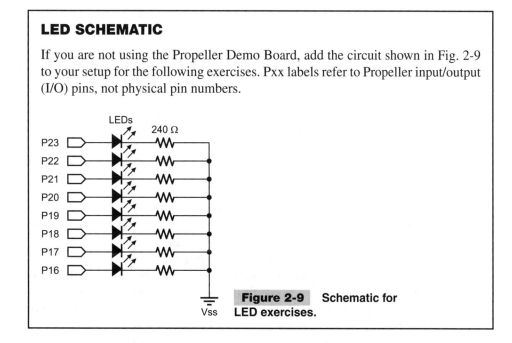

Figure 2-9   Schematic for LED exercises.

Figure 2-10   Communication dialog:
Verifying RAM.

You just wrote your first program using the Propeller's Spin language! It's a fully functional program called a Propeller Application. Go ahead and try it out!

✓ Download your application to the Propeller by pressing the <u>F10</u> key, or by selecting the <u>Run</u> → <u>Compile Current</u> → <u>Load RAM</u> menu item.
 ○ A message like Fig. 2-10 will appear briefly indicating the download status.

After downloading, the LED connected to the Propeller's I/O pin 16 should turn on. If you check the I/O pin with a voltmeter, you'll measure a little more than 3 volts DC.

EXPLANATION

As you may already realize, all this program does is make the Propeller set its I/O pin 16 to output a logic high (≈3.3 V). Don't worry—we'll do more exciting things in a moment, but first take a closer look at how the program works.

The PUB LED_On statement declares that the block of Spin code under it is a public method named LED_On. A *method* is a container that holds code of a specific purpose, and the name of the method indicates that purpose. Without testing it, you probably could have guessed what our LED_On method does. The term *public* relates to how we can use the method, which we'll discuss later. A Propeller Application usually contains multiple methods, and all executable Spin instructions must be grouped inside them to compile and execute properly.

All instructions below the PUB LED_On declaration are indented slightly to indicate that they are part of the LED_On method, like subitems in an outline.

> **Tip:** PUB is only one of many block designators that provide structure to the Spin language. There are also CON and VAR (constant and global variable declarations), OBJ (external object references), PRI (private methods), and DAT (data and assembly code). Look for "Block Designators" in Propeller Tool Help or the Propeller Manual.

The dira[16] := 1 and outa[16] := 1 statements set I/O pin 16 to an output direction and a logic high state, respectively. Both dira (directions) and outa (output states) are 32-bit variables whose individual bits control the direction (input/output) and output state (low/high) of each of the Propeller's corresponding 32 I/O pins.

The := "colon-equals" is an assignment operator that sets the variable on its left equal to the value of the expression on its right. We could assign full 32-bit values to each of these two variables; however, when working with I/O pins, it's often more convenient to target a specific bit. The number in brackets, [16], forces the assignment operator, :=, to affect only bit 16 of dira and outa, corresponding to I/O pin 16.

Tip: The Propeller's 32 I/O pins are general-purpose; each can be an input or output and each can drive the same voltage/current levels. All of the Propeller's eight processors can read any pin as an input, but a processor must set a pin's direction to output if it wants to use it as an output. To learn more, search for "I/O Pins" in Propeller Tool Help or the Propeller Manual.

The `repeat` instruction is a flexible looping mechanism that we'll learn more about as we experiment. As written in this example, `repeat` makes the Propeller "do nothing" endlessly.

What's the point of that? Without `repeat` our program would simply end, leaving nothing for the Propeller to do; it would terminate and reset the I/O pin to an input direction. In other words, the LED would light up for a small fraction of a second and then turn off forever, giving the appearance that our application did absolutely nothing!

Most applications have an endless amount of work to do in one or more loops. For simple tests like this one, however, we need to include a "do nothing" `repeat` loop in order to see the intended result.

Challenge: Try changing the application's numbers and downloading it again to see the effects. What happens when you change both occurrences of [16] to [17]? How about setting `outa` to 0 instead of 1?

A Blinking LED

Admittedly, our first example is an expensive alternative to a simple wire and light, but now you know a way to control the Propeller's I/O pins!

Here's a more exciting example using the techniques we learned, plus a little more.

✓ Create a new application with the following code.

Tip: Press Ctrl+N to start with a fresh edit pane.

```
PUB LED_Blink

    dira[16] := 1              'Set I/O pin 16 to output direction
    repeat                     'Loop endlessly...
      outa[16] := 1            '   Set I/O pin 16 to high
      waitcnt(clkfreq + cnt)   '   Delay for some time
      outa[16] := 0            '   Set I/O pin 16 to low
      waitcnt(clkfreq + cnt)   '   Delay again
```

Tip: This source is from: PCMProp/Chapter_02/Source/LED_Blink.spin.

Caution: Indention is important here; make sure to indent the lines under `repeat` by at least one space. Also, the mark (') that appears in some places is an apostrophe character.

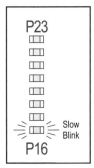

Figure 2-11   Demo Board LED blinking.

✓ Compile and download this application by pressing <u>F10</u>, or by using the <u>Run</u> → <u>Compile Current</u> → <u>Load RAM</u> menu item.

Now the LED on I/O pin 16 should blink on and off at roughly 1-second intervals. (see Fig. 2-11). No more is this a simple wire alternative!

EXPLANATION

Are you wondering what's to the right of the dira[16] := 1 statement yet? That's a comment. *Comments* describe the purpose of code; they mean absolutely nothing to the Propeller, but everything to the programmer and his or her friends. This one begins with an apostrophe ('), meaning it's a single-line code comment. There are other types of comments that we'll learn about soon, but for now, just remember that comments play a vital role in making your code understandable.

Impress your friends! Use comments generously!

Take a look at the rest of the comments in the program, and you should clearly see what each Spin statement does. We'll explain what is new to us.

We've seen repeat before, as an endless loop that did nothing, but now it's more useful. This repeat is a loop that endlessly executes a series of four statements within it. Did you notice that the statements under it are indented? That's important! It means they are part of the repeat loop. In fact, by default, the editor indicates that these are part of the loop by displaying little hierarchy arrows next to them, as in Fig. 2-12.

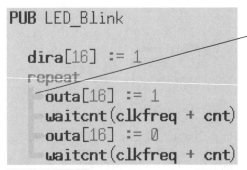

Hierarchy arrows, called "block group indicators," automatically appear to point out code that belongs to a group; in this case, the last four statements are part of an endless repeat loop.

Figure 2-12   A close-up of the LED_Blink method as it appears in the editor.

> ## INDENTATION IS IMPORTANT!!
>
> The Spin language conserves precious screen space by omitting begin and end markers on groups of code. Just as we rely on indention in outlines to show which subtopics belong to a topic, Spin relies on indention to know what statements belong to special control commands like `repeat`. You can toggle the block group indicators on and off by pressing Ctrl+I; Spin will understand the indention either way.

The `waitcnt` command is something we haven't seen before. It means, "Wait for System Counter," and serves as a powerful delay mechanism.

What do you do when you want to wait for five minutes? Well, of course, you check the current time, add five minutes, then "watch the clock" until that time is reached. The `waitcnt` command makes the Propeller "watch the clock" until the desired moment is reached.

The expression within `waitcnt`'s parentheses is the desired moment to wait for. Both `clkfreq` and `cnt` are built-in variables that relate to time. Think of `clkfreq` as "one second" and `cnt` as "the current time." So the expression `clkfreq + cnt` means "one second plus the current time" or "one second in the future."

Information: `Clkfreq` contains the current system clock frequency—the number of clock cycles that occur per second. `Cnt` contains the current System Counter value—a value that increments with every clock cycle. The value in `cnt` doesn't relate directly to the time of day; however, the difference between the value in `cnt` now and its value later is the exact number of clock cycles that passed during that time, which is a useful number for timing. For accurate timing, see "Timing Is Everything."

Challenge: Try changing the code so the LED is on for roughly one-eighth of a second and off for one second. Hint: '/' is the divide operator.

RAM versus EEPROM

If you followed the last example, you now have a happily blinking LED on your development board. Will it continue to blink after a reset or power cycle?

✓ Try pressing the reset button or switching power off and on again.

Did the LED ever light up again? No, it didn't, because we downloaded our application to random access memory (RAM) only. RAM contents are lost when power fails or a reset occurs, so when the Propeller started up again, our application was missing.

What if we want our application to start up again the next time the Propeller starts up? We need to download to the Propeller's external EEPROM to preserve our application even without power.

Figure 2-13   Communication dialog:
Programming EEPROM.

✓ Download your Propeller Application again, but this time by pressing the <u>F11</u> key
or by selecting the <u>Run</u> → <u>Compile Current</u> → <u>Load EEPROM</u> menu item.
    ○ Once again, a message will appear indicating the download status, but this time it
lasts longer as it programs the Propeller's external EEPROM chip (see Fig. 2-13).
✓ After successful download, press the <u>reset</u> button or switch power off and on
again.

This time, after a short delay, your application restarts and the LED blinks again.
This will continue to happen after every reset or power cycle. When you're ready for
your Propeller Application to "live" forever, you should download to EEPROM instead
of just to RAM.

WONDERING ABOUT THAT POWER-ON/RESET DELAY?

It takes about two seconds for the Propeller to complete its bootup procedure,
including the time to load your application from EEPROM.

    Don't worry—that delay won't increase with the size of your application!
The Propeller always loads all 32 KB from EEPROM, regardless of how many, or
how few, instructions your application contains. See "Boot Up Procedure" in the
Propeller Tool's Help to learn more.

A More Powerful Blink

Remember how we said to focus on simple problems first and tasks appropriate for
parallel processes will become apparent? Our blinking LED serves as one such example
process. Our application currently performs only one process in one specific way, but
it could perform this process in many different ways, either sequentially or in parallel.
Let's enhance our method to make it easier to do this.

    In this exercise, we'll create a more flexible version of our blinking LED method
and we'll demonstrate it in sequential operation. In the exercise immediately following,
we'll command multiple processes in parallel!

    We mentioned earlier that a method has a name and contains instructions. What might
not have been so apparent is that a method is like an action that can be called upon by
name, causing the processor to perform its list of instructions one at a time. As it turns
out, in the last example, we could have called our LED_Blink method by simply typing
its name elsewhere in the code, like this:

```
PUB SomeOtherMethod

  LED_Blink                        'Blink the LED
```

This is how methods are activated most of the time. We didn't need to do this in our previous examples because our applications had only one method and the Propeller naturally calls the first method in an application.

Not only can we call a method by name; we can also demand that it take on certain attributes. Suppose that we want *this* LED to blink *that* way and *for so long*. It's all possible with the same method as long as it's written to support those attributes.

Take a look at this new method we based on the previous example:

```
PUB LED_Flash(Pin, Duration, Count)
{Flash led on PIN for Duration a total of Count times.
 Duration is in 1/100th second units.}

  Duration := clkfreq / 100 * Duration    'Calculate cycle duration
  dira[16..23]~~                           'Set pins to output

  repeat Count * 2                         'Loop Count * 2 times...
    !outa[Pin]                             '  Toggle I/O pin
    waitcnt(Duration + cnt)                '  Delay for some time
```

Caution: Don't run this yet! Our application isn't ready until later in this exercise!

EXPLANATION

Our method declaration looks different, doesn't it?

```
PUB LED_Flash(Pin, Duration, Count)
```

Yes, we changed the name, but now there are also items in parentheses. The `Pin`, `Duration`, and `Count` items are *parameters* we made up to accept the attribute we spoke of earlier. Each of them is a placeholder for values (numbers or expressions) when the method is called. Figure 2-14 shows an example of this in action.

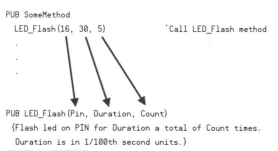

```
PUB SomeMethod
  LED_Flash(16, 30, 5)              'Call LED_Flash method
  .
  .
  .
PUB LED_Flash(Pin, Duration, Count)
{Flash led on PIN for Duration a total of Count times.
 Duration is in 1/100th second units.}
```

1. SomeMethod calls LED_Flash with parameters.

2. The parameter values are "copied" into LED_Flash's parameter variables.

3. LED_Flash references those values via their parameter names.

Figure 2-14   **Method call with parameters.**

To the method, the parameters are *local variables* for its own use. It can read them and manipulate them without affecting anything outside of itself.

In our new method, the two lines following the declaration are a different type of comment; a *multiline comment*. Multiline comments begin with an open brace ({) and end with a close brace (}) and can span more than one line of code.

The `Duration := clkfreq / 100 * Duration` statement means, "Make duration equal to `clkfreq` divided by 100 and multiplied by `Duration`'s original value." Remember, `clkfreq` is like "one second," so dividing it by 100 is "1/100th of a second," and multiplying that by `Duration` gives us a multiple of 100ths of a second. Note the comment above the line.

The `dira[16..23]~~` statement is a twist on an old theme. Recall that `dira` controls I/O pin directions and the number in brackets is the bit, or I/O pin, to affect. This statement is a clever way to affect all the bits from 16 to 23 at once. So what do they get set to? The trailing `~~` operator, when used this way, is a *set assignment* operator; it sets the bit(s) of the variable to which it is attached to high (1). It's shorthand, and without it we'd have to say: `dira[16..23] := %11111111`.

Note: We're setting all eight of these pins to outputs only because the Propeller Demo Board shares them with the VGA circuit, which causes certain LED pairs to light simultaneously when only one is actually activated. This would not happen if they were wired as in Fig. 2-9.

Did you notice that our `repeat` loop has changed? It now says `repeat Count * 2`. Previously, repeat was always an infinite loop, but now it is a finite loop that executes only `Count*2` times.

The contents of our loop changed as well. In the `!outa[Pin]` statement, the `!` is a *bitwise NOT assignment operator;* it toggles the bit(s) of the variable to which it is attached. It makes the I/O pin's output state toggle to the opposite state: high if it was low, low if it was high. It's shorthand for `outa[Pin] := NOT outa[Pin]`.

Tip: You can learn more about these and many other operators by searching for "Spin Operators" in Propeller Tool Help or the Propeller Manual.

So now our `LED_Flash` method takes three parameters (`Pin`, `Duration`, and `Count`), calculates the actual duration in clock cycles (in 100ths of a second units), sets the pin directions, and loops `Count*2` times, toggling `Pin` each time for the calculated duration.

But it really won't do anything for us if we don't call it properly!

Let's add another method to the top of our application so it appears as follows:

```
PUB Main

    LED_Flash(16, 30, 5)          'Flash led
    LED_Flash(19, 15, 15)         'Then another
    LED_Flash(23, 7, 26)          'And finally a third
```

```
PUB LED_Flash(Pin, Duration, Count)
{Flash led on PIN for Duration a total of Count times.
 Duration is in 1/100th second units.}

  Duration := clkfreq / 100 * Duration    'Calculate cycle duration
  dira[16..23]~~                          'Set pins to output

  repeat Count * 2                        'Loop Count * 2 times...
    !outa[Pin]                            '  Toggle I/O pin
    waitcnt(Duration + cnt)               '  Delay for some time
```

Tip: This source is from: PCMProp/Chapter_02/Source/LED_Flash.spin.

Now our application has two methods: Main and LED_Flash. Logically, our Main method is now our application's "director," giving commands to guide the nature of our application. The LED_Flash method is now a support method, obeying the Main method's commands.

Information: What makes Main so special? It's not its name—it's the position it holds. Main is special only because it's the first method in our application, the one the Propeller activates automatically when the application is started.

Main calls LED_Flash a number of times, each with different values for the parameters. What do you think is going to happen? Try it out and see!

✓ Download this application to the Propeller.

As you may have predicted, our application blinks pin 16's LED 5 times slowly, then pin 19's LED 15 times more quickly, and finally pin 23's LED 26 times very fast (see Fig. 2-15).

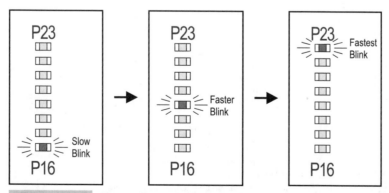

Figure 2-15   Demo Board LEDs blinking in sequence.

Specifically, when the application starts, the Propeller calls its first method, Main. The first line of Main is a call to LED_Flash, so the Propeller executes each statement in LED_Flash. When the finite loop finishes, there's no more code to execute in LED_Flash, so the Propeller "returns" to Main and executes the second line—another call to LED_Flash. This continues until it has executed all the statements of Main in sequence. Then, since there are no more statements in Main, it "returns." But to where? The Propeller called Main itself, so the "return" from Main causes the processor to terminate.

All Together Now

Now suppose the last application isn't quite what we needed. What if we need each blinking process to happen at the same time (in parallel) instead of one at a time (in sequence)?

On a single-core device, this request would be a nightmare because the timing of each individual process, as tested earlier, would negatively affect the timing of every process as a group. But on a multicore device like the Propeller, this is relatively easy! With minor changes to our code, we'll launch each instance of LED_Flash into a separate cog (processor) to execute in parallel.

Information: What's a cog? It's the Propeller's name for each of its processors. They are called cogs since they are simple and uniform, like the cogs on gears that mesh with others of their kind to induce change. Their simplicity assures reliability, and as a collective, they deliver powerful results.

Here's the updated code. Try it out now! We'll explain it in a moment.

```
VAR

    long  StackA[32]                    'Stack workspace for cogs
    long  StackB[32]
    long  StackC[32]


PUB Main

    cognew(LED_Flash(16, 30, 5), @StackA)   'Launch cog to flash led
    cognew(LED_Flash(19, 15, 15), @StackB)  'And another for different led
    cognew(LED_Flash(23, 7, 26), @StackC)   'And a third, all at same time


PUB LED_Flash(Pin, Duration, Count)
{Flash led on PIN for Duration a total of Count times.
 Duration is in 1/100th second units.}
```

```
Duration := clkfreq / 100 * Duration    'Calculate cycle duration
dira[16..23]~~                           'Set pins to output

repeat Count * 2                         'Loop Count * 2 times...
  !outa[Pin]                             '  Toggle I/O pin
  waitcnt(Duration + cnt)                '  Delay for some time
```

Tip: This source is from: PCMProp/Chapter_02/Source/LED_MultiFlash.spin.

✓ Download this application to the Propeller and note how it behaves differently from the last exercise.

Now all three LEDs blink simultaneously, at different rates and different counts, but each exactly the way they did before (see Fig. 2-16). There is no difference in each individual LED's behavior!

This is the beauty of a multicore device like the Propeller! You can focus your time on a simple implementation of a process with no regard to other parts of a final application and then simply connect the individual parts together in the end. And this can be done with many different types of processes, not just multiple instances of the same process.

EXPLANATION

Our application code didn't change much. In fact, the LED_Flash method didn't change at all.

First we added a new block at the top of the code—a VAR block—which declares a block of *global variables*. The statement long StackA[32] reserves 32 longs (128 bytes) of memory and calls it StackA. The declarations for StackB and StackC are similar. As the comment indicates, this memory is for cog workspace.

Information: When a cog executes Spin code, it needs *stack space*—that is, temporary workspace to process method calls and expressions. The Spin compiler automatically assigns the first cog's workspace (for the main application code), but additional cogs launched on Spin code need manually assigned workspace. Thirty-two longs is more than enough for this example. We show how to determine how much is enough later in "Sizing the Stack."

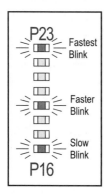

Figure 2-16   Demo Board LEDs blinking in parallel.

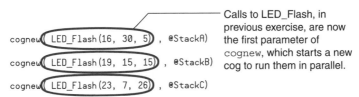

```
cognew( LED_Flash(16, 30, 5) , @StackA)
cognew( LED_Flash(19, 15, 15) , @StackB)
cognew( LED_Flash(23, 7, 26) , @StackC)
```

Calls to LED_Flash, in previous exercise, are now the first parameter of cognew, which starts a new cog to run them in parallel.

Figure 2-17   **Method call with parameters.**

Our Main method changed the way we call the LED_Flash method. Now each of our calls is the first parameter of cognew commands; as shown in Fig. 2-17. Cognew, as the name implies, starts a new cog to run the method indicated by its first parameter.

When the statement cognew(LED_Flash(16, 30, 5) , @StackA) is executed, a new cog starts up to run the LED_Flash method (with the parameters 16, 30, and 5). The second parameter of cognew, @StackA, directs the new cog to its assigned workspace. The @ operator returns the address of the variable it's attached to, so the new cog locates its workspace in the memory starting at the address of StackA.

As Main executes, it starts three other cogs, each running LED_Flash with different parameters and using different workspaces; then Main runs out of code, causing the first cog (the application cog) to terminate. As each of the remaining cogs finish executing their instance of LED_Flash, they individually terminate, leaving absolutely no cogs running and no LEDs flashing.

Wrapping It Up

So far we've created a nice method to perform our desired task: flashing an LED. It's been enhanced, tested in isolation, and even integration-tested as parallel processes. You didn't know it, but all this time we've been building towards this moment: the creation of a building block object.

An object is a set of code and data that is self-contained and has a specific purpose. Though we called our examples "Propeller Applications," that's only half the story. A *Propeller Application* is an *executable image* made from an object, which itself may be made up of one or more other objects. We've actually been designing an object all along. Now we want to transform it into a building block object.

Building block objects are meant to be subcomponents of other objects; they have a set of required inputs and deterministic outputs. Propeller users love building block objects because they can swiftly combine them into one object, or application, that performs with all the expert skills of the collective of objects. Figure 2-18 shows an analogy for this concept.

Take a look at the following code; we rewrote the top portions of our last exercise, but our core, the LED_Flash method, is only slightly different. Can you see how?

✓ Enter and save this code. Give it the name "Flash.spin."

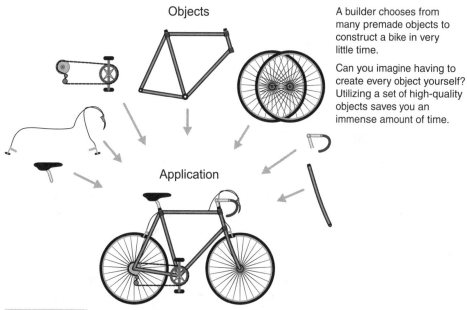

Objects

A builder chooses from many premade objects to construct a bike in very little time.

Can you imagine having to create every object yourself? Utilizing a set of high-quality objects saves you an immense amount of time.

Application

Figure 2-18   **Building block objects used in a bike application.**

```
VAR
  long  Cog                        'Holds ID of started cog
  long  Stack[32]                  'Stack workspace for cog


PUB Start(Pin, Duration, Count)
{{Start flashing led on Pin, for Duration, a total of Count times.}}

  Stop
  Cog := cognew(LED_Flash(Pin, Duration, Count), @Stack) + 1


PUB Stop
{{Stop flashing led.}}

  if Cog                           'Did we start a cog?
    cogstop(Cog~ - 1)              '  If so, stop it


PRI LED_Flash(Pin, Duration, Count)
{Flash led on PIN for Duration a total of Count times.
 Duration is in 1/100th second units.}

  Duration := clkfreq / 100 * Duration  'Calculate cycle duration
  dira[16..23]~~                        'Set pins to output
```

```
  repeat Count * 2              'Loop Count * 2 times...
    !outa[Pin]                  '  Toggle I/O pin
    waitcnt(Duration + cnt)     '  Delay for some time
```

Tip: This source is from: PCMProp/Chapter_02/Source/Flash.spin.

Caution: Don't run this yet! Our application isn't ready until later in this exercise.

EXPLANATION

The only change we made to LED_Flash is to declare it as PRI instead of PUB. Do you recall how PUB declares a "public" method? PRI declares a *private* method. The difference is that while public methods can be called by other objects, private methods cannot. This feature helps an object maintain its integrity. Generally, most object methods are declared as public since they are meant to be called by other objects at any time. Methods not designed for random calling from other objects should be declared as private, especially those that threaten the object's integrity if misused.

Since LED_Flash is our core method, we're making it private just as a matter of principle. We're making this object automatically handle cog launching, memory management, and LED blinking with one simple interface, so there's no need for outside objects to call LED_Flash directly.

In the VAR block we declared a variable, Cog, to hold the ID of the cog this object will launch. We'll use this to stop that cog later, if necessary. We also declared a single Stack variable of 32 longs, the same size as before.

We replaced our Main method from the previous example with two new methods: Start and Stop. The Start method launches a cog to flash our LED. The Stop method stops the cog launched by Start, if any.

Information: By convention, whenever you create an object that launches another cog, you should have methods named "Start" and "Stop" that manage the cog. This provides a standard interface that gives users of your object a clear indication of its intention.

Our Start method includes the same parameters as our LED_Flash method. Other objects will call our Start method to launch another cog to flash an LED with the given parameters.

Ironically, it seems, the first thing Start does is call Stop. This is because we want each instance of our object to maintain only one cog at a time. We don't want users calling our object's Start method multiple times and running out of cogs.

The next line, Cog := cognew(LED_Flash(Pin, Duration, Count), @Stack) + 1, is a combined expression and cognew instruction. If you look carefully, you'll see that the cognew instruction launches our LED_Flash method with the parameters originally given to Start and assigns it some workspace, @Stack. But why do we set the Cog variable equal to cognew() + 1? As it turns out, cognew returns a value equal to the ID (0 to 7) of the cog it actually started, or −1 if none were started. We'll use this to keep track of the cog we launched so we can stop it later, if desired. We add 1 to the ID to make the code in Stop more convenient, as you'll see in a moment.

In `Stop`, we check if a cog was started by us and, if so, we stop it. The `if` is a decision command. If its condition—`Cog` in this case—is "true," it executes the block of code within it: the indented `cogstop` command. The condition `if` statements evaluate can range from simple to elaborate expressions, but the result is always the same: it's either true or false. In this case, our decision is simple; it means, "If the value in the `Cog` variable is not zero, execute a `cogstop` statement." Remember that `Cog` was set to `cognew()` + 1, giving us a value of 0 (if no cog was started) or a value of 1 through 8 if a cog was started.

Our `cogstop(Cog~ - 1)` command, if executed, stops the cog whose ID is the value `Cog - 1`, and then post-clears (`~`) the `Cog` variable to zero. We clear it so an additional call to `Stop` does nothing.

Tip: You can find out more about the commands `cognew`, `cogstop`, and `if`, and the post-clear operator (`~`) in the Propeller Tool Help or Propeller Manual.

You may have noticed the first comment in `Start` and `Stop` begins with two brackets: `{{`. This is not a mistake; it's yet another type of comment—a *document comment*. Use it for embedding documentation right inside the object that can be seen using the Propeller Tool's Documentation view (see Fig. 2-19).

USING OUR BUILDING BLOCK OBJECT

Now that our Flash object is ready, others can use it by including its name in an `OBJ` block, like this one:

```
OBJ

    LED   : "Flash"        'Include Flash object
```

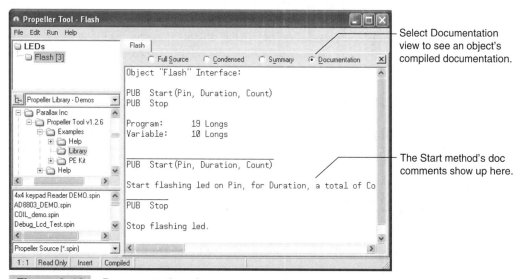

Figure 2-19   Documentation view.

This includes our Flash object that we saved in the previous steps and gives it the nickname "LED." Now we can refer to methods within the Flash object using *nickname.methodname* syntax, like this:

```
LED.Start(16, 30, 5)        'Blink LED 16 five times slowly
```

This statement calls LED's Start method. That method, in our Flash object, launches another cog to run the private LED_Flash method using the parameters given: 16, 30, and 5.

Remember the recent exercise where we launched our LED_Flash method on three separate LEDs at the same time? Now that we've neatly wrapped the critical code in our Flash object, other objects can achieve the same glory quite easily. Check out the following code.

✓ In a new edit pane, enter and save this code. Store it in the same folder as "Flash. spin" and name it "LEDs.spin."

```
OBJ

  LED[3]   : "Flash"              'Include Flash object


PUB Main

  LED[0].Start(16, 30, 5)         'Blink LED 16 five times slowly
  LED[1].Start(19, 15, 15)        'Blink LED 19 fifteen times faster
  LED[2].Start(23, 7, 26)         'Blink LED 23 twenty-six times fastest
```

Tip: This source is from: PCMProp/Chapter_02/Source/LEDs.spin.

✓ Download this application to the Propeller.

As this example shows, since our Flash object does the major work, our new application-level object is clean and simple, but can blink three LEDs at different rates simultaneously.

Tip: The LED[3] statement in the OBJ block declared an array of three Flash objects. Each one uses the same code but its own variable space. After compiling, you can explore the structure of your multiobject application in the Object View (upper-left pane of the Propeller Tool). Search for "Object View" in Propeller Tool Help to learn more.

Timing Is Everything

How fast has our Propeller been running all this time? We have a 5-MHz crystal connected (see Fig. 2-20), so it's reasonable to think it's running at 5 MHz, right?

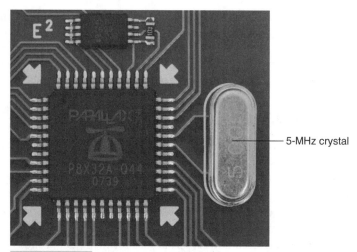

— 5-MHz crystal

Figure 2-20   Propeller, EEPROM, and crystal circuit on
Propeller Demo Board.

That is reasonable, but incorrect. The Propeller has some incredibly powerful clocking features, but as it turns out, none of our examples has set the clock mode. All this time our Propeller has been using its internal 12-MHz clock, leaving the external 5-MHz crystal dormant.

Some applications will never need an external crystal because the internal clock is just fine. For most applications, however, the internal clock is excessively inaccurate.

Information: The internal clock runs ideally at either 20 kHz or 12 MHz, but its frequency can vary by as much as ±66% from the ideal.

Since we've been using the internal 12-MHz clock, our waitcnt delays have not been very accurate; clkfreq contains the ideal frequency, not the actual frequency in this case.

To achieve much higher clock accuracy we need to use an external crystal. To make the Propeller use the external 5-MHz crystal in our circuit, our application needs to set some built-in constants.

```
CON
  _clkmode  =  xtal1 + pll16x      'Use low crystal gain, wind up 16x
  _xinfreq  =  5_000_000           'External 5 MHz crystal on XI & XO
```

This is a CON block with the typical clock configuration. The _clkmode constant is set for low crystal gain (xtal1) and a phase-locked loop (PLL) wind-up of 16 times. The _xinfreq tells the Propeller that the external crystal is providing it a 5-MHz clock signal. The combination of _clkmode and _xinfreq means we have an accurate 5-MHz clock multiplied by 16 (with the Propeller's internal PLL) for a total speed of 80 MHz. That's an amazing speed from such an inexpensive crystal.

Applying this to the previous examples may not appear to make a change, but if you looked at the difference on an oscilloscope it would be clear.

Tip: The clock mode constants can only be set in the application-level object. Clock mode constants in building block objects are ignored at compile time.

SYNCHRONIZED DELAYS

Despite the clock settings noted previously, the timing of events will still be slightly off unless we use waitcnt in a specific way. So far, our loops have performed an operation and then waited for an interval of time. Since that interval was our "ideal" delay time, it didn't account for the overhead of the rest of the instructions in the loop. The real delay between any two occurrences of our looped event is the time it took to start the loop iteration, plus the time to perform the event, plus our "idealized" loop delay.

For our application, timing accuracy isn't vital, but for many applications, accurate timing is a must. For example, code like the following causes a cumulative error in the moment the I/O pin toggles compared with the ideal moment in time, as seen in Fig. 2-21. Note that the Count, Duration, and Pin symbols are long variables.

```
dira[Pin]~~              'Set Pin to output
repeat Count * 2         'Loop Count * 2 times...
  !outa[Pin]             '  Toggle I/O pin
  waitcnt(Duration + cnt) '  Delay for some time
```

Tip: This source is from: PCMProp/Chapter_02/Source/Out_Of_Synch.spin.

In contrast, with just a slight rewrite of the code, it uses a single moment in time (Time 0 in Fig. 2-22) as an absolute reference for all future moments and perfectly synchronizes the I/O pin events to those moments. Note that the Count, Duration, Pin, and Time symbols are long variables.

```
dira[Pin]~~              'Set Pin to output
Time := cnt              'Determine reference moment
repeat Count * 2         'Loop Count * 2 times...
  waitcnt(Time += Duration) '  Delay for some time
  !outa[Pin]             '  Toggle I/O pin
```

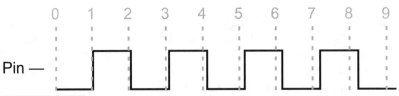

Time (units of Duration)

0 1 2 3 4 5 6 7 8 9

Pin —

Figure 2-21   I/O pin events out of synch with ideal duration.

Tip: This source is from: PCMProp/Chapter_02/Source/In_Synch.spin.

Time (units of Duration)

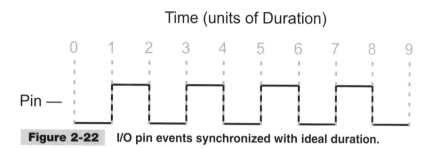

Figure 2-22   I/O pin events synchronized with ideal duration.

Just before the loop, Time is set to the current System Counter value. Inside the loop, we first wait for the moment indicated by Time + Duration, then perform the I/O event. The Time += Duration expression calculates that next moment and adjusts Time to that moment in preparation for the next loop iteration.

This technique works for any loop as long as Duration is greater than the longest possible loop overhead.

Tip: You will find an in-depth look at tackling timing-related bugs in Chapter 3. You can also find out more by looking up "Synchronized Delays" in Propeller Tool Help or the Propeller Manual.

Sizing the Stack

In "All Together Now," we mentioned the need for stack space when launching Spin code into another cog. Now we'll discuss how to size it.

The stack used by a cog running Spin code is temporary workspace. Within the stack, the amount of memory actually used changes with time. It grows while evaluating expressions and calling nested methods, and shrinks when returning expression results and returning from methods.

While developing objects, it is best to size the stack reserved for new cogs larger than necessary to avoid the strange results that can plague a program whose stack is too small. Until you get experienced with sizing the stack, we recommend reserving 128 longs initially and optimizing it later.

Why should you optimize, and when is the best time to do it? You should optimize the size of stack space so your object does not waste precious memory for all the applications that use it. Only when you're absolutely done with your object should you consider optimizing stack space; doing so beforehand could be disastrous as you make final tweaks to your code.

To determine the optimal stack size needed for your object, we recommend using the Stack Length object that comes with the Propeller Tool.

Tip: Many objects come with the Propeller Tool software. To find demonstration code for them, choose <u>Propeller Library</u>—<u>Demos</u> from the drop-down list above the Folder View. To find the actual building block objects, choose <u>Propeller Library</u> instead.

✓ Load the Stack Length Demo object (see the previous Tip) and read its comments.
   o If you want to learn even more, load the Stack Length object itself and read its top comments and its "Theory of Operation" section.

We'll use the Stack Length Demo object as a template for our test. As it suggests, we copy and paste its "temporary code" above the existing code in our Flash object, as shown here:

```
'-------- Temporary Stack Length Demo Code --------
CON
    _clkmode     = xtal1 + pll16x       'Use crystal*16 for fast serial
    _xinfreq     = 5_000_000            'External 5 MHz crystal XI & XO

OBJ
    Stk    :       "Stack Length"       'Include Stack Length Object

PUB TestStack
    Stk.Init(@Stack, 32)                'Init reserved Stack space
    Start(16, 30, 5)                    'Exercise object under test
    waitcnt(clkfreq * 2 + cnt)          'Wait ample time for max stack usage
    Stk.GetLength(30, 115200)           'Send results at 115,200 baud


'-------- Flash object code --------
VAR
    long   Cog                          'Holds ID of started cog
    long   Stack[32]                    'Stack workspace for cog


PUB Start(Pin, Duration, Count)
{{Start flashing led on Pin, for Duration, a total of Count times.}}

    Stop
    Cog := cognew(LED_Flash(Pin, Duration, Count), @Stack) + 1


PUB Stop
{{Stop flashing led.}}

    if Cog                              'Did we start a cog?
      cogstop(Cog~ - 1)                 '  If so, stop it
```

```
PRI LED_Flash(Pin, Duration, Count)
{Flash led on PIN for Duration a total of Count times.
 Duration is in 1/100th second units.}

  Duration := clkfreq / 100 * Duration    'Calculate cycle duration
  dira[16..23]~~                          'Set pins to output

  repeat Count * 2                        'Loop Count * 2 times...
    !outa[Pin]                            '  Toggle I/O pin
    waitcnt(Duration + cnt)               '  Delay for some time
```

Tip: This source is from: PCMProp/Chapter_02/Source/Flash_Stack_Test.spin.

Note that we carefully checked and modified the code in the temporary TestStack method so that:

1. "Stack," in the Stk.Init statement, is the actual name of our reserved stack space for the LED_Flash method.
2. "32," in the Stk.Init statement, is the actual number of longs we reserved for LED_Flash's stack space.
3. The Start statement contains valid parameters to our Flash object's Start method.
4. The waitcnt statement waits long enough for our LED_Flash method to be fully exercised (at least one iteration of its repeat loop).

RUNNING THE STACK TEST

Now it's time to run the test. The Stack Length object outputs its result serially on the Propeller's programming port. To receive the information, you need a simple terminal application. As it so happens, the Propeller Tool installer includes one called Parallax Serial Terminal.

✓ Start the Parallax Serial Terminal software.
   o An icon for it may have been placed on your desktop during installation, or you may find it in your computer's Start → All Programs → Parallax Inc → Propeller Tool . . . path.
   o A window should appear similar to Fig. 2-23.
✓ Select your Propeller's programming port from the Parallax Serial Terminal's Com Port field.
✓ Select 115200 from the Baud Rate field.
✓ Arrange the Parallax Serial Terminal and Propeller Tool windows so you can get to each one quickly with the mouse.
✓ Select the Propeller Tool software.

When you selected the Propeller Tool software, did you notice something happened with the Parallax Serial Terminal? The title bar (top of window) and the lower-right

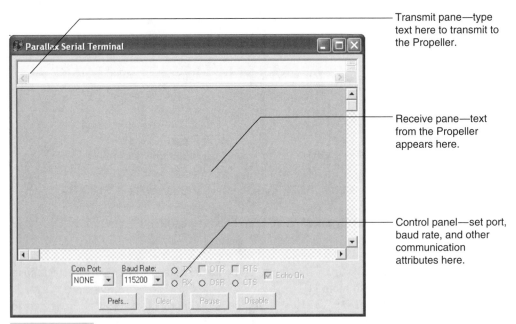

Transmit pane—type text here to transmit to the Propeller.

Receive pane—text from the Propeller appears here.

Control panel—set port, baud rate, and other communication attributes here.

Figure 2-23   **Parallax Serial Terminal: default display.**

button (control panel) change to let you know the Parallax Serial Terminal has closed the serial port (see Fig. 2-24). This is important because it lets the Propeller Tool software open the port to download our application.

✓ Start the download of our modified Flash object and, during the download, click the Enable button of the Parallax Serial Terminal.

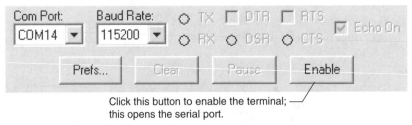

Click this button to enable the terminal; this opens the serial port.

Figure 2-24   **Control panel of Parallax Serial Terminal: waiting for enable.**

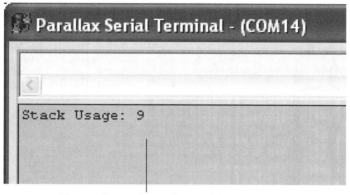

The stack test shows the resulting
utilization; 9 longs.

Figure 2-25   **Stack utilization message from Propeller.**

After our modified Flash object is downloaded, the Parallax Serial Terminal will
open the serial port and wait for input. The pin 16 LED will flash, as expected, and a
message will soon appear in the receive pane (see Fig. 2-25).

Now we know we only need to reserve 9 longs of space for the LED_Flash method's stack. We can adjust the stack size, remove the temporary code, and publish our
object.

```
VAR
  long  Cog                          'Holds ID of started cog
  long  Stack[9]                     'Stack workspace for cog
  .
  .
  .
```

Propeller Objects and Resources

We encourage you to learn more about the Propeller. Besides reading the rest of this book
and exploring outside resources we've shown you, study other developers' objects. Dozens
are included with the Propeller Tool software, and hundreds more are in a central location
called the Propeller Object Exchange—obex.parallax.com (see Fig. 2-26).

There's an active user forum full of Propeller users with ideas, questions,
answers, and genuine motivation to share (see Fig. 2-27). Visit the Parallax Forums

The Propeller Object Exchange is where developers go to share their objects with others. It's free! You can quickly download, try out, experiment with, and learn from other Propeller objects.

Objects are organized by category for easy locating. Drill down through lists for more detail including user ratings and reviews.

Figure 2-26 Propeller Object Exchange (obex.parallax.com).

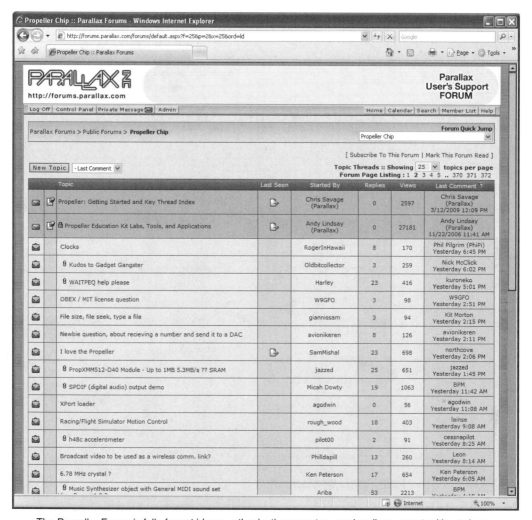

The Propeller Forum is full of great ideas, enthusiastic supporters, and endless opportunities to learn.

Figure 2-27   Parallax Propeller Forum.

at forums.parallax.com. The top threads contain links to many valuable Propeller resources.

More tips and examples can be found by joining a Propeller Webinar (a live web-based meeting) or viewing archived webinars at www.parallax.com/go/webinar (see Fig. 2-28). This is a way to connect to Parallax staff and get an inside look at how to use the multicore Propeller.

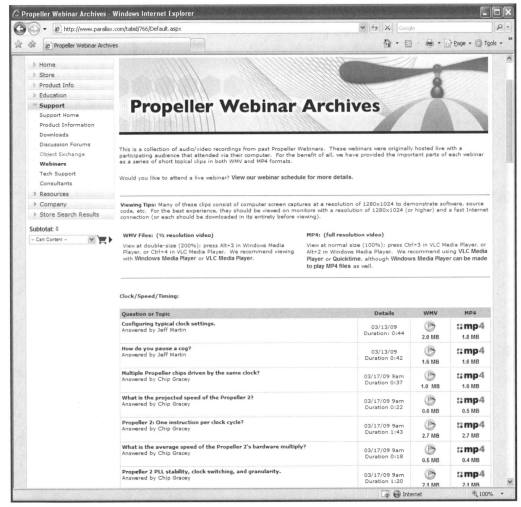

Figure 2-28   Propeller Webinars (www.parallax.com/go/webinar).

Summary

Together we built our own object that evolved into a building block other developers could use. Along the way, we learned about methods, I/O, loops, decisions, timing, and single-core versus multicore processing. The next chapter will build up your object debugging skills, and then you'll have the foundation you need to dive into the many fascinating projects that fill the remainder of this book.

Exercises

To further your learning experience, we recommend trying the following exercises on your own:

1 Explore the Propeller Library and Demos included with the Propeller Tool. Compile and run as many as possible, and take the time to study the code.
2 Modify the examples we built to perform tasks of your choice. Can you make an application that performs three or four different tasks at once?
3 Log on to the Propeller Forum and read through, and even post to, some of the recent threads.

3

DEBUGGING CODE FOR
MULTIPLE CORES

Andy Lindsay

The chapter title, "Debugging Code for Multiple Cores," might sound a little like a computer science or engineering course topic that students struggle through and then wax poetic about how hard it was later in life. If that's the kind of challenge you were looking for, sorry, you won't find it here. As with application development, debugging multicore applications with the Propeller microcontroller is typically easier than debugging equivalent single-core, time-sliced implementations. The Propeller microcontroller's architecture, programming language, and object design conventions all work together to minimize the likelihood of coding errors (aka bugs). They also help keep any coding errors that do sneak in on the surface where they are easier to spot. In addition, there are a number of healthy coding habits that help prevent multiprocessor coding mistakes, as well as software packages, useful objects, and techniques you can use to reduce the time it takes to find and correct coding errors. These preventative measures, software packages, and bug finding and correcting techniques are the focus of this chapter as it introduces the following:

- Propeller features that simplify debugging
- Object design guidelines for preventing multiprocessing bugs
- Common multiprocessor coding mistakes
- Survey of Propeller debugging tools
- Debugging tools applied to a multiprocessing problem

The most common root cause of coding errors that do make their way into Propeller multicore applications is our natural tendency to forget that segments of the application code get executed in parallel. Thinking in multiprocessing terms seems like it should be a simple thing to remember, but especially at first, it's all too easy to forget. Once forgotten, obvious bugs can start to seem subtle and difficult to find, at least until the results of some test provides the necessary reminder. So, start your parallel processing-think memory exercises as you go through this chapter by keeping in mind that segments of multicore Propeller application code get executed in parallel by more than one cog.

Resources: Demo code and other resources for this chapter are available for free download from ftp.propeller-chip.com/PCMProp/Chapter_03.

Propeller Features That Simplify Debugging

The Propeller microcontroller's architecture, programming language, and design conventions that object authors adhere to don't just simplify application development—they help prevent a myriad of coding errors that could otherwise plague application developers.

ARCHITECTURE THAT PREVENTS BUGS

While designing the Propeller microcontroller, one of Chip Gracey's first and foremost goals was to distill its design so that the rules for accomplishing any task would be simple and straightforward. Two examples where this design approach prevents a variety of bugs are in I/O pin and memory access.

There are some multicore microcontroller designs where each processor has direct access to only its own bank of I/O pins, so extra communication steps are necessary for a core to interact with an I/O pin outside of its bank. In contrast, with the Propeller chip, any cog or group of cogs can influence any Propeller microcontroller I/O pin or group of I/O pins at any time. Each cog has its own output and direction registers for all I/O pins, and to make I/O pin states follow a simple set of rules, all the cogs' I/O pin direction and output settings pass through the top set of OR gates shown in Fig. 3-1. With this arrangement, if one cog sets an I/O pin register bit to output, a different cog can still leave its I/O pin direction register bit set to input, and even monitor the state of the I/O pin to find out what signals the other cog is transmitting.

The rule for multiple cogs controlling outputs is also simple. If one or more cogs control the same I/O pin output, a cog sending a high (binary 1 in the output register bit) will win and the I/O pin will be set high, even if other cogs have binary 0s in the same output register bits. This scheme makes it possible for one cog to modulate higher-speed on/off carrier signals that another cog is transmitting. Whenever the modulating cog sends a high signal, the carrier signal's low signals don't make it through. When the modulating cog sends a low signal, the carrier's high and low signals can make it through. Figure 3-2 shows an example where the upper trace is the signal from two cogs

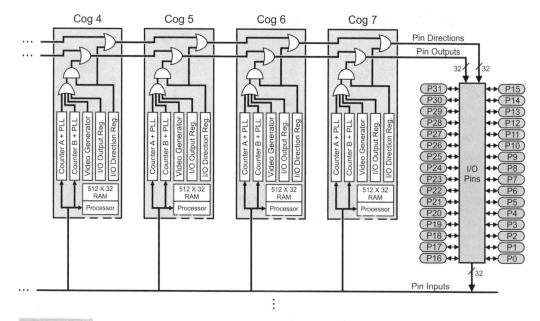

Figure 3-1   Block diagram excerpt—shared I/O pin access.

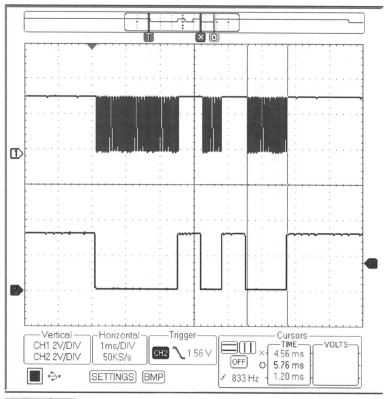

Figure 3-2   Signal modulation with two cogs sharing I/O pin access.

sharing an I/O pin and the lower trace is a copy of the modulator cog's signal, which stops the upper trace's carrier signal whenever it's high.

Low-level main memory access collisions are another example of a potential debugging problem that could have been left to the programmer to prevent. *Memory access collisions* can occur if two processors attempt to access the same 32-bit long in memory at the same instant. Low-level memory access collisions are prevented in the Propeller microcontroller by giving each cog Main RAM access in the round-robin fashion shown in Fig. 3-3. This completely eliminates the possibility of low-level memory collisions because each cog takes its turn accessing memory elements. This also frees the application from any concerns about taking turns at accessing individual memory elements and immensely simplifies the code. In addition to 32 K bytes (8 K longs) of Main RAM, each cog has its own 2 K bytes (512 longs) of Cog RAM. Each cog has exclusive access to its own Cog RAM, without taking turns, which can be useful for speed-optimized processes.

Even though the Propeller chip's architecture has eliminated the possibility of low-level main memory access collisions, there can still be timing issues with cogs reading from or writing to groups of memory elements during the same time period. In that case, one cog might get half old and half new values, as the other cog is busy updating the same group of memory elements. So, the Propeller microcontroller's main memory has eight semaphore bits, called *locks,* which simplify the task of making sure that one cog doesn't try to read a group of variables at the same time another cog is updating them.

Communication between cogs is another design puzzle that has a variety of solutions, some of which could have made coding complex and bug-prone. With the Propeller microcontroller, cogs can exchange information through the Propeller chip's Main RAM. Again, since each cog gets sequential access to individual memory elements, bugs as a result of low-level memory contention are not possible, and lock bits built into

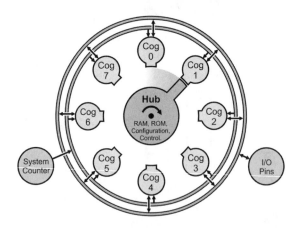

Hub and Cog Interaction

Figure 3-3   **Block diagram excerpt—round robin main memory access.**

main memory for updating groups of memory elements make this a simple, effective, and bug free means for cogs to exchange information.

One last but not-so-obvious characteristic of the Propeller chip's design that helps reduce bugs is its hardware symmetry. As mentioned in Chap. 1, all cogs are physically identical, rather than being specialized for certain functions. This allows any cog to be as useful as any other cog for any task, so it is not necessary to assign code to a specific chunk of hardware (unless desired, as the Spin language certainly provides for this). This allows the next available cog to handle whatever task is presented, and there is no need to determine if unexpected behavior is a result of code waiting for, or running in, a certain type of core.

LANGUAGE AND PROGRAMMING CONVENTIONS THAT HELP PREVENT BUGS

The Spin and Assembly languages incorporated into the free Propeller Tool software have a number of features that help prevent multiprocessing bugs. For example, the lock bits have a set of commands that help manage main memory. Likewise, there is a set of cog commands to give the developer more control over the hardware, if desired. For example, a cog can be selected and launched by number with coginit so the developer can know exactly which process is happening where. Code can also be launched into the next available cog with cognew, report where it landed with cogid, and a cog's ID used with cogstop will shut down that cog. An arrangement similar to this is incorporated into objects available from the Propeller Object Exchange because it allows building block objects to launch code into the next available cog without interfering with any cogs that the application might already be using.

The object-based nature of the language was introduced in the previous chapter, and one of the most important features of building block objects from the Propeller Library and Propeller Object Exchange is that their authors (usually) follow conventions established by Parallax to make their interfaces simple and trouble-free. Building block objects that launch code into other cogs take care of most of the multiprocessing grunt work. Good objects also contain methods that simplify configuring the process executed by the other cog and exchanging information with the top-level application object.

By convention, a building block object that launches a process into another cog has a Start method that receives configuration information and contains code that launches the new cog. It also has a Stop method for shutting the process down and freeing the cog. In many cases, these objects also have methods that provide a data exchange interface. In other cases, the parent object passes information about its variable addresses so that the object can directly write to and/or read from the parent object's variables. Regardless of whether a method interface, a memory sharing interface, or some combination of the two gets used, the building block object that manages the process keeps the interface simple.

For example, let's consider the Propeller Library's Keyboard object. After its Start method gets called, its assembly language code takes care of communication with the keyboard and buffers any key presses. The application object can then call the Keyboard object's Key method to get the latest buffered key press (or find out that there's nothing in the buffer) whenever it has time. Another example is the Sigma-Delta ADC object, which is designed to provide digitized analog voltage measurements and will be demonstrated in Chap. 4. This object's Start method is designed to receive a variable's memory address from the application object. After the Sigma-Delta ADC object's Start method launches its analog-to-digital conversion code into a new cog, that cog always copies the most recently measured voltage value into the parent object's variable that was set aside for receiving the measurements. In either case, the end result is a simple and easy-to-use interface, which, in turn, tends to be bug-free because all the application code has to concern itself with using the information it has received.

Object Design Guidelines

If you plan on designing a building block object that launches a process into another cog, either for an application or for the Propeller Object Exchange, the Start and Stop method conventions introduced in the previous chapter are crucial, and designing a bug-free interface that communicates with other objects through shared memory is equally crucial.

Figure 3-4 shows an example of one of the ways a building block object that has launched a process (either a method or some assembly code) into another cog can provide an information bridge between the two cogs. This figure shows a call to one of its methods after the object has launched a cog as a result of a call the application object made to the building block object's Start method. To exchange information

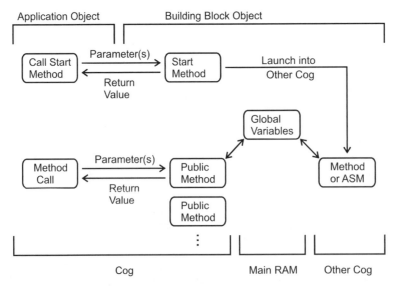

Figure 3-4   **Cog information exchanges with object.method calls.**

with the other cog, the application object calls one of the building block object's public methods. Code in those public methods is executed by the same cog that is executing the application object's method call. Those public methods can exchange information with the method(s) or assembly language (ASM) code executed by the other cog through the building block object's global variables. Propeller Library examples of objects that use this approach include the Parallax Serial Terminal, Keyboard, and Mouse objects.

Another common design for building block objects that manage processes in other cogs involves a `Start` method with one or more parameters that receive one or more memory addresses from the application object. The process that the building block object launches into another cog then uses those memory addresses to update and/or monitor one or more variables in the application object. Instead of `object.Start(value1, value2,...)`, the application object would use `object.Start(@varaible1, @variable2,...)` to pass addresses of variables that the application object expects the building block object to work with. Once the building block object knows these addresses, it passes them to a Spin or assembly language coded process that it launches into another cog. That process can use the memory addresses to read directly from and/or write directly to the application object's variables, as shown in Fig. 3-5. Code in the application object can then exchange information with the other cog by simply writing to or reading from those variables.

There are also other, less common variations and combinations of the two cog information exchange arrangements just discussed. In some cases, they are used to support a particular set of tasks the building block object is expected to perform. Regardless of the design, the building block object's documentation should be clear about how it exchanges information with the application object, and the documentation should also be clear about what its public methods do, the parameters they expect, and the values that return. The object should also be thoroughly tested to verify that it functions as advertised.

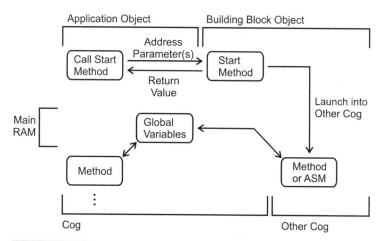

Figure 3-5   **Cog information exchanges through mutually agreed-upon memory addresses.**

Common Multiprocessor Coding Mistakes

Thanks to the Propeller microcontroller's architecture, programming languages, and object design conventions, the list of common multiprocessor-related coding mistakes is small. This section explains each potential coding mistake, its symptoms, and how to correct it.

- Missing call to a building block object's `Start` method
- Missing I/O pin assignments in a new cog
- Incorrect timing interval
- Code that missed the `waitcnt` boat
- Only one cog is waiting while the other has moved on
- Memory collisions
- Wrong address passed to a method
- Forgotten literal # indicator for assembly language code
- Method in a new cog outgrows stack space

STUCK ON A BUG?

If you get stuck on a bug with test code using multiple processors, go through this list because chances are, the bug will be one of these items. The community at http://forums.parallax.com can also help with finding bugs and correcting misunderstandings about how a given piece of code works.

MISSING CALL TO A BUILDING BLOCK OBJECT'S START METHOD

A building block object that executes code in more than one cog typically has a `Start` method, which has to be called to launch the code that does its job into another cog. The Test Float32.spin application object utilizes three building block objects to calculate and display the floating point tangents of integer-degree values entered into the Parallax Serial Terminal's transmit windowpane. Two of those building block objects have `Start` methods: Parallax Serial Terminal and Float32. The Parallax Serial Terminal object has assembly code that gets launched into another cog that maintains full duplex serial communication with the PC, and the Float32 object also has assembly code that gets launched into another cog to optimize the speed of its floating point calculations. In addition to `Start` and `Stop` methods, both of these objects have methods that take care of exchanging information with code running in the other cogs. These methods all use the scheme shown in Fig. 3-4, accepting parameters and passing them to the other cogs, or returning results they got from other cogs, or both. The third building block object is

FloatString. This object is a collection of useful methods for converting floating point values to their string representations, but it does not use any other cogs to make these conversions, so it does not have a `Start` method.

Each of these objects has documentation comments that explain how to use their methods.

✓ Load Test Float32.spin into the Propeller Tool software.
✓ Use <u>Run</u> → <u>Compile Current</u> → <u>View Info F8</u> to compile the application.
✓ Use the upper-left Object View pane to open the Parallax Serial Terminal, Float32, and FloatString objects.
✓ Click the <u>Documentation</u> radio button to display each of the building block objects in documentation view.

The Parallax Serial Terminal and Float32 objects both have `Start` and `Stop` methods listed first in their Object Interface sections, which is visible in the Propeller Tool software's documentation view. This is your clue that each of these objects launches code into another cog and that your application code will have to call each of their `Start` methods before calling any of their other methods. If an object's `Start` method is optional, its documentation comments should state that, and if it doesn't, a call to the object's `Start` method is probably required for the object to do its job(s).

```
" Test Float32.spin

CON

    _clkmode = xtal1 + pll16x          ' Crystal and PLL settings
    _xinfreq = 5_000_000               ' 5 MHz crystal x 16 = 80 MHz

OBJ

    pst : "Parallax Serial Terminal"   ' Serial communication object
    fp  : "Float32"                    ' Floating point object
    fs  : "FloatString"                ' Floating point string object

PUB Go | degrees, radians, cosine

    pst.Start(115200)                  ' Start Parallax Serial Terminal cog
    fp.Start                           ' Don't forget this!!!

    pst.Str(String("Calculate tangent...", pst#NL))
    repeat
      pst.Str(String(pst#NL,"Enter degrees: "))
      degrees := pst.DecIn             ' Get degrees
      ' Convert to floating point radians.
      radians := fp.Radians(fp.FFloat(degrees))
      ' Calculate tangent.
```

```
cosine  := fp.Tan(radians)
' Display result.
pst.Str(String("Tangent = "))
pst.Str(fs.FloatToString(cosine))
```

Test Float32.spin works correctly since both of the Parallax Serial Terminal and Float32 objects' Start methods were called, and the interactive testing for floating point tangent calculations is shown in Fig. 3-6. To enter numbers into the Parallax Serial Terminal, click the *Transmit windowpane* before typing. The Transmit windowpane is just above the Receive windowpane, and in Fig. 3-6, it has the number 30 entered into it, just above the first line in the Receive windowpane that reads "Calculate tangent..."

If this is your first time running the Parallax Serial Terminal, follow these instructions:

✓ Make sure your Propeller chip's power supply and programming cable are connected.
✓ In the Propeller Tool software, make sure Test Float32 is the active tab. In other words, the Test Float32.spin code should be displayed in the Propeller Tool software's Editor pane.
✓ In the Propeller Tool software, click <u>Run</u> → <u>Identify Hardware... F7</u> and make a note of which COM port the Propeller is connected to. Then, in the Parallax Serial

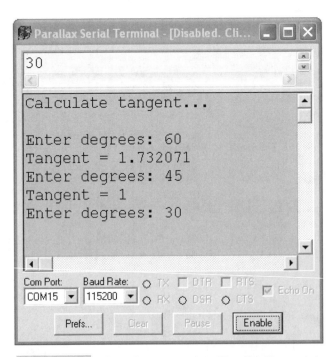

Figure 3-6   Calculate tangents with a floating point object.

Terminal software, set the <u>Com Port</u> drop-down menu to the COM port number you got from the Propeller Tool software.

✓ Set the <u>Baud Rate</u> drop-down to <u>115200</u> so that it matches the baud rate passed to the Parallax Serial Terminal object's `Start` method with the Test Float32.spin object's `pst.Start(115200)` method call.

When you load a program that exchanges messages with the Parallax Serial Terminal into the Propeller chip with the Propeller Tool, make sure to click the Parallax Serial Terminal's <u>Enable</u> button as soon as the Propeller Tool software's Communication window reports Loading… If you wait too long to click the Parallax Serial Terminal's <u>Enable</u> button, it might have already missed the Propeller chip's user prompt messages. If that's the case and you used the Propeller Tool's <u>Run</u> → <u>Compile Current</u> → <u>Load RAM F10</u> to load the program, you will have to reload it. If you instead used <u>Run</u> → <u>Compile Current</u> → <u>Load EEPROM F11</u>, you can restart the program by either pressing and releasing the Propeller board's Reset button or double-clicking the Parallax Serial Terminal's <u>DTR</u> check box.

✓ Make sure there is a check mark in the Parallax Serial Terminal's <u>Echo On</u> check box. This displays text you type in the Parallax Serial Terminal's Transmit windowpane in its Receive windowpane.

✓ In the Propeller Tool software, load the program into the Propeller chip, either with <u>Run</u> → <u>Compile Current</u> → <u>Load RAM F10</u> or with <u>Run</u> → <u>Compile Current</u> → <u>Load EEPROM F11</u>.

✓ As soon as the Propeller Tool software's Communication window reports Loading…, click the Parallax Serial Terminal's <u>Enable</u> button. Don't wait, or you might miss the user prompts.

The Parallax Serial Terminal's transmit windowpane is shown in Fig. 3-6 with the number 30 typed into it. It's just above the "Calculate tangent…" title in the Receive windowpane's display.

✓ Type integer-degree angles into the Parallax Serial Terminal's transmit windowpane, pressing the Enter key after each one. Try 30 first and verify that your results match Fig. 3-6.

The Propeller will reply by sending the string representation of the floating point result to the Parallax Serial Terminal.

Creating the "forgot to call a building block object's `Start` method" bug is easy. Just comment one or both of the `Start` method calls in Test Float32.spin by placing an apostrophe to the left. Then, load the modified application into the Propeller chip. Since the application depends on the processes the two objects run in other cogs, either one will create rather drastic bug symptoms. Commenting `pst.Start(115200)` will prevent any serial messages from being exchanged between the Propeller chip and Parallax Serial Terminal. Commenting `fp.Start` will cause all the floating point results to be 0.

AUTHOR'S NOTE

Forgetting to include `Start` method calls in application objects is my most common coding error. In fact, when I demonstrated on the fly floating point examples during Propeller seminars and trainings, I forgot the `fp.start` method call on several occasions. Each time, it took a couple of minutes to figure out, and it's amazing how time seems to slow to a standstill while trying to find and fix a bug in front of a large group.

Even after all that, I still almost forgot to include "Missing call to a building block object's `Start` method" in my list of most common multiprocessor coding mistakes.

MISSING I/O ASSIGNMENTS IN NEW COG

As mentioned in the "Architecture that Prevents Bugs" section, each cog has its own I/O direction and output registers. Although this solves a number of potential problems, there is still one coding error people tend to make: configuring the I/O from the wrong cog. The typical form of this error is code that makes I/O pin configurations in one cog, and then launches a new cog to control the I/O pins. If the new cog isn't also configured to work with those I/O pins, it won't be able to control their output states. Furthermore, if the cog that launched the new cog never needed to control the I/O pins, there isn't any reason for its code to make any I/O pin configurations at all. In that case, the code that the new cog executes is the only code that needs to configure its I/O registers. For example, IO Declaration Bug.spin launches a process that is supposed to make a light blink, but the light emitting diode (LED) in Fig. 3-7 won't blink because the I/O pin was set to output by the cog executing the `Go` method. The cog executing the `Blinker` method never sets its I/O pin to output, so it has no control over the I/O pin's output state.

LIGHT AND PUSHBUTTON CIRCUITS

The light and pushbutton circuits in this section are explained in more detail in "Propeller Education Kit Labs: Fundamentals"—4: I/O and Timing Basics Lab. A more basic introduction to these circuits and examples of building them from schematics is also included in early "What's a Microcontroller?" chapters. Both are free downloads from www.parallax.com.

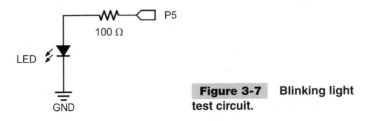

Figure 3-7 **Blinking light test circuit.**

```
' IO Declaration Bug.spin

VAR

    long stack[10]                      ' Array cog executing Blinker

PUB Go

    dira[5] := 1                        ' **BUG P5→output in the wrong cog
    cognew(Blinker, @stack)             ' Launch a new cog to control P5
    repeat                              ' Optionally keep cog running

PUB Blinker

    repeat                              ' Infinite loop
      !outa[5]                          ' Invert P5 output register bit
        waitcnt(clkfreq/4 + cnt)        ' Delay for 1/4 s
```

This problem can be corrected by moving the I/O pin direction setting to the method that is launched into the new cog.

```
' IO Declaration Bug (Fixed).spin

VAR

    long stack[10]                      ' Array cog executing Blinker

PUB Go

    cognew(Blinker, @stack)             ' Launch a new cog to control P5
    repeat                              ' Optionally keep cog running

PUB Blinker

    dira[5] := 1                        ' P5→output in the right cog
    repeat                              ' Infinite loop
      !outa[5]                          ' Invert P5 output register bit
        waitcnt(clkfreq/4 + cnt)        ' Delay for 1/4 s
```

As an aside, adding `pin` and `delay` parameters to the `Blinker` method gives the `Go` method some flexibility for setting I/O pin and delay.

```
' Other Cog Blinks Light.spin

VAR

    long stack[10]                      ' Array cog executing Blinker
```

```
PUB Go

  cognew(Blinker(5, clkfreq/4), @stack)       ' Launch new cog
  repeat                                      ' Optionally keep this cog running

PUB Blinker(pin, delay)                       " Blink light method

  dira[pin] := 1                              ' Set I/O pin to output
  repeat                                      ' Repeat loop
    waitcnt(delay + cnt)                      ' Delay 1/4 s
    !outa[pin]                                ' Invert I/O pin output state
```

TIMING INTERVAL ERRORS

Although precise timing was not required for the last three blinking light code examples, there are many other situations where a precise time interval is crucial. Examples include the Parallax Serial Terminal's signaling for serial communication with the PC and establishing a sample interval for taking sensor and signal measurements, which will be utilized in the next chapter's Sigma-Delta A/D conversion examples. The blinking light example programs have two potential sources of "bugs" that contribute to an inexact timing interval. First, they do not use a precise external clock, and second, code in their waitcnt commands does not compensate for the time it takes other commands in the repeat loops to execute. Both of these bugs are easy to spot and easy to fix.

Wrong Clock Frequency Settings   When a Spin top file does not specify the clock settings, the Propeller uses its internal RCFAST setting by default. Although this oscillator is nominally 12 MHz, the actual frequency can vary anywhere from 8 to 20 MHz. In contrast, crystal oscillators are much more precise, with variations measured in parts per million, or ppm, which indicates how many cycles per million the oscillator might vary. A typical value for the 5-MHz crystal that comes with the many of the Propeller kits and boards is +/- 30 ppm. Since this oscillator is only off by, at most, 30 signal cycles per million, the Propeller will be off by, at most, 30 clock ticks per million. At 80 MHz, that's, at most, 2400 clock ticks off, which is quite a bit better than +8 million or –4 million!

Applications that require accurate timing tend not to function correctly when the clock settings that are supposed to specify the external crystal included on most Propeller kits and demo boards are omitted. One example of symptoms this coding error can cause is garbled messages displayed by the Parallax Serial Terminal. Without the more precise timing provided by the external oscillator, the Propeller can end up communicating at a baud rate that's slightly different from the one specified in the code. Figure 3-8 shows the results with the correct system clock settings (above) and without them (below).

The incorrect clock settings bug is easy to create, observe, and fix. Try this:

✓ Make a copy of the Parallax Serial Terminal QuickStart object in the Propeller Library Examples folder, and rename it Test Missing Clock Settings.spin.

✓ Open Test Missing Clock Settings.spin with the Propeller Tool and change its pst. Start(115200) method call to pst.Start(9600).

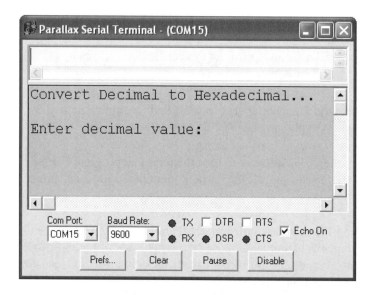

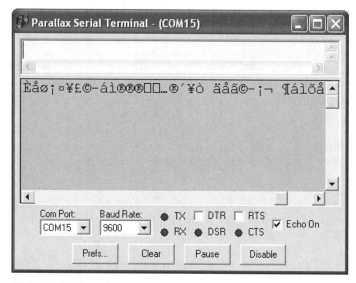

Figure 3-8   Serial communication with and without the external crystal settings.

This will change the baud rate the Parallax Serial Terminal uses from 115200 bps to 9600 bps.

✓ Change the Parallax Serial Terminal's Baud Rate drop-down menu to 9600 bps.
✓ Load the program into the Propeller chip, and remember to click the Parallax Serial Terminal's <u>Enable</u> button as soon as the Propeller Tool's Communication window displays the Loading… message.

✓ Verify that the Parallax Serial Terminal displays the "Convert decimal to hexadecimal…" message.

✓ Comment the _clkmode and _xinfreq system clock directives by placing an apostrophe to the left of each one.

This will make the Propeller Tool software's Spin compiler ignore these directives and instead use the default internal RCFAST clock settings.

✓ Use the Propeller Tool software to load the modified program (with a clock configuration bug that throws off the serial communication signal timing) into the Propeller.

While running the modified program, the Parallax Serial Terminal will probably display garbled messages similar to those on the right side of Fig. 3-8. To fix the bug, uncomment the _clkmode and _xinfreq system clock directives and load the corrected program back into the Propeller chip.

```
' Test Missing Clock Settings.spin

CON
  ' *** Comment these two declarations to create the timing bug.
  _clkmode = xtal1 + pll16x       ' Crystal and PLL settings
  _xinfreq = 5_000_000            ' 5 MHz crystal x 16 = 80 MHz

OBJ

  pst : "Parallax Serial Terminal"  ' Serial communication object

PUB go | value

  pst.Start(9600)                   ' Start Parallax Serial Terminal cog

'-------- Replace the code below with your test code ---------------

  pst.Str(String("Convert Decimal to Hexadecimal...")) ' Heading
  repeat                          ' Main loop
    pst.Chars(pst#NL, 2)          ' Carriage returns
    ' Prompt user to enter value
    pst.Str(String("Enter decimal value: "))
    value := pst.DecIn            ' Get value
    ' Announce output
    pst.Str(String(pst#NL,"Your value in hexadecimal is: $"))
    ' Display hexadecimal value
    pst.Hex(value, 8)
```

Incorrect Loop Interval Code   Incorrect Loop Interval.spin has the correct clock settings, but its timing is not yet precise because there is still a bug in the way the `repeat` loop is written. The `repeat` loop currently delays for one-quarter of a second, inverts the state of the I/O pin, and repeats the loop again. The loop is going slower than 4 Hz because the `repeat` and `!outa[5]` commands both take time to execute and there's nothing in the loop that compensates for it. Furthermore, adding more commands to the loop would cause it to take even longer to repeat.

```
' Incorrect Loop Interval.spin

CON

  _clkmode = xtal1 + pll16x          ' Crystal and PLL settings
  _xinfreq = 5_000_000               ' 5 MHz crystal x 16 = 80 MHz

VAR

  long stack[10]                     ' Stack for cog executing Blinker

PUB Go

  cognew(Blinker(5, clkfreq/4), @stack)    ' Blinker method to new cog

PUB Blinker(pin, delay)              " Blink light method

  dira[pin] := 1                     ' Set I/O pin to output
  repeat                             ' Repeat loop
    waitcnt(delay + cnt)             ' Delay 1/4 s
    !outa[pin]                       ' Invert I/O pin output state
```

Correct Loop Interval.spin shows a simple modification that keeps loop timing precise. The command `t := cnt` just before the `Blinker` method's `repeat` loop copies the current number of clock ticks stored in the `cnt` register to the variable `t`. Then, every time through the loop, the command `waitcnt(t += delay)` adds the number of clock ticks in `delay` to `t`, and then waits until the `cnt` register catches up. These synchronized delays were first introduced in Chap. 2.

```
' Correct Loop Interval.spin

CON

  _clkmode = xtal1 + pll16x          ' Crystal and PLL settings
  _xinfreq = 5_000_000               ' 5 MHz crystal x 16 = 80 MHz

VAR

  long stack[10]                     ' Stack for cog executing Blinker
```

```
PUB Go

    cognew(Blinker(5, clkfreq/4), @stack)      ' Blinker method → new cog

PUB Blinker(pin, delay) | t            " Blink light method

    t := cnt                           ' Get current cnt register value

    dira[pin] := 1                     ' Set I/O pin to output
    repeat                             ' Repeat loop
      waitcnt(t+=delay)                ' Synchronized delay
      !outa[pin]                       ' Invert I/O pin output state
```

The clkfreq register stores the number of clock ticks in 1 s, and, in the case of Correct Loop Interval.spin, that's 80,000,000 since the system clock is configured to run at 80 MHz. When the cognew command launches the Blinker method into a new cog, the Blinker method call passes 5 to the Blinker method's pin parameter and clkfreq/4 to its delay parameter. In this case, the delay parameter receives the result of clkfreq/4, which is 80,000,000/4 = 20,000,000. Next, the command t := cnt copies the current clock tick count stored by the cnt register into the Blinker method's local variable t.

Let's assume that when the code gets to t := cnt that the cnt register holds the value 600,000,025. After the I/O pin assignment, the code enters the repeat loop, and then waitcnt(t+=delay) waits for the cnt register, which increments with every clock tick, to accumulate to 620,000,025. The value that the waitcnt command waits for is t+=delay, which is equivalent to t := t + delay. Since t stores 600,000,025 and delay stores 20,000,000, the result of t+=delay stored in t is 620,000,025. Then the waitcnt command waits for the cnt register to accumulate to that value before allowing the cog to continue executing code. Next time through the loop, t+=delay adds another 20,000,000 to t, so the waitcnt commands waits for the cnt register to get to 640,000,025. The third time through the loop, the waitcnt command waits for 660,000,025, and the fourth time through it waits for 680,000,25, and so on. Each time through the loop, the waitcnt command waits for 20,000,000 ticks, which is the number of clock ticks in quarter-seconds. So long as the rest of the commands in the loop take less than one-quarter of a second to execute, it doesn't matter how long they take because the waitcnt command uses the t variable, which accumulates by the number of clock ticks in a quarter of a second each time through the loop.

Code that Missed the Waitcnt Boat Continuing from the previous example, now the cog executing the Blinker method is repeating the loop with a precise timing interval of every 20,000,000 clock ticks. If the commands in the loop take longer than 20,000,000 clock ticks to execute, the waitcnt command will just keep waiting until the cnt register eventually rolls over and gets back around to the target value. Since the cnt register is 32 bits, it rolls over every 2^{32} clock ticks. That's 4,294,967,296 clock ticks, which takes about 53.7 s when the Propeller's system clock is running at 80 MHz.

The symptoms of this bug seem drastic because the cog appears to stop responding, only to pick up where it left off almost 54 s later!

Delay Beyond Interval.spin has a bug that can easily miss the waitcnt boat. Let's say the pushbutton in Fig. 3-9 is supposed to control a second indicator light connected to P6. The nested repeat while (ina[21]==1) loop prevents the outer loop from repeating until the pushbutton connected to P21 is released. A brief tap on the pushbutton immediately after the LED changes state won't cause the cog to miss the waitcnt boat, but if you are still pressing the button one quarter of a second later, the P5 LED will stop flashing and the cog will appear to stop responding because its waitcnt command is waiting for a time that has already passed. Although the P5 LED will start blinking again within 54 s, which proves that the cog is still functioning, it's definitely a coding error that needs to be fixed.

```
' Delay Beyond Interval.spin

' This code has a bug that causes the cog to stop executing
' code for almost 54 seconds when the button connected to
' P21 is still pressed when the LED changes state.

CON

    _clkmode = xtal1 + pll16x        ' Crystal and PLL settings
    _xinfreq = 5_000_000             ' 5 MHz crystal x 16 = 80 MHz

VAR

    long stack[10]                   ' Stack for cog executing Blinker
```

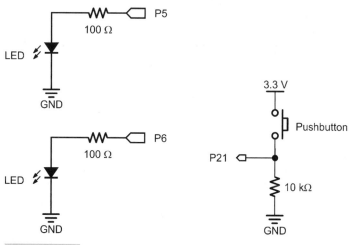

Figure 3-9   Test circuit.

```
PUB Go

    cognew(Blinker(5, clkfreq/4), @stack)        ' Blinker method → new cog

PUB Blinker(pin, delay) | t              '' Blink light method

    t := cnt                             ' Get current cnt register value
    dira[6]   := 1                       ' Set P6 to output
    dira[pin] := 1                       ' Set I/O pin to output
    repeat                               ' Repeat loop
      waitcnt(t+=delay)                  ' Synchronized delay
      repeat while (ina[21]==1)          ' *** BUG
        outa[6] := 1                     ' Turn on P6
      outa[6]:=0                         ' Turn P6 back off
      !outa[pin]                         ' Invert I/O pin output state
```

The application intended to have the light connected to P6 turn on when the
pushbutton connected to P21 was pressed. This is easy to do without disrupting
the cog's timing because the ina[21] register will store 1 when the pushbutton is
pressed or 0 when it's not pressed. To make the P6 light turn on, outa[6] needs to
store 1; to turn the light off, outa[6] should store 0. So the assignment outa[6] :=
ina[21] will turn the light on whenever the button is pressed and leave it off when
the button is released.

```
    repeat                               ' Repeat loop
      waitcnt(t+=delay)                  ' Synchronized delay
      outa[6] := ina[21]                 ' *** BUG -- fixed
      !outa[pin]                         ' Invert I/O pin output state
```

There are also situations where a loop should monitor a process until just before the
allotted time runs out. For example, let's say the application needs the P6 LED to turn
off immediately when the button is released, without a potential quarter-second delay.
The previous code could take up to a quarter of a second before turning the light back
off, but maybe only a couple of milliseconds delay is acceptable. Here is an example
that monitors the P21 pushbutton and mirrors the state with the P6 I/O pin until about
a millisecond before the loop has to repeat.

```
    repeat
      waitcnt(t+=delay)
      !outa[pin]
      ' Keep repeating button LED loop until time is almost up.
      repeat until cnt-t => (delay - clkfreq/1000)
        outa[6] := ina[21]               ' *** BUG -- fixed
```

The inner repeat loop continues its outa[6]:=ina[21] process until the current value
of the cnt register minus the value stored by the cnt register at the end of the previous

time interval is greater than or equal to some number of clock ticks that's slightly less than the outer loop's time interval. In other words,

current cnt – previous cnt >= value slightly less than time interval

The statement repeat until cnt-t => (delay - clkfreq/1000) follows this form. When the waitcnt command is done, the variable t stores the cnt register value at the end of the previous time interval, which was the instant the waitcnt command allowed the cog to continue to the next command. This value is subtracted from the current value in the cnt register and then compared to the number of clock ticks in delay - clkfreq/1000. That's the number of clock ticks in a quarter of a second minus the number of clock ticks in 1 ms, which is a value slightly less than the time interval stored in the delay parameter.

WAITCNT TRICKS WITH RC DECAY

The RC Decay measurements in the next chapter rely on this approach to prevent the cog that's taking the measurement from getting stuck when the RC Decay measurement takes longer than the allotted time.

COGS GET SENT TO DIFFERENT MEMORY ADDRESSES TO EXCHANGE INFORMATION

When an application needs two or more cogs to exchange information, the application code should be consistent about what memory address or addresses the cogs access. Cogs Not Sharing Info (Bug).spin is an example where the cog executing the Go method assumes the cog executing the Blinker method can see its delay local variable. Meanwhile, the cog executing the Blinker method takes its own delay parameter and uses it in a repeat loop. Even though each method is relying on a variable named delay, they are two separate local variables. A local variable can only be accessed by the method that declares it, so the Go and Blinker methods are accessing different instances of a variable with the same name, and each instance resides at a different memory address. So, regardless of the value entered into the Parallax Serial Terminal's Transmit windowpane, the Blinker method will continue to toggle the LED at the rate it received when the Go method launched it into a new cog.

✓ Remember to set the Parallax Serial Terminal software's Baud Rate back to 115200 before running this example program.

```
' Cogs Not Sharing Info (Bug).spin
' This program demonstrates a bug where different cogs fail
' to exchange information because they are both using local variables
' named delay.
```

```
CON

    _clkmode = xtal1 + pll16x          ' Crystal and PLL settings
    _xinfreq = 5_000_000               ' 5 MHz crystal x 16 = 80 MHz

VAR

    long stack[10]                     ' Stack for cog executing Blinker

OBJ

    pst    : "Parallax Serial Terminal"

PUB Go | delay                         ' Go method

    pst.Start(115200)                  ' Start Parallax Serial Terminal cog

    cognew(Blinker(5, clkfreq/4), @stack)    ' Blinker method → new cog

    repeat                             ' Main loop
      pst.Str(String("Enter delay ms: "))    ' Prompt user input
      delay := pst.DecIn * (clkfreq/1000)    ' User input → delay var

PUB Blinker(pin, delay) | t            ' Blinker method

    t := cnt                           ' Current cnt register → t variable

    dira[pin] := 1                     ' Set I/O pin to output
    repeat                             ' Repeat loop
      waitcnt(t+=delay)                ' Synchronized delay
      !outa[pin]                       ' Invert I/O pin output state
```

Since all methods in an object can access global variables, the solution to this problem is to modify the program so that both methods exchange information through a single global variable named delay. The code in Cogs Sharing Info (Bug fixed).spin corrects the problem by declaring a global variable named delay. The Go method updates the delay variable whenever a new value gets entered into the Parallax Serial Terminal, and the Blinker method uses the delay variable's value to control the rate of the loop that toggles the LED circuit on and off.

```
' Cogs Sharing Info (Bug fixed).spin

CON                                    ' Constant declarations

    _clkmode = xtal1 + pll16x          ' Crystal and PLL settings
    _xinfreq = 5_000_000               ' 5 MHz crystal x 16 = 80 MHz

VAR                                    ' Variable declarations
```

```
    long stack[10]              ' Stack for cog executing Blinker
    long delay                  ' Global var for cogs to share

  OBJ                           ' Object declarations

    pst    : "Parallax Serial Terminal"   ' Serial communication object

  PUB Go                        ' Go method

    pst.Start(115200)           ' Start Parallax Serial Terminal cog

    delay := clkfreq/4          ' Initialize delay variable

    cognew(Blinker(5), @stack)  ' Launch Blinker method into new cog

    repeat                           ' Repeat loop
      pst.Str(String("Enter delay ms: "))   ' Prompt user for input
      delay := pst.DecIn * (clkfreq/1000)    ' User input → delay var

  PUB Blinker(pin) | t          ' Blinker method

    t := cnt                    ' Current cnt register → t variable

    dira[pin] := 1              ' Set I/O pin to output
    repeat                      ' Repeat loop
      waitcnt(t+=delay)         ' Synchronized delay
      !outa[pin]                ' Invert I/O pin output state
```

In building block objects, this is the very same approach that makes it possible for a method call to pass a value to another cog running different code inside the object (see Fig. 3-4). When the building block object's method gets called, that method sets a global variable so that the method executed by the other cog (or assembly language code) can access it.

MEMORY COLLISIONS

The Propeller chip's architecture eliminates the possibility of memory collisions for any single element in main memory. However, cogs sharing a group of memory elements can still encounter problems if code is not in place that gives each cog exclusive access to those elements. Figure 3-10 shows a typical memory collision scenario for cogs that share more than one main memory element. In this scenario, one cog has updated two of three variables while another cog jumped in and fetched all three. The one cog may have been waiting for a sensor measurement for the third variable, so the other cog got two up-to-date values and one old one. As mentioned earlier, the Propeller has built-in lock bits that cogs can set when they are working on memory and clear when they are done. Provided the code in all the cogs that are working with a particular group of shared memory elements follows the same rules and waits for the lock bit to clear before setting

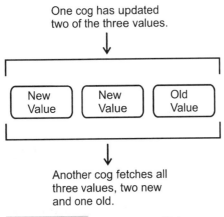

One cog has updated
two of the three values.

| New Value | New Value | Old Value |

Another cog fetches all
three values, two new
and one old.

Figure 3-10   Memory collision.

it and working with the memory, this bug will never happen. The upcoming "Debugging Tools Applied to a Multiprocessing Problem" section features an example of testing for memory collisions and applying locks to resolve them.

WRONG ADDRESS PASSED TO A METHOD

Let's say that the previous object, Cogs Sharing Info (Bug fixed).spin, has been separated into a top-level file and a building block object: Top File.spin and Building Block. spin. Top File.spin gives Building Block.spin the nickname blink in its OBJ block. The Building Block object has a Blinker method that will run in another cog and is expected to monitor the Top File object's global variable (as in Fig. 3-5). Following is the working example of such an application.

Blink.Start(5, @delay) in Top File.spin passes 5 to the pin parameter in Building Block's Start method and passes the *address* of its delay variable to the delayAddr parameter. Building Block's Start method then launches the Blinker method into a new cog and passes along both parameters for the Blinker method to work with. The most common bug here is omission of the @ sign to the left of the delay variable in Top File.spin's Start method call. If the @ sign were missing, the blink.start method call would pass the current value of the delay variable—in this case, clkfreq/4 from the line of code just above it, instead of the Main RAM *address* of the delay variable.

```
'' Top File.spin

CON                          ' Constant declarations

  _clkmode = xtal1 + pll16x   ' Crystal and PLL settings
  _xinfreq = 5_000_000        ' 5 MHz crystal x 16 = 80 MHz

VAR                          ' Variable declarations
```

```
    long delay                       ' Delay variable

OBJ                                  ' Object declarations

    pst   : "Parallax Serial Terminal"    ' Parallax Serial Terminal.spin
    blink : "Building Block"              ' Building Block.spin object

PUB Go                               ' Go method

    pst.Start(115200)                ' Start Parallax Serial Terminal cog

    delay := clkfreq/4               ' Initialize delay variable
    blink.Start(5, @delay)           ' Call start method, P5 blink 2 Hz

    repeat                               ' Repeat loop
      pst.Str(String("Enter delay ms: "))   ' User prompt
      delay := pst.DecIn * (clkfreq/1000)   ' User input → delay var
```

Building Block.spin has a `Start` method with a `cognew` command to launch its `Blinker` method, passing along the `pin` and `delayAddr` parameters. Each time through the `Blinker` method's `repeat` loop, it uses `delayAddr` to fetch the current value of Top File's `delay` variable. Since `delayAddr` stores the *address* of the application object's `delay` variable, the expression `Long[delayAddr]` returns the value stored at that location in main memory. So, instead of `waitcnt(t+=delay)`, the `Blinker` method's `repeat` loop uses `waitcnt(t+=long[delayAddr])`. If `@` had been omitted from `@delay` in Top File. spin's `Start` method call, `Blinker` would have received and attempted to use the value 20,000,000 for the address of the Top File object's `delay` variable and any update the Top File object makes to the value of its `delay` variable would go unnoticed, as `Blinker` executed its `repeat` loop in another cog.

```
'' Building Block.spin


VAR                                  ' Variable declarations

    long stack[20], cog              ' Stack array & cog variable

PUB Start(pin, delayAddr) : success
{{Start blinking process in a new cog.
  Parameters:
    pin         - I/O pin number to send the high/low signal
    delayAddr   - Address of the long variable that stores the delay
  Returns    : zero if failed to start or nonzero if it succeeded}}

    ' Launch process into new cog and return nonzero if successful
    success := (cog := cognew(blinker(pin, delayAddr), @stack) + 1)
```

```
PUB Stop                              "Stop blinking.

  if cog                              ' If cog is not zero
    cogstop(cog~ - 1)                 ' Stop cog & set cog to zero

PUB Blinker(pin, delayAddr) | t       " Method blinks light
" Updates rate based on value in parent objects variable at
" delayAddr.

  t := cnt                            ' Current cnt register → t variable
  dira[pin] := 1                      ' Set I/O pin to output
  repeat                              ' Repeat loop
    waitcnt(t+=long[delayAddr])       ' Delay from parent object's var
    !outa[pin]                        ' Invert I/O pin output state
```

FORGOTTEN LITERAL # INDICATOR FOR ASSEMBLY LANGUAGE CODE

In Propeller assembly language, a line of code can have a label, and then a line of code elsewhere in the program can use that label to make the program "jump to" that label and continue executing code from there. The most rudimentary example of this is to label one line of code with :loop, then later in the code, another line uses the jump-to address (jmp) instruction to send the program back to the :loop label. The command should read jmp #:loop, and the # literal indicator causes program execution to jump to the instruction at the Cog RAM address of :label. A common mistake is to simply type jmp :loop instead. If the literal indicator gets left out, the contents of the actual machine language instruction (a binary value that corresponds to add count, #1 in the following example) instead becomes the Cog RAM jump target address, and the program wanders off somewhere unexpected.

```
:loop               add      count, #1
                    wrlong   count, addr
                    jmp      #:loop
```

Assembly Language Example   As an aside, Asm Cog Example.spin has an assembly language routine that counts and rapidly updates a Spin variable named asmCt. When cognew launches an assembly routine into a cog, the first argument is the address of a label where the cog should start executing assembly language code. The address of the first line of assembly code that should be executed in Asm Cog Example.spin is @AsmCounter. The second parameter is typically the address of a Spin variable in Main RAM that will be used for information exchange between the assembly code and the Spin code. The value in the second parameter gets copied to the cog's par register, which is accessible to the assembly code. In this case, the second parameter is @asmCt. The Spin code expects the assembly code to update its asmCt variable with each repetition of its counting loop, and the cog executing the Go method displays the value once per second.

```
'' Asm Cog Example.spin

CON                                  ' Constant declarations

  _clkmode = xtal1 + pll16x          ' Crystal and PLL settings
  _xinfreq = 5_000_000               ' 5 MHz crystal x 16 = 80 MHz

VAR                                  ' Variable declarations

  long asmCt                         ' Spin variable

OBJ                                  ' Object declarations

  pst   : "Parallax Serial Terminal"    ' Parallax Serial Terminal.spin

PUB Go | t, dt                       ' Go method

  cognew(@AsmCounter, @AsmCt)        ' Launch ASM code into new cog
  pst.Start(115200)                  ' Start Parallax Serial Terminal cog

  t := cnt                           ' Current cnt register → t variable
  dt := clkfreq                      ' Set up 1 second delay interval

  repeat                             ' Repeat loop
    waitcnt(t += dt)                 ' Synchronized delay
    pst.Home                             ' To top-left of terminal
    pst.Str(String(pst#NL, "AsmCt = "))  ' Display text
    pst.Dec(asmCt)                       ' Display deciaml asmCt val

DAT                                      ' DAT block
                    org                  ' ASM address reference
'
' Entry
'
AsmCounter          mov     addr, par    ' Parameter reg → addr
                    mov     count, #0    ' count := 0
:loop               add     count, #1    ' count += 1
                    wrlong  count, addr  ' asmCt:=count (see note)
                    jmp     #:loop       ' Jump up 2 lines
'
' Uninitialized data
'
addr                res     1            ' Cog RAM registers
count               res     1

''NOTE: wrlong count, addr copies the cog's count register to asmCt,
''which is a variable at the address @asmCt in Main RAM.
```

Inside ASM Cog Example.spin The DAT block in Spin programs can contain data and/or assembly code. The org directive gives the assembly code a reference inside the cog for the addresses of labels such as AsmCounter (address 0), :loop (address 2), and so on. As mentioned previously, the cognew command passed the address of the Spin asmCt variable as the second argument in the cognew command, and that address gets copied to the cog's par register. The command mov addr, par copies the contents of the par register to a register named addr, which was declared in the uninitialized data section at the bottom of the DAT block. The instruction mov count, #0 copies the value 0 to a register named count, which was also declared in the uninitialized data section. Without the literal indicator, the command would instead copy the contents of address 0 to the register named count. Since the contents of address 0 in this example would be the machine language code for the mov addr, par assembly language command, the jmp command would send the program to the wrong address.

The command add count, #1 adds 1 to the register named count. Next, wrlong count, addr copies the value of count (a cog RAM register) to an address in Main RAM. Since addr stores a copy of par, which in turn stores the address of the Spin asmCt variable, the command wrlong count, addr copies the contents of the cog's register named count to the object's asmCt Spin variable in Main RAM. The command jmp #:loop causes the next instruction that gets executed to be add count, #1. Again, remember to use the literal indicator with the jmp command to make the program jump to a particular line of assembly code.

For more information on assembly language, see the Propeller Manual's "Assembly Language Reference."

METHOD IN A NEW COG OUTGROWS STACK SPACE

This topic was already discussed in the previous chapter. If you add code to a method that was launched into a new cog and you can't see anything wrong with the code that was added, check to make sure you allotted enough stack space.

Survey of Propeller Debugging Tools

This section focuses on four commonly used software tools for debugging Propeller applications:

- TV Terminal
- Parallax Serial Terminal
- ViewPort
- Propeller Assembly Language Debugger

Some of these tools are simple text displays for checking variable values and text messages at certain points in the program, while others are more full-featured debuggers.

In addition to these tools, numerous debuggers of varying functionality and cost are available. Some are objects; others are objects combined with software. The best place to find the latest Propeller application debugger links, along with other software and applications, is in the Propeller Chip forum's Getting Started and Key Thread Index at http://forums.parallax.com.

TV TERMINAL

The *Propeller Demo Board* in Fig. 3-11 was the first Parallax board available with a built-in Propeller chip, and it highlighted the Propeller chip's multiprocessing capabilities with microphone, PS/2 mouse, and PS/2 keyboard inputs along with stereo headphone, VGA, and RCA video outputs. The two PS/2 connectors are shown in the upper-right of Fig. 3-11, and the microphone is in the middle, to the right of the P3 and P4 labels. Along the bottom, the stereo headphone, RCA, and VGA connectors are shown from left to right. The Propeller Tool software that accompanied this board included example objects to demonstrate all these features, and early Propeller designers made use of the RCA video output to debug their code with a TV using the TV_Terminal object. More recent Propeller boards and kits also have built-in audio/video connectors; kits and boards that do not have them typically offer inexpensive adapter options so that just about any Parallax Propeller kit can run demonstration objects that utilize the TV_Terminal object.

A number of Propeller Object Exchange (Obex) objects utilize the TV_Terminal object and hardware to demonstrate the object's functionality. To evaluate these objects, either the Propeller chip can be connected to a TV or the demonstration object can be ported to use the Parallax Serial Terminal. The Propeller Library's Keyboard_Demo. spin makes use of both the Keyboard and TV_Terminal objects to demonstrate a PS/2 keyboard; this requires both circuits shown in Fig. 3-12.

Figure 3-11   Propeller Demo Board.

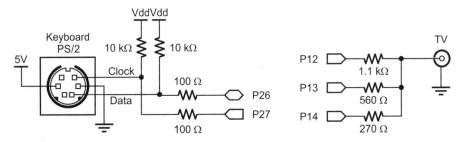

Figure 3-12   **TV terminal and keyboard circuits.**

The left side of Fig. 3-12 shows the Propeller Demo Board's 3-bit D/A converter (DAC) circuit that the application uses to generate baseband video with the TV_Terminal object. This circuit is also built into the Propeller Professional Development Board, and it can be constructed on the Propeller Education Kit breadboard and the Propeller Proto Board using the Parallax RCA–to–Breadboard adapter. The right side of Fig. 3-12 shows the PS/2 Keyboard circuit from the Propeller Demo Board. This circuit is also built into the Propeller Professional Development Board and can be constructed on the Propeller Education Kit breadboard with the Parallax PS/2 to Breadboard Adapter, or on the Propeller Proto Board with either the PS/2 to Breadboard Adapter or the Propeller Proto Board Accessory Kit.

The Keyboard_Demo.spin object's OBJ block declares the TV_Terminal and keyboard objects, giving them the nicknames term and kb. Both of these objects have Start methods, so both the TV terminal and keyboard processes get launched into other cogs. The application's repeat loop then takes the result of the keyboard object's getkey method and passes it to the TV_Terminal object's hex method, which displays the hexadecimal value of the keyboard key that was pressed on the TV.

```
' Keyboard_Demo.spin

CON

        _clkmode        = xtal1 + pll16x
        _xinfreq        = 5_000_000


OBJ

        term    : "tv_terminal"
        kb      : "keyboard"


PUB start | i

  'start the tv terminal
  term.start(12)
  term.str(string("Keyboard Demo...",13))
```

```
'start the keyboard
kb.start(26, 27)

'echo keystrokes in hex
repeat
  term.hex(kb.getkey,3)
  term.out(" ")
```

After a quick look at the TV_Terminal object's methods, we see it has an out method that will cause TV_Terminal to echo the actual keystroke on the display. Here is a modified repeat loop for displaying the keyboard characters instead of their hexadecimal values. Figure 3-13 shows the TV_Terminal output after some typing on the PS/2 keyboard.

```
'echo keystrokes
repeat
  term.out(kb.getkey)
```

PARALLAX SERIAL TERMINAL

The Parallax Serial Terminal introduced in Chap. 2 and featured in earlier examples in this chapter also provides a convenient first line of defense for catching program bugs and circuit errors. By adding a few lines of code at key points in a program, the Parallax Serial Terminal can display variable values and I/O pin states that bring simple mistakes to light with minimal hair-pulling. There is also an object named *PST Debug LITE*, available through the Propeller Chip forum's Propeller Education Kit Labs, Tools, and

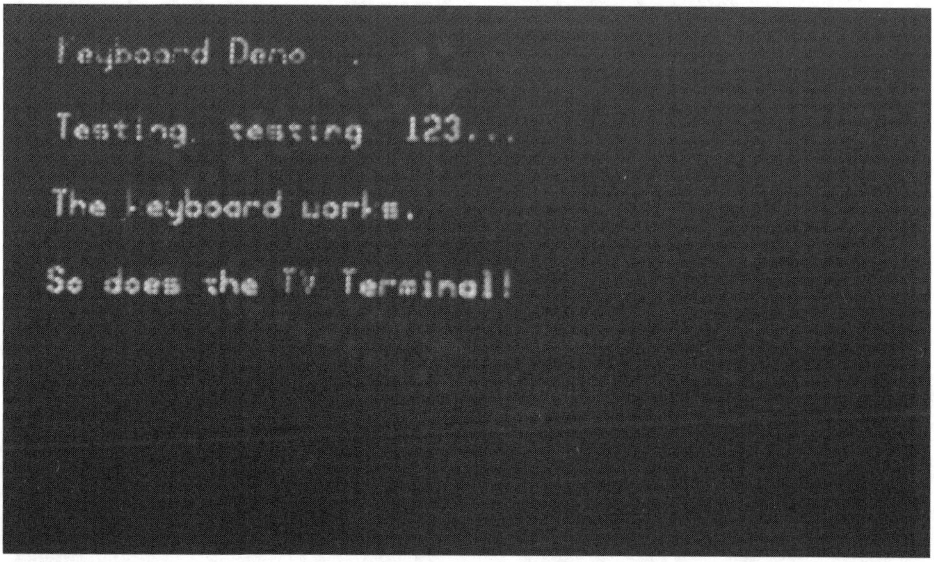

Figure 3-13   TV terminal output.

Applications thread at http://forums.parallax.com. This object extends the functionality of the Parallax Serial Terminal object to provide some additional debugging options. Figure 3-14 shows an example of a display you might see if you incorporate the PST Debug LITE object into your application.

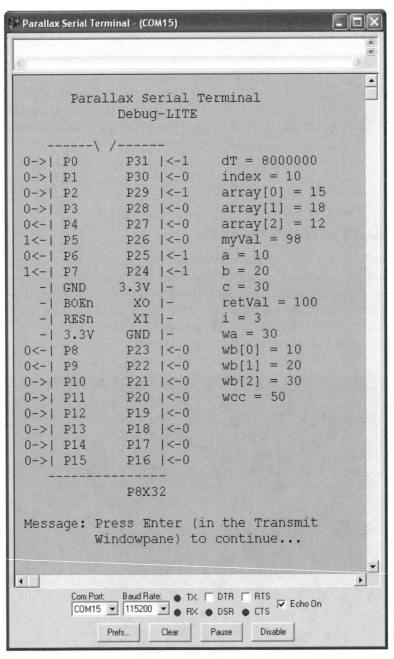

Figure 3-14   PST Debug LITE Display in the Parallax Serial Terminal.

PST Debug LITE allows you to add breakpoints and commands that display I/O pin states and their directions, as well as lists of variables and their values, with a few simple commands sprinkled into your code. This is one example in a set of free resources available for debugging; others are available through the Propeller Chip forum's *Getting Started and Key Thread Index* at http://forums.parallax.com.

VIEWPORT

For more elusive bugs, as well as for folks who are accustomed to IDE-style PC programming environments, Hanno Sander's *ViewPort* software provides a powerful set of tools for developing Propeller applications. ViewPort's Spin language Debugger offers familiar features like runtime stepping, breakpoints, and flyover variable display and update options, shown in Fig. 3-15*a*.

VIEWPORT—WHERE TO GET IT

The latest version of ViewPort is available for a 30-day free trial from www. parallax.com. Just type "ViewPort" in the Search field and click Go. ViewPort was developed by Hanno Sander, and extensive application information and tutorials are available from Hanno's www.mydancebot.com web site.

ViewPort also takes advantage of the Propeller chip's multicore design to provide a more advanced set of instrumentation and diagnostic tools with no extra hardware required. These tools are listed in Table 3-1 and shown in Fig. 3-15*b*, and they make it possible to analyze Propeller chip programs and application circuits with instruments that transform your Propeller chip into its own electronic workbench.

PASD—THE PROPELLER ASSEMBLY LANGUAGE DEBUGGER

Propeller Assembly Language can be especially useful for applications that require higher-speed processing, precise timing, or both. There are already many published objects with assembly code and Spin language interfaces for common tasks like communication, sensor monitoring, motor control, a variety of math algorithms, video display, and more. The *Propeller Assembly Source Code Debugger (PASD)* software shown in Fig. 3-16 is a free yet invaluable tool for anyone who needs to modify existing assembly code or write a Spin + assembly language object from scratch.

Later in this chapter we will see a brief PASD example that demonstrates setting breakpoints, stepping through assembly code, checking variable values in the Propeller chip's processor (cog) and shared (Main) memory areas, and viewing I/O pin states.

TABLE 3-1 VIEWPORT TOOLS		
DEBUGGING	**INSTRUMENTATION**	**MORE**
Text Terminal	Logic analyzer	Indicators and controls
IDE-style debugger	Oscilloscope	Fuzzy logic
	Spectrum analyzer XY-plot	Vision analysis

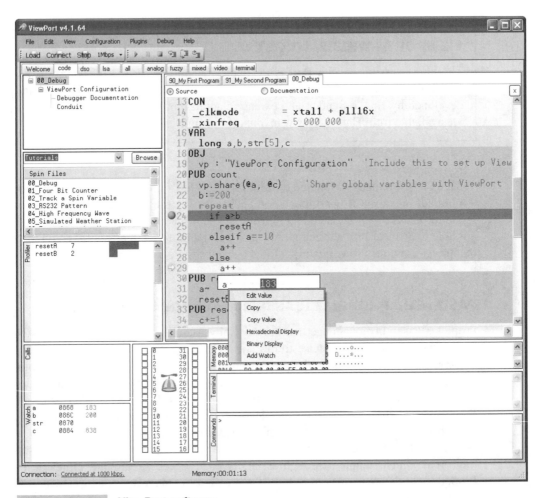

Figure 3-15a   ViewPort software.

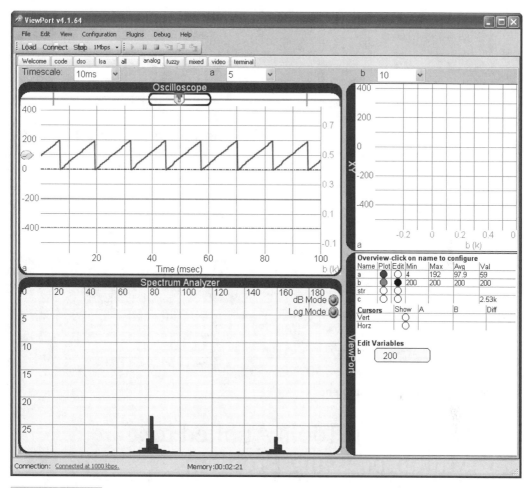

Figure 3-15b   ViewPort tools.

PASD—WHERE TO GET IT

The latest version of PASD is available as a freeware download from the PASD (debugger) Project page on Andy Schenk's Insonix company site:

www.insonix.ch/propeller/prop_pasd.html

This is a German language site with a link you can click for automatic Google translation to English. The accolades and thank-you messages on the Parallax Forum thread where Andy first unveiled this software contribution to the Propeller community are indeed well deserved.

http://forums.parallax.com/forums/default.aspx?f=25&m=214410

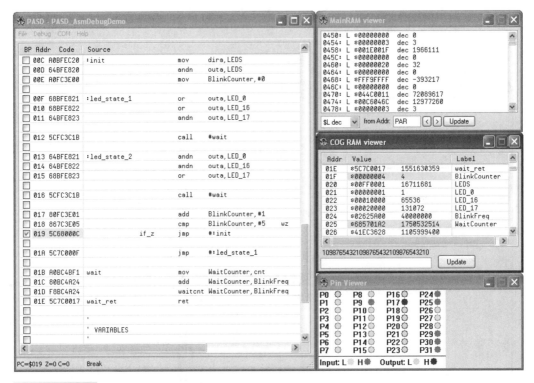

Figure 3-16 PASD environment.

Debugging Tools Applied to a Multiprocessing Problem

This section chronicles the development of an object that manages a timekeeping process in a separate cog and makes use of the software packages introduced in the previous section to test and debug at various stages. For the sake of keeping the example programs short, the timekeeping object will only track minutes, seconds, and milliseconds. An expanded version of the object that provides full calendar information can be obtained from http://obex.parallax.com.

DEVELOPMENT WITH THE PARALLAX SERIAL TERMINAL

The previous chapter and earlier examples in this chapter utilized the Parallax Serial Terminal object to make the Propeller chip communicate with the Parallax Serial Terminal software running on the PC. So the Parallax Serial Terminal object's and software's usefulness for quick tests at various stages of object development has already been demonstrated. This section will extend those examples and use the PST Debug LITE object, which has method calls that add convenience to setting breakpoints and displaying lists of variable values.

Step 1: Test in the Same Cog Let's say your Propeller application needs an object to make periodic timestamps of the minute, second, and millisecond when a series of events occurs, and a quick scan of http://obex.parallax.com didn't yield any candidates. Why not take a stab at writing a timestamp object from scratch? One way to get started would be to test and make sure the code that's counting time gets the correct sequence for minutes, seconds, and milliseconds.

Test Time Counting.spin uses the PST Debug LITE object's features to verify that the code counts time increments in the correct sequence. The object does not yet have millisecond pacing; it's just going as fast as it can in Spin. The object declares PST Debug LITE in the OBJ block and gives it the nickname debug. At the beginning of the TimerMs method, debug.Style(debug#COMMA_DELIMITED) configures PST Debug LITE to display comma-delimited lists. Since PST Debug LITE is really just the Parallax Serial Terminal object with some additional features, its Start method needs to be called in the same way the Parallax Serial Terminals object's Start method is called. Test Time Counting.spin uses debug.Start(115200). Near the end of the TimerMs method's repeat loop, debug.KeyCheck checks the PST Debug LITE object's serial receive buffer to find out if the Parallax Serial Terminal sent a character that was typed into its Transmit windowpane. If the Propeller chip receives a character from the Parallax Serial Terminal, the debug.Vars(@m, String("long m, s, ms")) method call passes the address of the m variable along with a string copy of the variable declaration that starts with the m variable.

```
" Test Time Counting.spin
" First test of timestamp code verifies proper counting sequence.


CON

    _clkmode = xtal1 + pll16x            ' Crystal and PLL settings
    _xinfreq = 5_000_000                 ' 5 MHz crystal x 16 = 80 MHz

OBJ

    debug  : "PST Debug LITE"            ' Debug object

VAR

    long m, s, ms                        ' Timekeeping variables


PUB TimerMs

    debug.Style(debug#COMMA_DELIMITED)   ' Configure debug display
    debug.Start(115200)                  ' Start debug cog

    repeat                               ' Infinite loop
```

```
ms++                          ' Add 1 to millisecond count
if ms == 1000                 ' If milliseconds = 1000
  ms := 0                     '  Set milliseconds to 0
  s++                         '  Increment seconds
if s == 60                    ' If seconds = 60
  s := 0                      '  Set seconds to 0
  m++                         '  Increment minutes
if m == 60                    ' If minutes = 60
  m := 0                      '  Set minutes to 0


' If key pressed, display variable list
if debug.KeyCheck
  debug.Vars(@m, String("long m, s, ms"))
```

Since the PST Debug LITE object was configured to display comma-delimited lists, calls to debug.Vars result in the variable lists shown in Fig. 3-17a. After loading the

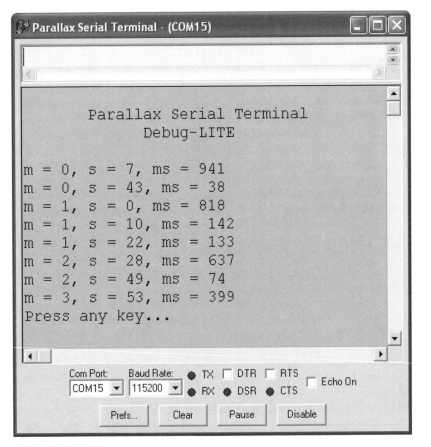

Figure 3-17a   Variable display with Parallax Serial Terminal.

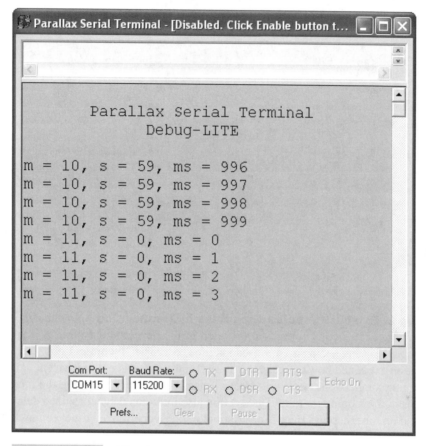

Figure 3-17b   (*Continued*)

code and placing the cursor in the Parallax Serial Terminal's Transmit windowpane, a few quick keypresses to check the timestamp results look encouraging. The display in Fig. 3-17b was obtained with some modifications to the code to verify that the variable values roll over and increment in the correct order.

✓ Before trying the example programs in this section, uncheck the Parallax Serial Terminal's <u>Echo On</u> check box.

The modified Test Time Counting 2.spin code initializes the m, s, and ms variables to 995, 59, and 10, and the repeat loop is changed from unconditional to 8 repetitions. The if debug.KeyCheck condition is commented, and the debug.Vars method call is outdented so that it is part of the repeat loop block but not part of any if conditions preceding it. With these modifications, it displays eight steps in the sequence, from 10 minutes, 59 seconds, 996 milliseconds through 11 minutes, 0 seconds, 3 milliseconds.

```
' Test Time Counting 2.spin
'
'  ...

   debug.Start(115200)                              ' Start debug cog

   ms := 995                                        ' **Add
   s  := 59                                         ' **Add
   m  := 10                                         ' **Add

   repeat 8                                         ' ** Modify to 8 reps
     ms++                                           ' Add 1 to millisecond count

'
'  ...

     if m == 60                                     ' If minutes = 60
       m := 0                                       ' Set minutes to 0

    'if debug.KeyCheck                              '**Comment
    debug.Vars(@m, String("long m, s, ms"))         '**Outdent
```

Step 2: Establish Precise Time Base in Another Cog   Now that we know the counting sequence is correct, a next step in the development process might involve establishing and testing a 1-millisecond time base. The "Incorrect Loop Interval Code" section explained how waitcnt(clkfreq/1000 + cnt) is not precise enough for time-keeping applications because it doesn't take into account how long the other instructions in the loop might take. That section also introduced synchronized delay code for accurate timekeeping. Here is a variation on the synchronized timekeeping loop:

```
   t  := cnt                                        ' Current clock counter
   dt := clkfreq/1000                               ' Ticks in 1 ms

   repeat                                           ' Infinite loop
     waitcnt(t+=dt)                                 ' Wait for the next ms
     '  ...
```

Before entering the repeat loop, timekeeping code uses t := cnt to copy the current value of the cnt register to the t variable. Remember that the cnt register increments with every clock tick, so t := cnt copies the time at that moment into the t variable. Next, the code copies the time increment clkfreq/1000 to a time interval variable named dt. Inside the loop, waitcnt(t+=dt) adds dt, which is the number of ticks in 1 ms, to the previously recorded value of t, and then waitcnt waits until cnt register reaches that time. The next time the loop repeats, it again adds the number of clock ticks in 1 ms to the previous value of t and then waits for the cnt register to get to this next value. As long as the other commands in the loop do not take longer than 1 ms to execute, this technique ensures that the 1-millisecond time base will be maintained even though the other commands in the loop might take varying amounts of time.

Step 3: Test the Timekeeping Code in Another Cog The "Code That Missed the Waitcnt Boat" section demonstrated how the `waitcnt(t+=dt)` approach can be unforgiving if the commands inside the loop exceed the time interval—1 ms in this case. If debugging commands in the loop take too long to display the values, it can cause the next `waitcnt` command to miss the value of the `cnt` register that it's waiting for. In that case, the `cnt` register will eventually get back to that target after it rolls over. One way of preventing the debugging code from interfering with the time base is to launch the time counting code with the precise delay into another cog. A separate cog can then monitor the time counting values. Test Timestamp from Other Cog.spin launches a version of the `TimerMs` method that has been modified for precise timing into another cog. Meanwhile, the cog executing the `Go` method allows you to periodically sample timestamps through the Parallax Serial Terminal by placing the cursor in its Transmit windowpane and then periodically pressing a key on your computer keyboard.

```
" Test Timestamp from Another Cog.spin
" This example program demonstrates an initial test of some
" timestamp prototype code that fails to expose a memory bug.

" IMPORTANT: This code has a hidden bug!!!

CON

  _clkmode = xtal1 + pll16x              ' Crystal and PLL settings
  _xinfreq = 5_000_000                   ' 5 MHz crystal x 16 = 80 MHz

OBJ

  debug  : "PST Debug LITE"              ' Serial communication object

VAR

  long stack[40]                         ' Ample stack for prototyping
  long m, s, ms, t, dt                   ' Timekeeping variables

PUB Go | minutes, seconds, milliseconds

  debug.Style(debug#COMMA_DELIMITED)     ' Configure debug display
  debug.Start(115200)                    ' Start debug cog

  cognew(TimerMs, @stack)                ' Launch timekeeping cog

  repeat
    longmove(@minutes, @m, 3)            ' Copy TimerMs vars
    debug.ListHome                       ' Display values
    debug.Vars(@m, String("| minutes, seconds, milliseconds"))
    debug.break
```

```
PRI TimerMs

  t  := cnt                              ' Current clock counter
  dt := clkfreq/1000                     ' Ticks in 1 ms

  repeat                                 ' Infinite loop
    waitcnt(t+=dt)                       ' Wait for the next ms
    ms++                                 ' Add 1 to millisecond count
    if ms == 1000                        ' If milliseconds = 1000
      ms := 0                            '  Set milliseconds to 0
      s++                                '  Increment seconds
    if s == 60                           ' if seconds = 60
      s := 0                             '  Set seconds to 0
      m++                                '  Increment minutes
    if m == 60                           ' If minutes = 60
      m := 0                             '  Set minutes to 0
```

Judging from the Parallax Serial Terminal in Fig. 3-18, it's working great, isn't it? Even with modifications similar to the previous code that test at times like 10 minutes, 59 seconds, 999 milliseconds, it is unlikely to expose a hidden memory collision bug in Test Timestamp from the Other Cog.spin.

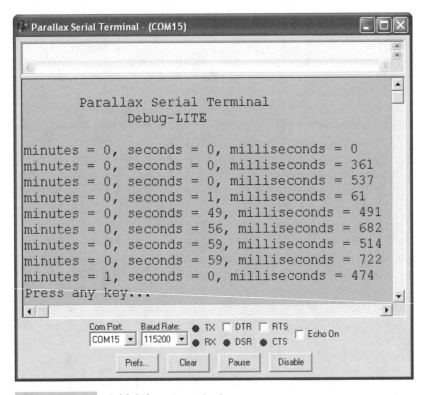

Figure 3-18   Initial timestamp test.

Step 4: Remember that Multiple Processors Are Involved and Test! Going back to the "Common Multiprocessor Coding Mistakes" list, we can check off a few possible bugs already. The timing interval is correct, and it doesn't miss the `waitcnt` boat. The two cogs are using global variables to exchange values and appear to be doing so correctly. However, memory collisions have not been ruled out. Remember from the "Memory Collisions" section that cogs using groups of variables to exchange information can end up exchanging a combination of new and old values if locks are not used to give commands executed by different cogs exclusive access to the group of memory elements. Since Test Timestamp from Another Cog.spin is now executing code in two separate cogs accessing the same group of variables, it stands to reason that the `longmove` command might copy values at unexpected instants, such as when the `ms` variable stores 1000 or when the `s` variable stores 60. Since the `ms` variable should only store 0...999 and the `s` variable should only store 0...59, this would be a bug that could, in turn, cause other bugs in the application.

If there really is a potential memory collision problem in this code, the best way to expose it is to initialize the values to a point where they are about to roll over and then run the code in slow motion, with long delays. The long delays make windows of opportunity for memory collisions to occur. The cog making copies of the group of variables can also sample it more quickly than the timekeeping cog updates it, which should expose the bug, if there is one. In Find Hidden Bug in Timestamp.spin, the `dt` variable is assigned the number of clock ticks in one second instead of the number of ticks in a millisecond, and there are quarter-second delays after each timekeeping variable is incremented. The Go method's `repeat` loop checks the timestamp eight times, one-third of a second apart.

```
'' Find Hidden Bug in Timestamp.spin
'  ...
'


PUB Go | minutes, seconds, milliseconds

'  ...
'

  cognew(TimerMs, @stack)                    ' Launch timekeeping cog

'  ...
'

  m   := 10                                  ' **Add
  s   := 59                                  ' **Add
  ms  := 999                                 ' **Add

  repeat 8                                   ' **Modify to 8 reps Main loop
    waitcnt(clkfreq/3+cnt)                   ' **Add
    longmove(@minutes, @m, 3)                ' Copy timestamp vars
    'debug.break                             ' **Remove
    debug.Vars(@m, String("| minutes, seconds, milliseconds"))
```

```
PRI TimerMs

  t  := cnt                          ' Current clock counter
  dt := clkfreq'/1000                ' **Modify Ticks in 1 ms

  repeat                             ' Infinite loop
    waitcnt(t+=dt)                   ' Wait for the next ms
    ms++                             ' Add 1 to millisecond count
    waitcnt(clkfreq/4+cnt)           ' **Add
    if ms == 1000                    ' If milliseconds = 1000
      ms := 0                        '   Set milliseconds to 0
      s++                            '   Increment seconds
      waitcnt(clkfreq/4+cnt)         ' **Add
    if s == 60                       ' if seconds = 60
      s := 0                         '   Set seconds to 0
      m++                            '   Increment minutes
      waitcnt(clkfreq/4+cnt)         ' **Add
    if m == 60                       ' If minutes = 60
      m := 0                         '   Set minutes to 0
```

With this modified program, the results of possible memory collisions are now exposed. Take a look at the third and fourth lines in Fig. 3-19 Parallax Serial Terminal

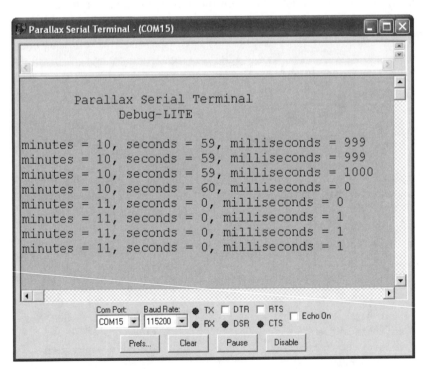

Figure 3-19   Timestamp test—memory collisions exposed.

display. The timestamp should never contain 1000 milliseconds or 60 seconds. The previous keypress example program might have exposed this same bug over time, but it would have taken a lot of keypresses!

Step 5: Fix the Bug Exposed by the Testing   Locks provide an effective remedy for the memory collisions that can occur when two cogs exchange memory using a group of variables. The Propeller microcontroller's main memory has eight semaphore bits, called locks or lock bits, and the Spin language has a set of commands for using them. The locknew command checks out a lock bit from the pool. Once a lock bit has been checked out, lockset and lockclr can set and clear it, and lockret can return it to the pool. Cogs that access common memory then use code that checks a given lock bit before accessing memory. If the lock bit is set, the code makes the cog wait for the other cog accessing the memory to clear it first. If the lock bit is not set, the cog has to set it before accessing the memory and then clear the bit when it's finished accessing the memory.

LOOKING UP LOCKS

The lock management commands are available in both Spin and Propeller Assembly. The Propeller Manual discusses the use of locks in detail and has code examples for both languages in their respective reference sections. The Propeller Manual is available in print and is included as a tagged PDF in the Propeller Tool software's Help.

Using locks in Spin programs essentially boils down to the following code excerpt. This code first checks out a lock bit using locknew, and it stores the result (0...7 or -1 if no locks are available) in a variable named semID. Next, code in any cog that accesses the memory has to wait for that lock bit to be clear before setting it and modifying the Main RAM. The repeat until not lockset(semID) loop does this by repeatedly setting lockset. Since lockset returns the previous state of the lock, it will return true if it was already set by another cog. In that case, the repeat loop will keep trying until lockset returns 0. When it returns 0, the fact that lockset was called means it was set to 1, this time by the cog waiting for access. Since the condition of the repeat loop is not lockset(semID), when it returns 0, the NOT of 0 is TRUE, so the repeat loop will allow the code to move on and access the shared memory. As soon as the code is done with the shared memory, it should immediately use lockclr(semID) to clear the lock and minimize the amount of time other cogs have to wait for memory access.

```
' Code in one cog
'
if (semID := locknew) == -1          ' If lock, check one out
  pst.Str(String("Error, no locks!"))  ' Else display error
'
' ...
'
```

```
    repeat until not lockset(semID)     ' Wait till clear, set lock
    '" Shared memory operations
    lockclr(semID)                      ' Clear lock
    '
     '...code in anther cog
    '
    repeat until not lockset(semID)     ' Wait till clear, set lock
    '" Shared memory operations
    lockclr(semID)                      ' Clear lock
```

Step 6: Test the Bug Fix   Test Timestamp Bug Fix.spin uses the lock approach and tests for memory collisions by leaving in the extra delays to keep that memory collision window wide open. The semID variable declaration and `locknew`, `lockset`, and `lockclr` commands that were added to fix the bug are labeled with ***Fix in the comments. In the Go method, the lock is set before the `longmove` command and is cleared immediately afterwards. This is all that's needed in the Go method for memory access, and it happens quickly. In the `TimerMs` method, which gets executed by another cog, the lock gets set before the code starts modifying the ms variable, and it continues through the s and then m variables. When it's done with the m variable, it clears the lock again before the loop repeats.

```
" Test Timestamp Bug Fix.spin
" This program tests to verify that semaphores prevent memory
" memory collisions.

CON

  _clkmode = xtal1 + pll16x           ' Crystal and PLL settings
  _xinfreq = 5_000_000                ' 5 MHz crystal x 16 = 80 MHz

OBJ

  debug  : "PST Debug LITE"           ' Serial communication object

VAR

  long stack[40]                      ' Ample stack for prototyping
  long m, s, ms, t, dt                ' Timekeeping variables
  long semID                          ' ***Fix Semaphore ID variable

PUB Go | minutes, seconds, milliseconds

  debug.Style(debug#COMMA_DELIMITED)  ' Configure debug display
  debug.Start(115200)                 ' Start debug cog

  if (semID := locknew) == -1         ' ***Fix If no locks in pool
```

```
      debug.Str(String("Error, no locks!"))  ' ***Fix Display error message
      cognew(TimerMs, @stack)                 ' Launch timekeeping cog

      m  := 10                                ' **Add
      s  := 59                                ' **Add
      ms := 999                               ' **Add

      repeat 8                                ' **Modify to 8 reps Main loop
        waitcnt(clkfreq/3+cnt)                ' **Add
        repeat until not lockset(semID)       ' ***Fix Wait for lock, set
        longmove(@minutes, @m, 3)             ' Copy timestamp vars
        lockclr(semID)                        ' ***Fix Clear lock
        'debug.break                          ' **Remove
        debug.Vars(@m, String("| minutes, seconds, milliseconds"))

PRI TimerMs

      t  := cnt                               ' Current clock counter
      dt := clkfreq'/1000                     ' **Modify Ticks in 1 ms

      repeat                                  ' Infinite loop
        waitcnt(t+=dt)                        ' Wait for the next ms
        repeat until not lockset(semID)       ' ***Fix Wait for lock, set
        ms++                                  ' Add 1 to millisecond count
        waitcnt(clkfreq/4+cnt)                ' **Add
        if ms == 1000                         ' If milliseconds = 1000
          ms := 0                             ' Set milliseconds to 0
          s++                                 ' Increment seconds
          waitcnt(clkfreq/4+cnt)              ' **Add
        if s == 60                            ' if seconds = 60
          s := 0                              ' Set seconds to 0
          m++                                 ' Increment minutes
          waitcnt(clkfreq/4+cnt)              ' **Add
        if m == 60                            ' If minutes = 60
          m := 0                              ' Set minutes to 0
        lockclr(semID)                        ' ***Fix Clear lock
```

Judging by the Parallax Serial Terminal display in Fig. 3-20, the bug has been fixed. As this program runs, you'll be able to see a visible pause between the second and third lines as the code waits for the cog running the TimerMs method to clear the lock.

When all the extra waitcnt commands are removed and dt is restored to the number of ticks in a millisecond (clkfreq/1000), the code should now resemble Test Timestamp Bug Fix Full Speed.spin. Placement of the lockclr command immediately after longmove in the Go method will be especially important. If the lockclr command were instead placed after the debug.Vars call, the cog running the TimerMs method would wait so

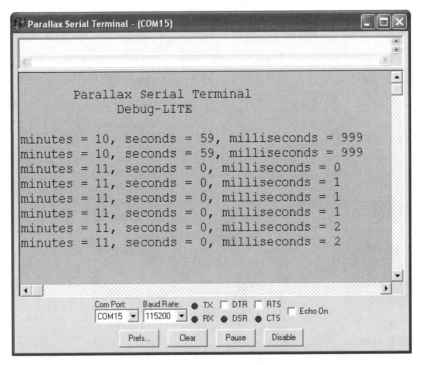

Figure 3-20   Timestamp test—memory collisions corrected.

long for the lock to clear that its `waitcnt` command would miss the target `cnt` register value in the next iteration of its `repeat` loop (missing the `waitcnt` boat bug).

Step 7: Remove Test Code   The next step is to remove the delays that were added for the sake of exposing the memory collisions. In other words, all the lines of code with comments labeled ** need to be reversed, and the best way to make sure you don't miss any is with a Search…Replace. All lines with **Add comments can be removed. Likewise, all the lines of code with **Remove can be uncommented, and all the lines with **Modify can be unmodified. It's important that you don't forget to restore `dt := clkfreq` back to `dt := clkfreq/1000`. After one last test to make sure the code still correctly timestamps events, it's ready to get incorporated into a building block object.

```
'' Test Timestamp Bug Fix Full Speed.spin
'' This program tests to verify that semaphores prevent memory
'' memory collisions at full speed, no slow motion delays.

CON

  _clkmode = xtal1 + pll16x          ' Crystal and PLL settings
  _xinfreq = 5_000_000               ' 5 MHz crystal x 16 = 80 MHz
```

```
OBJ

  debug  : "PST Debug LITE"                      ' Serial communication object

VAR

  long stack[40]                                 ' Ample stack for prototyping
  long m, s, ms, t, dt                           ' Timekeeping variables
  long semID                                     ' ***Fix Semaphore ID variable

PUB Go | minutes, seconds, milliseconds

  debug.Style (debug#COMMA_DELIMITED)            ' Configure debug display
  debug.Start (115200)                           ' Start debug cog

  if (semID := locknew) == -1                    ' ***Fix If no locks in pool
    debug.Str (String("Error, no locks!"))       ' ***Fix Display error message
  cognew (TimerMs, @stack)                        ' Launch timekeeping cog

  repeat                                         ' Main loop
    repeat until not lockset (semID)             ' ***Fix Wait for lock, set
    longmove (@minutes, @m, 3)                   ' Copy timestamp vars
    lockclr (semID)                              ' ***Fix Clear lock
    debug.break                                  ' Breakpoint
    debug.Vars (@m, String("| minutes, seconds, milliseconds"))

PRI TimerMs

  t  := cnt                                      ' Current clock counter
  dt := clkfreq/1000                             ' Ticks in 1 ms

  repeat                                         ' Infinite loop
    waitcnt (t+=dt)                              ' Wait for the next ms
    repeat until not lockset (semID)             ' ***Fix Wait for lock, set
    ms++                                         ' Add 1 to millisecond count
    if ms == 1000                                ' If milliseconds = 1000
      ms := 0                                    '   Set milliseconds to 0
      s++                                        '   Increment seconds
    if s == 60                                   ' if seconds = 60
      s := 0                                     '   Set seconds to 0
      m++                                        '   Increment minutes
    if m == 60                                   ' If minutes = 60
      m := 0                                     '   Set minutes to 0
    lockclr (semID)                              ' ***Fix Clear lock
```

Step 8: Incorporate into a Building Block Object Now that the code is known to work, it's time to move it into a building block object. Test Timestamp Object.spin is an example of a top-level object that utilizes the new Timestamp Object. Notice it does not have to concern itself with any locks. All it has to do is (1) declare the Timestamp Object, (2) call its Start method and pass the start time, and (3) call its GetTime method to get a timestamp. From this application object's standpoint, the code is simple and it can be oblivious to all the cog and lock bookkeeping the Timestamp object takes care of. This is a typical hallmark of a Propeller Library or Propeller Object Exchange object that manages a process in another cog. It takes care of any cog and lock details, and provides method calls and/or a memory interface for exchanging information with the application-level object.

```
" Test Timestamp Object.spin
" Wait for key press and display ms since TimerMs was launched into a cog.

CON

  _clkmode = xtal1 + pll16x                ' Crystal and PLL settings
  _xinfreq = 5_000_000                     ' 5 MHz x 16 = 80 MHz

OBJ

  debug  : "PST Debug LITE"                ' Serial COM object
  time   : "Timestamp Object"              ' TimeStamp object

PUB Go | minutes, seconds, milliseconds

  debug.Style(debug#COMMA_DELIMITED)       ' Configure debug display
  debug.Start(115200)                      ' Start debug cog
  time.Start(10, 59, 999)                  ' Start timestamp cog

  repeat                                   ' Main loop
    time.GetTime(@minutes)                 ' Get a timestamp
    debug.ListHome                         ' Display timestamp vars
    debug.Vars(@minutes, String("| minutes, seconds, milliseconds"))
    debug.break                            ' Set breakpoint
```

In addition to all the cog bookkeeping and stack declaration, when the TimerMs method gets moved into its own building block object, it should take care of all the locks. Timestamp Object.spin is one approach to incorporating the TimerMs method into a building block object. Before launching the TimerMs method into a new cog, the Start method checks out a lock. Before shutting down the cog, the Stop method returns the lock. As with earlier examples, commands in the object's other methods that get called have to wait for the lock to clear and then set it before performing any operations on the m, s, and ms variables. Those are the object's three global variables shared by the cog executing the TimerMs method and the cog executing other methods such as GetTime and SetTimer.

This design allows the application object's code to concern itself with just one thing: getting the timestamp whenever it needs it. As you examine the Timestamp Object, keep in mind that all the methods except `TimerMs` are executed by the same cog that's executing the Test Timestamp Object's methods. In other words, the same cog that's executing the `repeat` loop in the Test Timestamp object makes a call to `time.GetTime`, executes the code in the Timestamp Object's `GetTime` method and returns. Meanwhile, the Timestamp Object's `TimerMs` method is executed by a different cog. The Timestamp Object's `GetTime` and `SetTimer` methods both use the memory access approach shown in Fig. 3-4, and they rely on the `TimerMs` method executed by another cog to keep the `m`, `s`, and `ms` variables updated with the current time.

```
" Timestamp Object.spin
" Supplies minutes, seconds, milliseconds timestamp upon request.

VAR

  long stack[40]              ' Ample stack for prototyping
  long t, dt, m, s, ms        ' Timekeeping variables
  long semID, cog             ' Semaphore ID and cog variables

PUB Start(minutes, seconds, milliseconds) : success
{{Start timekeeping in a new cog.
  Parameters:
    minutes      - starting minutes count
    seconds      - starting seconds count
    milliseconds - starting milliseconds count
  Returns      : 0 if failed to start or nonzero if it succeeded}}

  longmove(@m, @minutes, 3)       ' Copy local → global variables

  ' If checked out lock and launched new cog, return nonzero, else
  ' return zero.
  if not semID := locknew         ' If checked out a lock
    Stop                          ' Stop if already running
      ' Launch process into new cog and return nonzero if successful
    success := (cog := cognew(TimerMs, @stack) + 1)

PUB Stop                          "Stop timekeeping.

  if cog                          ' If cog is not zero
    lockret(semID)
    cogstop(cog~ - 1)             '  Stop cog & set cog to zero

PUB SetTimer(minutes, seconds, milliseconds)
{{Set the timer.
  Parameters:
    minutes      - minutes count
    seconds      - seconds count
    milliseconds - milliseconds count}}
```

```
    repeat until not lockset(semID)      ' Set lock on time variables
    longmove(@m, @minutes, 3)            ' Copy values received
    lockclr(semID)                       ' Clear lock on time variables

PUB GetTime(minAddr)                     '' Get current timestamp
{{Get the current timestamp.
  Parameters:
    minAddr      - address of the caller's minutes variable

      IMPORTANT:   Caller must declare long variables in this sequence:
                   minutes, seconds, milliseconds}}

    repeat until not lockset(semID)      ' Set lock on time variables
    longmove(minAddr, @m, 3)             ' Copy times to caller's vars
    lockclr(semID)                       ' Clear lock on time variables

PRI TimerMs

    t  := cnt                            ' Current clock counter
    dt := clkfreq/1000                   ' Ticks in 1ms

    repeat                               ' Infinite loop
      waitcnt(t+=dt)                     ' Wait for the next ms
      repeat until not lockset(semID)    ' Set lock
      ms++                               ' Add 1 to millisecond count
      if ms == 1000                      ' If milliseconds = 1000
        ms := 0                          '  Set milliseconds to 0
        s++                              '  Increment seconds
      if s == 60                         ' if seconds = 60
        s := 0                           '  Set seconds to 0
        m++                              '  Increment minutes
      if m == 60                         ' If minutes = 60
        m := 0                           '  Set minutes to 0
      lockclr(semID)
```

DEVELOPMENT WITH VIEWPORT

ViewPort has a code view that provides an IDE-style debugger for the top-level object. You can use ViewPort to monitor variable values, set breakpoints, run, pause, and step. You can also use it to modify variable values before continuing from a breakpoint. In addition, ViewPort makes it possible to look at variable behavior graphically with oscilloscope, logic analyzer, spectrum analyzer, and other tools. For analysis of the timekeeping application, the oscilloscope can provide graphical verifications that the timekeeping object functions as intended.

Note: New ViewPort functions and features are on the horizon at the time of this writing. To check for updated versions of code examples from this section that are compatible with the latest version of ViewPort, go to ftp.propeller-chip.

com/PCMProp/Chapter_03/Source/. As mentioned earlier, the latest version of ViewPort is available for a 30-day free trial from www.parallax.com.

The ViewPort software communicates with the Propeller through an object named conduit that launches a variable monitoring and communication process into another cog. For debugging, the conduit object has to be configured by the top level object's Spin code to work with the ViewPort debugger. The .spin files that will be debugged also have to be placed in the same directory with certain objects that are packaged with the ViewPort installation. Below is a list of general steps for preparing a .spin application for debugging with ViewPort:

- Copy objects into C:\Program Files\ViewPort...\mycode.
- Open the top object with ViewPort by clicking the <u>code</u> tab and then using <u>File</u> → <u>Open</u> to find the file.
- Remove any object that might try to communicate with the PC through the Propeller chip's programming connection (for example, Parallax Serial Terminal or PST Debug LITE).
- For debugging, make sure the Propeller's system clock is set to 80 MHz.
- Declare the conduit object, typically by adding `vp:"conduit"` to the OBJ block.
- Call the conduit object's `config` method and pass it a list of the variable names.
- Pass the start and end addresses of the contiguous list of variables that you want to be able to monitor and update to the ViewPort object's share method. The variable names passed to the `config` method must correspond with these variables.

Test Timestamp Object with ViewPort.spin started out as the program in the previous example: Test Timestamp Object.spin. With the modifications just discussed, it's now ready for ViewPort debugging. The call to the conduit object's `config` method sends a list of variable names: `"var:minutes,seconds,milliseconds"`. The code also passes the addresses of the start and end variables in the list with a call to `vp.share`, so ViewPort will monitor `@minutes` through `@milliseconds`. Make sure to keep the names in the `config` method call the same as the names of the variables in your code. This will ensure that you can use all of ViewPort's debugging features with those variables.

```
" Test Timestamp Object with ViewPort.spin
" Check minutes, seconds, and milliseconds in code tab's Watch
" windowpane and signal timing in the dso tab's Oscilloscope.

CON

  _clkmode = xtal1 + pll16x               ' Crystal and PLL settings
  _xinfreq = 5_000_000                    ' 5 MHz x 16 = 80 MHz

OBJ

  vp      : "conduit"                     ' ViewPort conduit object
  time    : "TimeStamp Object"            ' TimeStamp object
```

```
PUB Go | minutes, seconds, milliseconds
  ' Configure ViewPort display & share @minutes through @milliseconds.
  vp.config(String("var:minutes,seconds,milliseconds"))
  vp.share(@minutes, @milliseconds)

  time.Start(10, 59, 999)                    ' Start timestamp cog

  repeat                                     ' Main loop
    time.GetTime(@minutes)                   ' Get a timestamp
```

To run the program in ViewPort:

✓ Copy the following objects to the C:\Program Files\ViewPort...\mycode directory:
  o Test Timestamp Object with ViewPort.spin
  o TimeStamp Object
✓ Open ViewPort.
✓ Click the <u>code</u> tab.
✓ Set the drop-down menu above the <u>Spin Files</u> list to the <u>mycode</u> directory.
✓ Open the program (Test Timestamp Object with ViewPort.spin).
✓ Click the <u>Start Debugging</u> button. The <u>Start Debugging</u> button looks like the triangular play button on most music and video players.
✓ Examine the variables listed in the lower-left Watch windowpane.

The Watch windowpane in Fig. 3-21 will show the `minutes`, `seconds`, and `milliseconds` variables and their values. While the program is running, the `repeat` loop repeatedly updates the `minutes`, `seconds`, and `milliseconds` variables. Meanwhile, the conduit object uses another cog to stream those variable values to the PC, where the ViewPort software updates its display with their values. As you watch these variables counting upwards, it provides a quick verification that the Timestamp Object is providing the current times as expected.

ViewPort has a variety of graphical analysis tools, and one that's exceedingly useful for evaluating this object is the digital storage oscilloscope (<u>dso</u>) view. This view can graphically display the `minutes`, `seconds`, and `milliseconds` variables. The <u>dso</u> view shown in Fig. 3-22 also has measurement tools for quickly verifying that the `minutes`, `seconds`, and `milliseconds` variables really are keeping time according to minutes, seconds, and milliseconds.

Follow these instructions to perform the test shown in Fig. 3-22:

✓ If it's not currently active, click the <u>Start Debugging</u> button in the <u>code</u> tab.
✓ Click <u>dso</u> tab.
✓ In the <u>Overview</u> table in the lower-right area, click the <u>Plot</u> buttons next to <u>milliseconds</u> and <u>seconds</u>.
✓ Adjust the <u>Horizontal</u> dial to <u>500 ms/div</u>.
✓ In the <u>Trigger</u> tab below the <u>Oscilloscope</u> display, click <u>Continuous</u> → <u>Ch1</u> → <u>Fall</u>, and <u>Normal</u>.

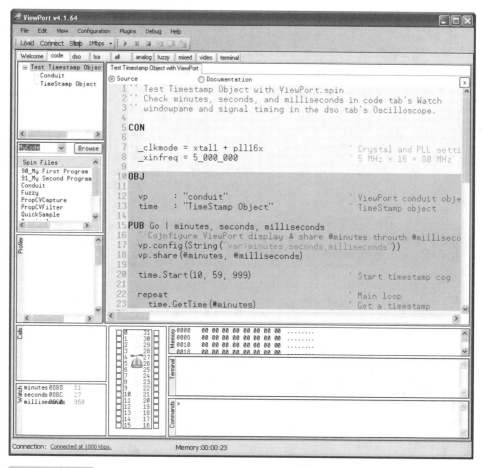

Figure 3-21   ViewPort timestamp variable monitoring.

✓ Adjust the <u>trigger</u> threshold along the left of the display so that it is in the middle of the <u>milliseconds</u> triangle wave.

✓ Click the <u>stop</u> button.

✓ Click the <u>Cursor</u> tab and then click the <u>Horizontal</u> and <u>Vertical</u> cursor buttons.

Horizontal and vertical cursors will appear as dotted lines on the <u>Oscilloscope</u> display. The cursors can be clicked and dragged, and the <u>Cursors</u> section in the lower-right ViewPort <u>Cursors</u> table displays the time difference between the vertical cursors and the amplitude difference between the horizontal cursors.

✓ Drag the vertical and horizontal cursors so that they frame one of the triangles in the <u>milliseconds</u> trace in Fig. 3-22.

✓ Examine the <u>Cursors</u> table.

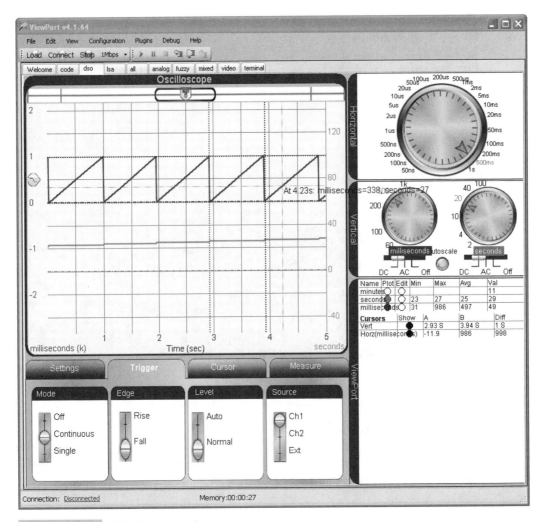

Figure 3-22 ViewPort dso view.

The <u>Cursors</u> table should indicate that the difference between the vertical cursors is about 1 s and the difference between the horizontal cursors is about 1000 ms. The cursor mode slider also has <u>Pan</u> and <u>Zoom</u> features for getting a closer look at a given waveform.

✓ Try it.

Adding Terminal Functionality to ViewPort Let's say the next task at hand is to count events and send an alarm message when 11 events have occurred. The ViewPort debugger has a Terminal pane that can be used to send character "events" to the Propeller chip. The corresponding terminal object that the Propeller application code uses has all the same functionality as conduit, but it also supports terminal communication, and even includes familiar method calls like Char, CharIn, Dec, DecIn, and more from the Parallax Serial Terminal object. In general, additional changes that have to be made for debugging with ViewPort's terminal object include:

- Replace the conduit object declaration with a terminal object declaration.

```
vp      : "terminal"         ' ViewPort conduit object
```

- Add two longs for sending characters to, and receiving characters from, the terminal immediately before the contiguous list of long variables.

```
long vptx, vprx
  long minutes, seconds, milliseconds, count
```

- Add a config method call with a string that configures the terminal.

```
vp.config(String("start:terminal::terminal:1"))
```

- Update your vp.config and vp.share calls to include the transmit and receive variables that were added to the variable declarations in a previous step.

```
vp.config(String("var:vptx,vprx,minutes,seconds,milliseconds,count"))
vp.share(@vptx, @count)
```

Test Timestamp Object with ViewPort Events.spin incorporates those changes into the test program, along with some code that waits to receive a character from ViewPort's Terminal. The Test Timestamp... object's Go method's repeat loop checks repeatedly for a character in the terminal object's input buffer. When the RxCheck method returns something other than -1, it indicates that there's a character waiting in the buffer, which in turn indicates that a character was typed into ViewPort's Terminal. The Test Timestamp... object waits for this event with if vp.RxCheck <> -1. When the event occurs, code in the if statement block clears the input buffer by calling vp.RxFlush, and then it gets a timestamp from the Timestamp object. Next, if count++ == 10 compares the count variable to 10 and then post-increments it. If the count variable is equal to 10, code in the nested if block calls the Alarm method. The Alarm method sends the string "Alarm!" to the terminal. While the actual event might be triggered by a sensor and the alarm might be a signal to a speaker, this application is useful for prototyping. For example, if the sensor sample is still on its way in the mail, you can still make progress and have application code under way before the sensor arrives.

```
" Test Timestamp Object with ViewPort Events.spin
" Use ViewPort to get events from the keyboard and test to make sure
" the alarm occurs after eleventh event.
```

```
CON

  _clkmode = xtal1 + pll16x              ' Crystal and PLL settings
  _xinfreq = 5_000_000                   ' 5 MHz x 16 = 80 MHz

OBJ

  vp      : "terminal"                   ' ViewPort conduit object
  time    : "Timestamp Object"           ' TimeStamp object

VAR

  long vptx, vprx
  long minutes, seconds, milliseconds, count

PUB Go                                   ' Go method
  'Configure ViewPort for Terminal communication, variable display,
  'and share @vptx through count.
  vp.config(String("start:terminal::terminal:1"))
  vp.config(String("var:vptx,vprx,minutes,seconds,milliseconds,count"))
  vp.share(@vptx,@count)
  waitcnt(cnt+clkfreq)
  vp.clear

  time.Start(10, 59, 999)                ' Start timestamp cog

  repeat                                 ' Main loop
    if vp.rxcheck <> -1                  ' If key in buffer
      vp.RxFlush                         ' Clear buffer
      time.GetTime(@minutes)             ' Get a timestamp
      vp.Dec(Count)                      ' Count to Terminal
      if count++ == 10                   ' Call alarm when count=10
        Alarm

PUB Alarm                                ' Alarm method

  vp.Str(String("Alarm!"))               ' Display alarm string in Terminal
  count~                                 ' Clear count variable
```

The Test Timestamp… object makes it possible to effectively utilize ViewPort's debugging features, including terminal, stepping, and variable modification, shown in Fig. 3-23.

To debug the Test Timestamp Object with ViewPort Events.spin object, follow these steps:

✓ Copy Test Timestamp Object with ViewPort Events.spin to C:\Program Files\ ViewPort...\mycode.

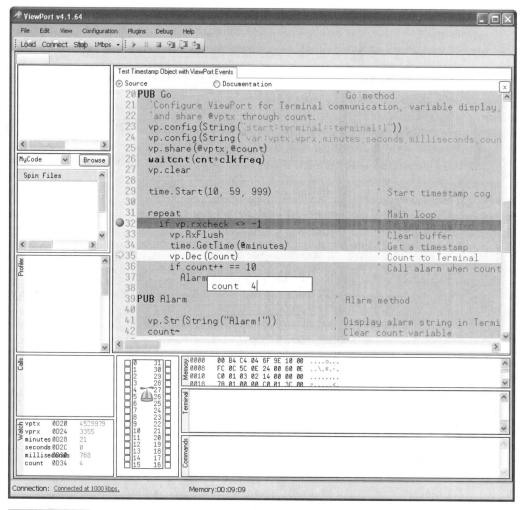

Figure 3-23   Terminal, stepping, and variable modification.

✓ With ViewPort in <u>code</u> view, open Test Timestamp Object with ViewPort Events. spin.

✓ Click the <u>Start Debugging</u> button.

✓ Place your cursor in the <u>Terminal</u> field, and press the ENTER key a few times. Each time, the <u>Watch</u> window should update the variable values.

✓ Hover your mouse cursor over the buttons to the right of the <u>Start Debugging</u> button to make their flyover labels appear. Use them to locate the <u>Pause</u>, <u>Step Over</u>, and <u>Step Into</u> buttons.

✓ Next, click the <u>Pause</u> button, and repeatedly click the <u>Step Over</u> button.

The yellow highlighter indicating the line that gets executed will not step into the `if vp.RxCheck <> 1` statement's code block because there is no character waiting in the terminal object's receive buffer.

✓ Click ViewPort's <u>Terminal</u> field and press your keyboard's ENTER key. Then, click the <u>Step Over</u> button three more times.

This time, the yellow highlighter that indicates the active line of code should advance into the code under the `if` block. The third step will place the active line at the `vp.Dec (count)` line. The next line will only call the `Alarm` method if `count` is equal to 10.

✓ Hover the cursor over the `count` variable in the `if count++ == 10` line and wait for the flyover with the variable value to appear.
✓ Click the value in the flyover and change it to 10.
✓ Click the <u>Step Into</u> button.

The active line highlighter should jump into the `Alarm` method. In the <u>Calls</u> window, it will indicate that the code being executed is in the `Alarm` method. After the `vp.str` call, the "Alarm!" text should appear in the <u>Terminal</u>.
  Also, try this:

✓ Set a breakpoint on `vp.Dec (count)` by clicking the line number next to that line of code. Then, click the Start Debugging button.
✓ A red *breakpoint marker* should appear on that line. Since no character is in the buffer, the code still can't get to it.
✓ Click the <u>Terminal</u> and press the ENTER key.

The active line highlighter will appear over the breakpoint.

✓ Change the value of count to 10 again, but this time, try the <u>Step Over</u> button.
✓ Stepping over the `Alarm` method call should cause the "Alarm!" message to appear in the <u>Terminal</u> pane, but the active line of code should jump back to the if `vp.rxcheck...` line.

DEVELOPMENT WITH THE PROPELLER ASSEMBLY DEBUGGER

PASD is a must if you plan on writing assembly code for your objects. Assembly language is great for making high-performance versions of Spin objects, and it can also be used to tackle certain tasks that the interpreted Spin language might not be fast enough for. An example of the speed improvements that can be attained with assembly language is the SimpleSerial Spin object's top baud rate of 19.2 kbps compared to ViewPort's conduit object, which uses assembly language to support rates up to 2 Mbps.
  Like ViewPort and the Parallax Serial Terminal, the PASD software relies on an object (PASDebug.spin) running in another cog to send it information. The code being debugged needs to declare the object and receive some slight modifications before it can be examined with PASD. Here are the general steps for setting up an object for debugging with PASD:

- Copy the test code and any objects it relies on into the same directory with the PASD software and PASDebug object.
- Verify that the object's system clock settings are at least 5 MHz with an external crystal. (The next example and the one included with PASD uses the same 80-MHz system clock settings from the Parallax Serial Terminal and ViewPort examples.)
- Comment any other code that might try to access the programming port (ViewPort, Parallax Serial Terminal, etc.).
- Add a declaration for the PASDebug object.

```
'{'
OBJ
   dbg    :         "PASDebug"                '<--- Add for Debugger
'}'
```

- After the cognew command that launches the assembly routine into a cog, add a call to the PASDebug object's Start method that passes the address where the assembly code starts. (31 and 30 are the Propeller I/O pins used for COM port communication through the Propeller programming tool.)

```
cognew(@entry, @m)
'{'
   dbg.start(31,30,@entry)          'debugger
'}'
```

- At address 0 in the code, just after org and the entry point label, add the PASD Debugger Kernel.

```
DAT
                              org
   '
   ' Entry
   '
entry

'{'
'
'   --------- Debugger Kernel add this at Entry (Addr 0) ---------
      long $34FC1202,$6CE81201,$83C120B,$8BC0E0A,$E87C0E03,$8BC0E0A
      long $EC7C0E05,$A0BC1207,$5C7C0003,$5C7C0003,$7FFC,$7FF8
'   -------------------------------------------------------------
'}'
```

Note that each of these PASD additions starts with '{' and ends with '}'. Since, at the time of this writing, the Propeller Tool software does not perform conditional compilations, this saves time adding and removing the assembly debugging functionality. The left-most apostrophes can be removed to comment out the debugging code, or added back to uncomment them. For example, this code is not commented:

```
'{'
  dbg.start(31,30,@entry)          'debugger
'}'
```

Now, it's commented:

```
{'
  dbg.start(31,30,@entry)          'debugger
}'
```

Also, an apostrophe following each brace makes it possible to use a couple of find or replace sessions to quickly uncomment and comment the debugger code.

TimeStamp Dev (ASM).spin is a test program for an assembly routine that performs the same function as the `TimerMs` method in Timestamp Object.spin. In fact, the required changes to substitute the assembly routine for the `TimerMs` method in the Timestamp Object would be minimal. The `cognew` command would have to be modified to launch the assembly code, and the `stack` variable array could be removed since assembly language does not require stack space in Main RAM that cogs executing Spin code use for expression calculations, method parameters, return addresses, and return values.

```
''  Timestamp Dev (ASM).spin
''  Supplies minutes, seconds, milliseconds timestamp upon request.

CON                                  ' Constant declarations

  _clkmode = xtal1 + pll16x          ' System clock settings
  _xinfreq = 5_000_000

VAR                                  ' Variable declarations

  long  cog, m, s, ms, dt, semID     ' Counting variables

OBJ                                  ' Object declarations
'{'
  dbg   :           "PASDebug"       '<--- Add for Debugger
'}'

PUB TestStart                        ' TestStart method

  m  := 10                           ' Initialize Spin Vars
  s  := 59
  ms := 999
  dt := clkfreq/1000
```

```
    if not semID := locknew          ' Check out a lock
      cognew(@entry, @m)
'{'
    dbg.start(31,30,@entry)          '<--- Add for Debugger
'}'
    Repeat                           ' Keep cog running

DAT                                  ' DAT block
                      org            ' ASM address reference
'
' Entry
'
entry

'{'
'   --------- Debugger Kernel add this at Entry (Addr 0) ---------
    long $34FC1202,$6CE81201,$83C120B,$8BC0E0A,$E87C0E03,$8BC0E0A
    long $EC7C0E05,$A0BC1207,$5C7C0003,$5C7C0003,$7FFC,$7FF8
'   -------------------------------------------------------------
'}'
                      mov     addr, par      ' Copy par → addr
                      rdlong  _m, addr       ' Copy m → _m
                      add     addr, #4       ' Point at next long
                      rdlong  _s, addr       ' s → _s
                      add     addr, #4       ' Point at next long
                      rdlong  _ms, addr      ' ms → _ms
                      add     addr, #4       ' Point at next long
                      rdlong  _dt, addr      ' dt → _dt
                      add     addr, #4       ' Point at next long
                      rdlong  sID, addr      ' semID → sID
                      mov     t, cnt         ' Copy cnt → t
                      add     t, _dt         ' Add _dt to t
timekeeper            waitcnt t, _dt         ' Wait for t+=dt
                      add     _ms, #1        ' _ms++
                      cmp     _ms, k     wz  ' if _ms==1000
             if_z     mov     _ms, #0        '   _ms:=0
             if_z     add     _s, #1         '   _s++
                      cmp     _s, #60    wz  ' if _s==60
             if_z     mov     _s, #0         '   _s:=0
             if_z     add     _m, #1         '   _m++
                      cmp     _m, #60    wz  ' if _m==60
             if_z     mov     _m, #0         '   _m:=0
:loop                 lockset sID        wz  ' if not lockset sID
             if_nz    jmp     #:loop
                      mov     addr, par      ' Copy par → addr
                      wrlong  _m, addr       ' Copy _m → m
                      add     addr, #4       ' Point at next long
                      wrlong  _s, addr       ' Copy _s → s
```

```
                    add      addr, #4          ' Point at next long
                    wrlong   _ms, addr         ' Copy _ms → ms
                    lockclr  sID               ' Clear lock bit
                    jmp      #timekeeper       ' Goto timekeeper

'
' Initialized data                            ' Cog RAM ASM vars
'
k                   long     1000             ' k := 1000
'
' Uninitialized data
'
_m                  res      1                ' More ASM variables
_s                  res      1
_ms                 res      1
_dt                 res      1
sID                 res      1
t                   res      1
addr                                                  res              1
```

The code between the Debugger Kernel and the timekeeper labels is responsible for copying initial values from Main RAM to Cog RAM. One important first test to run on assembly code is to make sure these values have been copied correctly. To run this test with PASD:

✓ Download PASD from www.insonix.ch/propeller/prop_pasd.html.
✓ Read the documentation and try the example code included in the download.
✓ Make sure that Timestamp Dev (ASM).spin is in the same folder with PASDebug.spin.
✓ Open Timestamp Dev (ASM) with the Propeller Tool software.
✓ If you don't know which COM port the Propeller chip is connected to for programming, press F7 in the Propeller Tool software and make a note of it.
✓ Run PASD.
✓ In PASD, click the COM menu and set the COM port to the Propeller chip's programming port.
✓ In PASD, click File and select Upload Code F11.

F11 VERSUS F2 IN PASD

File → Upload Code F11 in PASD loads the code into the Propeller chip and copies the ASM code to PASD. In contrast, Get ASM Code F2 just copies the ASM code over. It can be used if you manually uploaded the code with the Propeller Tool using either F10 or F11. You can then just use F2 in PASD to make a copy of the ASM code since it has already been loaded into the Propeller Chip. Each time you make a change to the ASM code with the Propeller Tool, make sure to load the modified code into the Propeller chip and get the latest copy into PASD, either with F11 from PASD or a combination of F10 or F11 in the Propeller Tool and F2 in PASD.

PASD then finds the Propeller Tool software, loads its code into the Propeller, makes a copy of the assembly language code, and loads it into PASD. When PASD reappears in the foreground, the address to the left of the first assembly instruction at Addr 00C near the upper-left corner of Fig. 3-24 should be highlighted.

✓ Use the Debug menu to open the Main RAM, Cog RAM, and Pin viewers.
✓ In the Main RAM viewer, select $L dec to display the long value stored at each address as hexadecimal and then as decimal.
✓ In the Cog RAM Viewer, scroll to Addr 02C. It should show the label k. You may also need to make the Value column wider to see the decimal equivalent to the right of its hexadecimal value.
✓ In the PASD window, select the check box to the left of the timekeeper label (Addr 018) to set a breakpoint there.
✓ Click the Debug menu and select Run F5.

The highlighting should advance from Addr 00C to Addr 018, indicating that the code ran until it stopped at the breakpoint.

✓ Arrange the windows as shown in Fig. 3-24.

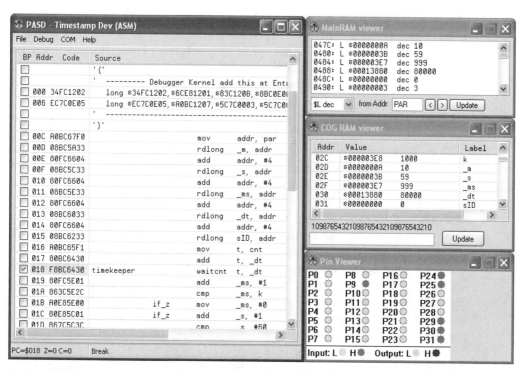

Figure 3-24   PASD assembly debugging session.

The assembly code between the `Debugger Kernel` and `timekeeper` label copies the contents of the `m`, `s`, and `ms` Spin variables in Main RAM to Cog RAM registers named `_m`, `_s`, and `_ms`. PASD lets you examine both the Propeller chip's Main RAM and Cog RAM to make sure the values were copied correctly. The command `cognew(@entry, @m)` launches the assembly code starting at the `entry` label into a new cog, and the address of the Spin variable `m` is passed to the cog's parameter register. The assembly language keyword for this register is `par`. The <u>Main RAM viewer</u> automatically starts displaying Main RAM from the `par` address. Notice that the initial values of 10, 59, and 999 are the first three values in the viewer. These correspond to the values that were loaded into the Timestamp Dev (ASM) object's global `m`, `s`, and `ms` Spin variables. Down in the <u>Cog RAM viewer</u>, the second line from the top shows that `_m` stores the value 10, and the third and fourth lines from the top show that `_s` and `_ms` store 59 and 999, respectively. So the assembly code passed the initialization test.

The PASD software's <u>Debug</u> menu also has <u>Run</u>, <u>Stop</u>, <u>Step</u>, <u>Step over</u>, and <u>Set address</u> for navigating your code while debugging. This software is a truly an invaluable tool for debugging Propeller assembly language code. For the price of a small download, this software can help save significant amounts of development time.

Summary

This chapter focused on debugging code for multiple cogs. The best form of debugging is bug prevention. Parallax addressed bug prevention well in the Propeller chip's design by providing features in the chip's architecture and programming languages that prevent many multiprocessing bugs. Utilizing objects to manage code that runs in other cogs is also an important way to prevent bugs. Prewritten objects that perform many common tasks are available from http://obex.parallax.com and advice on how to incorporate them into applications is also readily available from the Propeller forum at http://forums.parallax.com. Building block objects with code that runs in more than one cog follow a set of conventions for including `Start` and `Stop` methods to launch and shut down code that runs in another cog, along with any methods needed for configuration or information exchange between cogs.

One of the most common root causes of multiprocessing bugs is our natural tendency to forget that different segments of code are executed by different processors. Other common bugs relate to I/O and timing. For I/O, the most common bug is forgetting to add I/O assignments at the start of a method that gets launched into a new cog. If code in a cog is working with certain I/O pins, the method that gets launched into a new cog should initialize the I/O pin settings. Three common coding errors related to timing include forgetting to declare the clock settings, using `waitcnt(delay+cnt)` instead of synchronized delays with `t:=cnt...waitcnt(t+=delay)`, and writing code in a loop that takes longer than the synchronized `waitcnt` delay.

Although contention over an individual memory element is impossible with the Propeller chip, thanks to the fact that its Hub gives each cog main memory access in a round-robin fashion, two different processors accessing a *group* of longs during the same time can still result in corrupted data. This typically happens when one cog is partially done updating a group of values that another cog is reading. The cog doing the reading might read two new values and one old value, for example. The Propeller has built-in memory semaphore bits, called locks, that can be incorporated into code to prevent these memory collisions. Other common memory bugs include forgetting to use the @ operator to pass an address of a variable to a method instead of its value. For Spin methods that get launched into new cogs, make sure to allocate enough stack space for them, because another common bug happens when a method running in another cog doesn't have enough stack for all the calculations it needs to do.

Liberal use of debugging tools with each step of application development will help keep bugs out and reduce overall development time. The TV_Terminal object can be used to display messages from the Propeller chip as well as to test certain demonstration objects posted to the Parallax Object Exchange. The Parallax Serial Terminal software and its accompanying object are also excellent for simple tests, and the PST Debug LITE object can further automate some of the debugging tasks. ViewPort provides an IDE-style debugger along with graphical analysis tools for debugging, testing, and verification. For assembly language debugging, the PASD is a must. All of these software packages make use of objects that reside in one of the Propeller microcontroller's cogs and communicate with PC software to provide variable information and status.

Remember that if you run into a bug that's a real brain teaser, check back with the list at the start of the "Common Multiprocessor Coding Mistakes" section. Also, keep in mind that different segments of code in a multicore application are executed simultaneously. This will help prevent bugs as well as make the ones that do appear in your code easier to spot.

Exercises

1 Incorporate the ASM timestamp code into a building block object.
2 Design a full timekeeping object.
3 Expand the Keyboard_Demo object so that it allows you to enter numeric values that can be placed into variables. Hint: Save time by borrowing and modifying DecIn code from the Parallax Serial Terminal object.

4

SENSOR BASICS AND MULTICORE SENSOR EXAMPLES

Andy Lindsay

Introducing Sensors by Their Microcontroller Interfaces

The Propeller microcontroller's multicore architecture greatly simplifies getting information from many sensors and sensor arrays. The techniques for monitoring run-of-the-mill sensors like switches and resistive elements with the Propeller are similar to those used with most microcontrollers. However, thornier problems such as collecting measurements in parallel can be overcome with programs that make different cores (cogs) monitor different sensors or arrays of sensors. Other cogs can also be programmed to process and store the sensor measurements, and, in most designs, there will still be cogs left over for communication, display, and other application-specific tasks.

This chapter breaks sensors down into their most common microcontroller interfaces and then demonstrates how to measure an example sensor from each interface. With this approach, the chapter introduces a wide variety of sensors with just a few examples. For instance, a *photoresistor* is a sensor whose resistance varies with light. This light sensor can be replaced easily with sensors whose resistances vary with other quantities, such as rotation, salinity, temperature, or surface reflectivity. These four examples are still just the first few entries in a long list of resistive sensors. It turns out that the list gets even longer because the same circuit that is used to measure resistive sensors can also be used to measure capacitive sensors, and the technique used to measure those sensors also applies to certain diode, transistor, and conductivity sensors. So, the simple light sensor really represents an entire class of resistive, capacitive, and other sensors that can be used to measure a wide variety of physical quantities. With that in mind, here is a list of the common microcontroller/sensor interfaces introduced in this chapter:

- On/off
- Resistive, capacitive, diode, transistor, and other
- Pulse and duty cycle outputs
- Frequency outputs
- Voltage outputs
- Synchronous serial
- Asynchronous serial

To further simplify sensor measurements, the Propeller Object Exchange web site (obex.parallax.com) shown in Fig. 4-1 has prewritten objects for many sensors that can serve as "building blocks" for an application. These building block objects can take care

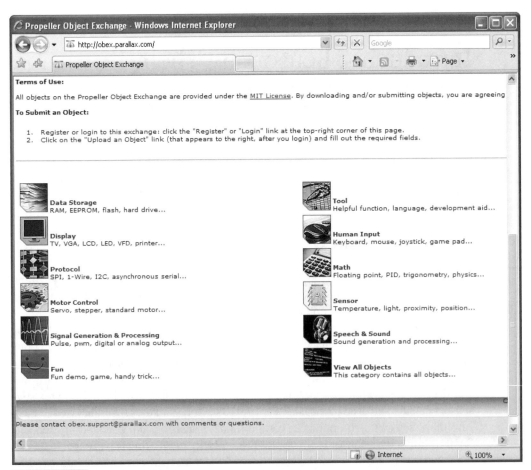

Figure 4-1   Propeller Object Exchange.

of many difficult programming tasks. They can also boil down sensor monitoring to a simple method call that returns the measurement, or, in some cases, a Start method call that arranges for an object to keep one or more variables in an application updated with the latest measurement(s).

Sensor-intensive applications tend to examine measurements and react to them on-the-fly or store them for later analysis, or sometimes both. Figure 4-2 shows examples of audio-spectrum analysis and a prototyping adapter for SD card data storage. These are two examples that address processing and storage included in this chapter. The treatment of processing and storage is not covered in depth here because examples will be plentiful in the project-oriented chapters that follow.

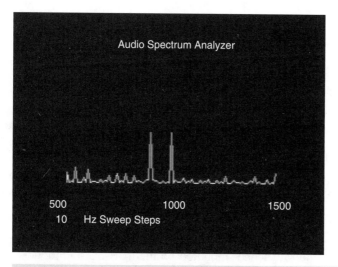

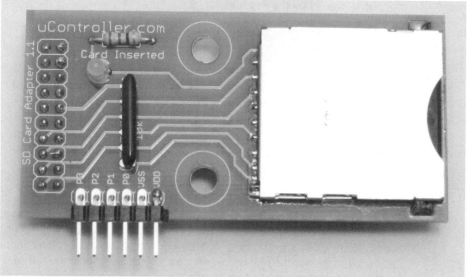

Figure 4-2   Audio signal frequency analysis and SD card adapter.

The examples in this chapter provide some basics on how a variety of sensors work, how the Propeller microcontroller interacts with different sensors to get measurements, and how to write programs that orchestrate the microcontroller's interactions with the sensors. Wherever possible, pointers to more information are included. In general, www.parallax.com has a page for any given sensor that Parallax manufactures or distributes, and the first place to look for more information about a given sensor would be in the Downloads section on the sensor's product page.

Resources: Demo code and other resources for this chapter are available for free download from ftp.propeller-chip.com/PCMProp/Chapter_04.

On/Off Sensors

On/off sensors send high or low signals, depending on whether a certain physical property was detected. The most common on/off sensors are pushbuttons and switches; other examples include infrared and beam-break detectors. Many more sophisticated sensor modules include an onboard microcontroller and/or onboard electronics to provide an on/off output for reduced prototyping effort at a higher price. Examples of this kind of sensor include the Parallax passive infrared (PIR) sensor for motion detection, sound level sensors, and gas sensors such as carbon monoxide, methane, and Propane gas. Figure 4-3 shows examples of some of these sensors, including the pushbutton switches, PIR, and carbon monoxide sensors.

Tip: With some extra work, or in some cases with an object from the Propeller Object Exchange, the raw sensor mounted on the PCB-based modules can be monitored directly by the Propeller chip.

PUSHBUTTONS

Figure 4-4 shows two example pushbutton contact sensor circuits, one with a pull-up resistor, and the other with a pull-down resistor. When a Propeller I/O pin is set to input, pressing the button in the circuit on the left causes a cog's ina register to store a binary-1 in the bit that holds that I/O pin's input state. For example, if a pushbutton circuit is connected to P21, ina[21] would return a 1 if the button is pressed or a 0 if it is released. The circuit on the right will result in the opposite values: 1 if released, 0 if pressed.

Caution: Always include that pull-up or pull-down resistor; do not leave it out. If the pull-up or pull-down resistor is left out, the I/O pin becomes an antenna when the button is not pressed. So the binary-1/0 result would depend on nearby electric fields or, in some cases, nearby I/O pin states, which can fluctuate. In other words, without the pull-up or pull-down resistors, there's no telling what the I/O pin will detect when the button is not pressed. This is why pull-up and pull-down resistors are incorporated into circuits involving electrical contact.

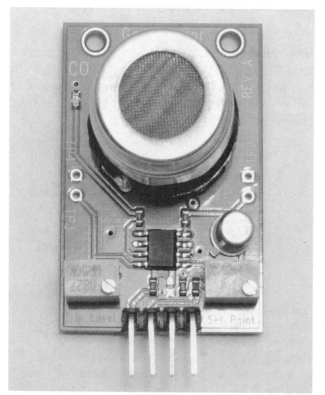

Figure 4-3   Examples of sensors with on/off outputs.

When the button on the left side of Fig. 4-4, is pressed, the I/O pin detects the short circuit to the 3.3 V connection and causes the input register bit for that pin to store a 1. When the button is released, the I/O pin detects ground = 0 V through the 10 kΩ resistor. The resistor "pulls down" the voltage to 0 V when the button is not pressed, and thus the name pull-down resistor.

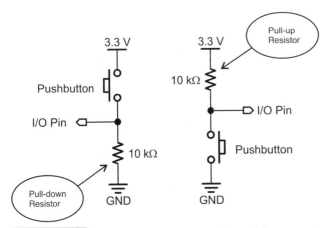

Figure 4-4   **Pushbutton circuits with pull-down and pull-up resistors.**

On/Off Sensor Example: Single Pushbutton   Figure 4-5 shows a single pushbutton connected to Propeller I/O pin P16 on the Propeller Education Kit platform, also known as the PE platform. This pushbutton circuit utilizes a pull-down resistor to keep the voltage applied to the I/O pin at ground when the button is not pressed, and when it gets pressed, the pushbutton connects the I/O pin to 3.3 V.

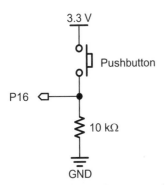

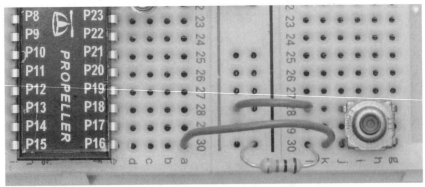

Figure 4-5   **Single pushbutton.**

Information: For explanations of how to build circuits in breadboards, download "What's a Microcontroller?" from www.parallax.com. For more information about the PE Kit, PE Platform, and PE Kit Labs textbook, type "Propeller Education" into the Search field at www.parallax.com, and then click the Go button.

The P16 Input States.spin object displays the state of I/O pin P16 in the Parallax Serial Terminal. When the pushbutton is pressed, the terminal should display "ina[16] = = 1"; and when it's not pressed, the terminal should instead display "ina[16] = = 0." The state of the incoming on/off signal is stored in bit 16 of the cog's ina register, which is `ina[16]`. When the pushbutton is pressed, `ina[16]` returns 1 because the pushbutton circuit applies 3.3 V to I/O pin P16. When the button is not pressed, `ina[16]` returns 0 because the pushbutton applies 0 V to the I/O pin through the pull-down resistor.

```
'' P16 Input States.spin

CON

  _clkmode = xtal1 + pll16x        ' Crystal and PLL settings.
  _xinfreq = 5_000_000             ' 5 MHz crystal x 16 = 80 MHz

OBJ

  pst : "Parallax Serial Terminal"  ' Serial communication object

PUB Go

  pst.Start(115200)                ' Start Parallax Serial Terminal

  repeat
    pst.Str(String(pst#HM, "ina[16] == "))
    pst.Bin(ina[16], 1)            ' Display 1 binary digit
    waitcnt(clkfreq/20 + cnt)      ' Wait 1/20 second
```

Figure 4-6 shows the Parallax Serial Terminal display while the pushbutton is not pressed. Of course, when the pushbutton is pressed, the display will instead read "ina[16] = 1." Assuming you have already tried programs in earlier chapters, you now know the steps for loading a program into the Propeller chip and displaying messages from the Propeller chip in the Parallax Serial Terminal. To test P16 Input States.spin, follow these steps:

✓ Use the Propeller Tool software to load the application into the Propeller chip.
✓ Click the Parallax Serial Terminal's <u>Enable</u> button.
✓ Watch the display as you press and release the pushbutton, and verify that it displays the correct binary values stored by `ina[16]` for the pressed and released button states.

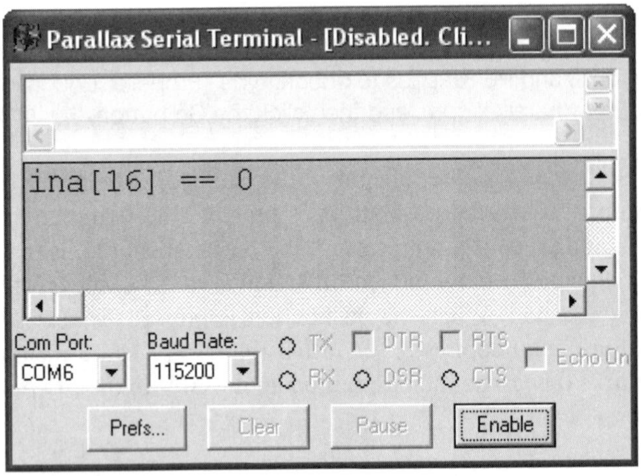

Figure 4-6   P16 input state display.

Keep in mind that the pushbutton is just one example of an on/off sensor. Also keep in mind that once the code has acquired that 1 or 0 measurement, it's up to the application designer to decide what to do with the information and write code to make it so. The Spin language has lots of flexibility. Pushbutton Decisions.spin shows a simple next step that calls different methods, depending on whether the sensor is sending the on/off signal. In Fig. 4-5, the pushbutton is the circuit sending those signals, but again, some completely different sensor could be sending on/off signals. While the program's Pressed and NotPressed methods just display some simple messages, an application might instead acquire a timestamp, move an actuator, update a display, start a mechatronic calibration sequence, sound an alarm, make a phone call, or maybe all of the above. With the Propeller microcontroller, all the tasks might even be performed in parallel by separate cogs.

```
'' Pushbutton Decisions.spin

CON

  _clkmode = xtal1 + pll16x       ' Crystal and PLL settings.
  _xinfreq = 5_000_000            ' 5 MHz crystal x 16 = 80 MHz

OBJ

  pst : "Parallax Serial Terminal"  ' Serial communication object

PUB Go

  pst.Start(115200)               ' Start Parallax Serial Terminal
```

```
repeat
  pst.Str (String(pst#HM, "ina[16] == "))
  pst.Bin (ina[16], 1)              ' Display 1 binary digit
  waitcnt (clkfreq/20 + cnt)        ' Wait 1/20 second
  if (ina[16] == 1)                 ' If ina[16] stores 1
    Pressed                         '   call pressed method
  else                              ' otherwise
    NotPressed                      '   call NotPressed method

PUB Pressed

  pst.Str (String(pst#NL, "Button pressed!", pst#CE))


PUB NotPressed

  pst.Str (String(pst#NL, "Button not pressed.", pst#CE))
```

On/Off Sensor Example: Pushbutton Array   Figure 4-7 shows a pushbutton array that includes two additional copies of the P16 pushbutton circuit. These copies are connected to adjacent I/O pins P17 and P18. This provides a small example of an on/off sensor array.

The Spin language has a simple feature to check multiple inputs on any range of contiguous I/O pins in a port. Instead of ina[16], which would just return the state of P16, an application can use ina[18..16] to check all three pushbuttons with one command. Ina[18..16] returns a three-digit binary number, with each digit indicating the state of one of the pins. For example, if the P18 and P16 pins were pressed, ina[18..16] would return %101. If the P18 and P17 I/O pins were pressed instead, ina[18..16] would return %110. Or, if just P16 were pressed, it would return %001.

✓ Build the circuit shown in Fig. 4-7.
✓ Load P18 to P16 Input States.spin into the Propeller chip.
✓ Test the pushbuttons by displaying their states with the Parallax Serial Terminal.

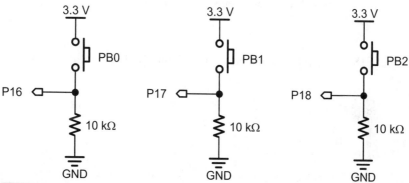

Figure 4-7   Pushbutton example of an on/off sensor array.

```
'' P18 to P16 Input States.spin

CON

   _clkmode = xtal1 + pll16x          ' Crystal and PLL settings.
   _xinfreq = 5_000_000               ' 5 MHz crystal x 16 = 80 MHz

OBJ

   pst : "Parallax Serial Terminal"   ' Serial communication object

PUB Go

   pst.Start(115200)                  ' Start Parallax Serial Terminal

   repeat
     pst.Str(String(pst#HM, "ina[18..16] == %"))
     pst.Bin(ina[18..16], 3)          ' Display 3 binary digits
     waitcnt(clkfreq/20 + cnt)        ' Wait 1/20 second
```

With three pushbuttons controlling three different binary digits, there are now 2^3 different combinations of binary results, from %000 to %111, which is decimal 0 to 7. Case statements can be useful for picking out certain conditions. For example, the case statement in the object P18 to P16 Input Decisions.spin first stores the value of ina[18..16] in a local variable named states. Later, in the Go method, a case statement checks if the P18 pushbutton has been pressed. The first condition the case statement checks for is %100..%111. This is all the values from %100 to %111, or decimal 4 to 7, which includes %100, %101, %110, and %111. If any of those conditions is true, the TopButton method gets called. Upon return, a waitcnt command executes, and then, the loop repeats. The case statement does not continue down through the other conditions; after the first true condition is detected, the code executes commands in the case block and then exits the case statement. Also, case statement conditions do not necessarily have to contain single-method calls. The other condition in the case statement has a code block with two method calls, though it could also contain expressions, commands, and so on.

✓ Test P18 to P16 Input Decisions.spin with the Parallax Serial Terminal.

```
' P18 to P16 Input Decisions.spin

CON

   _clkmode = xtal1 + pll16x          ' Crystal and PLL settings.
   _xinfreq = 5_000_000               ' 5 MHz crystal x 16 = 80 MHz
```

```
OBJ

  pst : "Parallax Serial Terminal"   ' Serial communication object
PUB Go | states

  pst.Start(115200)                   ' Start Parallax Serial Terminal

  repeat
    states := ina[18..16]
    pst.Str(String(pst#HM, "ina[18..16] == %"))
    pst.Bin(states, 3)                ' Display 3 binary digits
    case states                       ' Evaluate case by case
      %100..%111: TopButton           ' Method calls for buttons
      %010..%011: MiddleButton
      %001      : BottomButton
      other     :                     ' Otherwise, clear line
        pst.Newline
        pst.ClearEnd

    waitcnt(clkfreq/20 + cnt)         ' Wait 1/20 second

PUB TopButton
  pst.Str(String(pst#NL, "Emergency, top button is pressed!", pst#CE))

PUB MiddleButton
  pst.Str(String(pst#NL, "Warning, middle button is pressed!", pst#CE))

PUB BottomButton
  pst.Str(String(pst#NL, "Note, bottom button is pressed.", pst#CE))
```

MULTIPROCESSING EXAMPLE

With objects that take care of the multiprocessing grunt work, multicore application code becomes exceedingly simple. Consider P16 Multiprocessing Example.spin. If it weren't for the convention that a call to an object's Start methods results in one or more launched cogs, it would be difficult to tell that this program is making use of three cogs. The top object executes in one cog, and both the PST Debug LITE and Timestamp objects have code that other cogs execute. Since the two building block objects have methods that take care of information exchanges with the code running in the other cogs, P16 Multiprocessing Example.spin just looks like it's a single loop that makes a few method calls.

```
'' P16 Multiprocessing Example.spin

CON

  _clkmode = xtal1 + pll16x         ' Crystal and PLL settings.
  _xinfreq = 5_000_000              ' 5 MHz crystal x 16 = 80 MHz
```

```
OBJ

  debug : "PST Debug LITE"          ' Serial communication object
  time  : "TimeStamp Object"        ' TimeStamp object
PUB Go | minutes, seconds, milliseconds

  debug.Style(debug#COMMA_DELIMITED)' Configure debug display
  debug.Start(115200)               ' Start Parallax Serial Terminal
  time.Start(0, 0, 0)               ' Start timestamp cog

  repeat                            ' Main loop
    debug.Str(String("ina[16] == "))
    debug.Bin(ina[16], 1)           ' Display 1 binary digit
    debug.NewLine                   ' New line
    if ina[16] == 1                 ' If button pressed
      time.GetTime(@minutes)        ' Get timestamp & display
      debug.Vars(@minutes, String("| minutes, seconds, milliseconds"))
    waitcnt(clkfreq/20 + cnt)       ' Wait 1/20 second
    debug.Position(0, 4)
```

Figure 4-8 shows the Parallax Serial Terminal display, which gives the most recent value from the Timestamp object after a button press. Note that it also displays the state of P16 in the same way that the P16 Input States.spin object did. The P16 Multiprocessing Example object used the PST Debug LITE object instead of the Parallax Serial Terminal object to simplify displaying the list of time variables. Since the PST Debug LITE object is just the Parallax Serial Terminal object with some extra features, all of the Parallax Serial Terminal

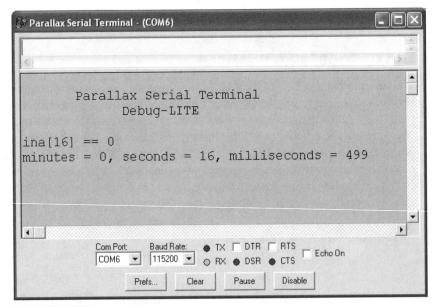

Figure 4-8   Customized PST Debug LITE display.

methods are still available to the application. The application uses debug.Position to place the cursor at zero spaces over and four spaces down between each repetition of the loop. Along with the .str method call that displays "ina[16] ==," this application demonstrates how one of the PST Debug LITE's display modes can be customized for different debugging tasks. As an aside, removing the debug.Style method call could be more beneficial here since it would then display the I/O pin direction and states along with the variable values.

CONVERTING ANALOG SENSORS TO ON/OFF SENSORS

Many analog sensors can be incorporated into circuits that behave as on/off sensors by virtue of the fact that Propeller microcontroller I/O pins have a 1.65 V input threshold voltage. Provided the voltage applied to an I/O pin set to input is in the 0 to 3.3 V range, a cog interprets voltages above 1.65 V as binary-1 and voltages below 1.65 V as binary-0. For example, the voltage output at the potentiometer's W terminal in Fig. 4-9 varies from 0 to 3.3 V as the knob is turned. Half of its range of motion will result in voltages above 1.65 V and the other in voltages below 1.65 V. So, as the potentiometer's knob is turned from one end of its range to the other, the voltage at its W terminal will pass the 1.65 V threshold at about the halfway point, and the Propeller can detect this as a change from 0 to 1 or in the other direction from 1 to 0.

✓ Test the circuit shown in Fig. 4-9 with P18 to P16 Input States.spin. Keep an eye on the middle of the three binary digits; it's the one that displays the state of P17.

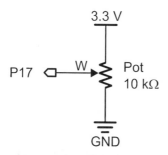

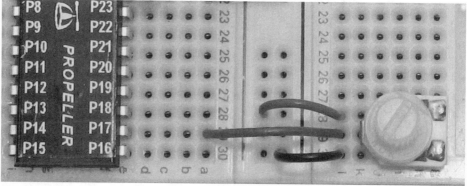

Figure 4-9   Potentiometer in an on/off circuit.

A photoresistor's resistance varies with light intensity on its cadmium sulfide element. This analog sensor can be converted to a digital sensor for the Propeller chip by placing it in series with a fixed-value resistor and then connecting the node between the two resistors to an I/O pin, as shown in Fig. 4-10. As the light levels change, the photoresistor's resistance changes and the voltage between the photoresistor and the fixed resistor in Fig. 4-10 also changes. For that particular circuit, brighter light levels reduce the photoresistor's resistance, which in turn makes the voltage between the two resistors increase. Dimmer light levels cause an increase in the photoresistor's resistance, so the voltage between two resistors decreases. To set the threshold, simply measure the resistance of the photoresistor at the light level that should cause the circuit's output to be 1.65 V, and then pick a fixed resistor of the same value. For example, if the photoresistor's resistance is measured at 10 kΩ, with the light level applied that should cause the transition, the fixed resistor should also be 10 kΩ. Reason being, the voltage between two equal resistors in series is half the voltage applied across both of them. Since 3.3 V is applied across both resistors, the voltage between them when they are equal will be 1.65 V, which is the Propeller chip's I/O pin threshold voltage.

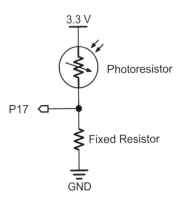

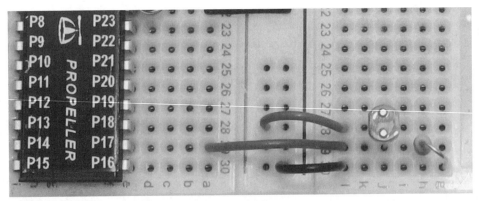

Figure 4-10   **Photoresistor as an on/off circuit.**

Information: The Cadmium Sulfide (CdS) cell or photoresistor was one of the most common ambient light sensors built into devices. With the advent of the European Union's Restriction of use of certain Hazardous Substances (RoHS) directive, cadmium sulfide photoresistors can no longer be built into devices imported into or manufactured in Europe. This has given rise to a number of photoresistor replacement products, including certain phototransistors and linear light sensors. Many of these devices are designed to be drop in replacements for photoresistors, making it possible for manufactures to comply with RoHS without redesigning all their circuit boards to accommodate new parts and circuits.

ON/OFF SENSORS THAT DEPEND ON SIGNALING

Some on/off sensors require special signals to make them work properly. One example is the Parallax CO sensor, which has a control pin that requires a sequence of high/low voltage levels for reliable readings from its alarm pin. Another example, shown in Fig. 4-11, is an infrared object detection circuit. The tube on the right houses an infrared light emitting diode (IR LED) like the ones found in common TV remotes, and the sensor on the left is a PNA4602 infrared sensor, which can be found in many TV sets and other entertainment system components. If this sensor detects infrared flashing on/off in the 38 kHz range, it sends a low signal; otherwise, it sends a high signal. In the circuit in Fig. 4-11, it sends a low signal if the infrared light is reflected by a nearby object; otherwise, it sends a high signal.

It might seem like another cog would be required for sending the 38 kHz signal, but that's not the case. Each of the Propeller microcontroller's eight cogs has two counter

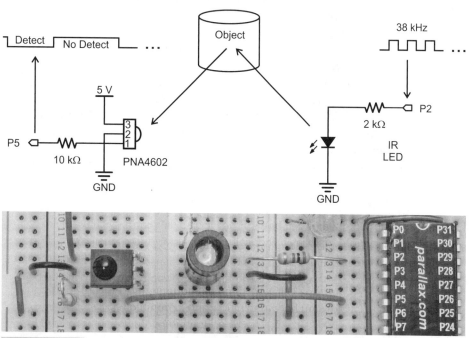

Figure 4-11 Infrared object detector.

modules. A counter module is a configurable state machine that's capable of performing a variety of tasks independently for its cog. One of those tasks is square wave generation, and counters are capable of generating frequencies that range from 0 Hz to 128 MHz.

Information: Other examples of independent tasks that counter modules can perform include pulse generation, pulse and decay measurement, and duty modulation for digital to analog (D/A) conversion, and that's still just scratching the surface. For more information about counter modules, get the Propeller *Education Kit Labs: Fundamentals* PDF textbook from www.parallax.com, and consult the Counter Modules and Circuit Applications lab. Additionally, an extensive list of links to counter module documentation and applications for the Propeller is available from http://www.parallax.com/go/counters.

The Propeller Education Kit Labs: *Fundamentals* book and the PE Kit Tools section in Propeller Forum's Propeller Education Kit Labs sticky-thread make use of an object named SquareWave as a tool for generating square waves utilizing a cog's counter modules. SquareWave is similar to the Synth object in the Propeller Library, and both SquareWave and Synth were developed from example code from the Propeller Library's CTR.spin object. Instead of launching another cog, these objects configure a counter module to generate the square wave in the background so that the cog's code can move on to other tasks in the foreground.

The Test IR Detect.spin application uses the SquareWave object to transmit the 38 kHz square wave signal to the IR LED circuit connected to P2. The SquareWave object does not have a Start method since it's not launching a process into another cog. So, the main loop calls the SquareWave object's Freq method and passes it I/O pin 2, channel 0, and a frequency of 38,000 Hz. The Freq method's channel parameter can be either 1 or 0 because each cog has two counter modules: A and B. The SquareWave object's Freq method expects a channel parameter of 0 to select counter module A, or 1 to select counter module B. This example selects counter module A to generate the 38 kHz square wave, but it could just as easily have selected counter B with a channel argument of 1. After transmitting the 38 kHz signal for 1 ms, the code copies the state of ina[5] to a variable named detect. Before displaying the detect variable's value and repeating the loop, the application code makes a second call to the SquareWave object's Freq method, passing a frequency of 0 Hz to stop the 38 kHz signal to the IRLED.

```
'Test IR Detect.spin
CON

  _clkmode = xtal1 + pll16x        ' Crystal and PLL settings.
  _xinfreq = 5_000_000             ' 5 MHz crystal x 16 = 80 MHz

OBJ

  pst : "Parallax Serial Terminal"  ' Serial communication object
  sqw : "SquareWave"                ' Square wave object
```

```
PUB go | detect

  pst.Start(115200)                    ' Start Parallax Serial Terminal

  repeat
    sqw.Freq(2, 0, 38000)              ' P2 IR LED flicker at 38 kHz
    waitcnt(clkfreq/100 + cnt)         ' Wait 10 ms
    detect := ina[5]                   ' Check IR receiver
    sqw.Freq(2, 0, 0)                  ' Turn off IR LED

  ' Cursor home, display label, detector state, clear to end of line.
  pst.Str(String(pst#HM, "IR Detector = "))
  pst.Dec(detect)
  pst.ClearEnd
```

Information: This example is an adaptation of Test IR Detect.spin in *PE Kit Tools: Transmit Square Wave Frequencies*. This post is part of the PE Kit Tools series, published in the Propeller Education Kit Labs sticky-thread in the Propeller forum at http://forums.parallax.com.

The IR receiver is an active-low device, meaning it sends a low signal to indicate that it is receiving a 38 kHz infrared signal, or a high signal to indicate that it is not receiving the signal. To test this program and circuit:

✓ Build the IR LED + Detector circuit shown in Fig. 4-11.
✓ Load Test IR Detect.spin into the Propeller chip.
✓ Enable the Parallax Serial Terminal.
✓ Verify that the Parallax Serial Terminal displays IR Detector = 1 when an object is not detected (Fig. 4-12) and IR Detector = 0 when an object is detected.

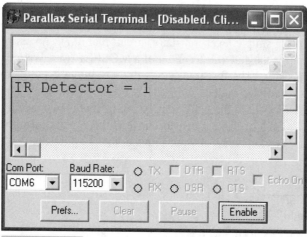

Figure 4-12 Object not detected.

SENSORS WITH OUTPUTS GREATER THAN 3.3 V

Sensors that transmit voltages outside the 0 to 3.3 V range should never be connected directly to a Propeller I/O pin. Lots of circuits are available for making a sensor's higher voltage level output compatible with the Propeller chip's 3.3 V input. Some examples of circuits to protect a Propeller I/O pin from higher voltage signal levels include series resistor circuits, transistor circuits, level translator integrated circuits (ICs), optoisolators, and solid state relays.

Series resistors are the quickest and easiest solution, especially for prototyping. Fig. 4-13 shows an example of a 3.9 kΩ resistor in series between a 5 V sensor output and a 3.3 V Propeller I/O pin. This circuit provides adequate protection, and a 10 kΩ resistor would also be fine for most applications. Note that the 5 V output infrared detector in Fig. 4-11 has a 10 kΩ resistor between its output and the I/O pin to provide this protection.

The reason a series resistor suffices is because Propeller I/O pins have protection diodes that drain current to the Propeller chip's positive supply if an incoming signal is above 3.3 V, or to ground if an incoming signal is less than 0 V. Assuming an incoming 5 V signal through a 3.9 kΩ series resistor, there will be a 1.25 V drop across the series resistor and a 0.45 V drop across the built-in protection diode, as the 3.3 V supply absorbs the current (Voltage drop values are approximate and will vary slightly). The resulting voltage applied to the I/O pin is 3.75 V, which is still safe for the I/O pin, provided the current into the diode is under the 500 μA limit specified in the *Propeller Datasheet*. Using voltage = current · resistance, or V = IR, with V = 1.25 V and R = 3.9 kΩ, the current through the resistor (and the diode) will be about 321 μA, which is well under 500 μA limit.

Note: Although the Propeller Datasheet specifies 3.6 V as the absolute maximum allowable voltage applied to an I/O pin, that value assumes no series resistance. With series resistance, it's the protection diode's job to make sure the I/O pin can tolerate the applied voltage.

If a sensor is adversely affected by the current load resistor protection puts on its output, Fig. 4-14 shows a simple level-translator circuit that draws no current thanks to that fact that its input is the gate (G) terminal of a MOSFET (metal oxide semiconductor field effect transistor). The first transistor's drain terminal is pulled up to Vcc, the supply for the higher voltage system, with a 10 kΩ resistor. The second transistor's drain (D) terminal is pulled up to the 3.3 V supply, so the highest voltage it will send to the I/O pin is 3.3 V. Each of these transistors inverts the signal. Since the two are cascaded, the signal is inverted twice, and the signal at the I/O pin is the 3.3 V version

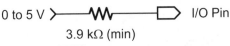

0 to 5 V   I/O Pin

3.9 kΩ (min)

Figure 4-13   Protect an I/O pin with a series resistor.

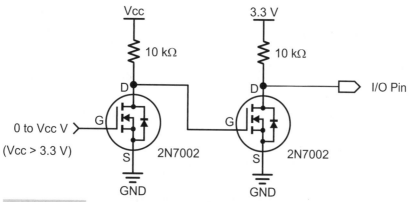

Figure 4-14   Level shifter with two FET transistors.

of the input signal V(Vcc). In other words, if V(Vcc) is high—6 V, for example—the voltage at the I/O pin will be 3.3 V. On the other hand, if V(Vcc) is low—0 V, the voltage at the I/O pin will be 0 V.

Other interfaces for higher voltages include level-translator ICs, optoisolators, and solid-state relays. Level-translator ICs are sometimes referred to as *level shifters,* and they provide voltage translation in a single chip. Some level translators boast extra protection circuitry, as well as maximum switching speeds that are significantly faster than the circuit in Fig. 4-14, which is only good for switching speeds up to 20 or so kHz. *Optoisolators* also come in IC packages, and they provide the I/O pin with a high degree of protection because the high and low signals are transmitted optically instead of electrically. With an optoisolator, the sensor's output gets connected to an infrared LED that's inside a small enclosure. If the sensor sends a high signal, the infrared LED emits light. The infrared LED is right next to an infrared transistor in the enclosure with a pull-up resistor that is connected to the lower voltage system's supply. So the transistor's output will pass either 3.3 or 0 V to the Propeller I/O pin, depending on whether the sensor's output turned the LED on or off. Since an optoisolator circuit uses light to carry the signal, the two systems can be kept electrically separate, not even requiring a common ground. *Solid-state relays* (SSRs) are designed to either pass a signal from a higher voltage system to a lower voltage system, or take a signal from a lower voltage system to control a higher voltage system. One of the low-voltage options is 3.3 V; some of the more common higher voltage options include 12 or 24 VDC, 110 or 220 VAC.

Resistive, Capacitive, Diode, Transistor, and Other

This category features lots of different analog sensors that can all be measured with the same technique, commonly referred to as *RC decay* or *RC time* measurement. The list of different sensors that can be measured with this technique is extensive. For example, three common resistive sensors are the photoresistor, potentiometer, and thermocouple. The photoresistor's resistance varies with light level. The potentiometer's resistance varies with the position of its adjusting knob or screw, and the thermocouple's resistance varies with temperature. You can find devices for many different physical properties, including gas concentration, force, humidity, temperature, fluid level, and salinity, and that's still just a sampling of the myriad of resistive sensors available. Examples of capacitive sensors include another type of humidity sensor, as well as touchpad and displacement sensors. Photodiodes and phototransistors are also compatible with this technique and tend to be selective for particular colors of light, and some are available with optical filters to make them even more selective.

HOW RC DECAY MEASUREMENTS WORK

RC decay circuits have resistive (R) and capacitive (C) elements connected in parallel to an I/O pin. If the R element is a variable sensor, the C element has to be a fixed value. Conversely, if the C element is a variable sensor, the R element has to be a fixed value. The example in Fig. 4-15 uses a variable resistor, which means the resistive element is the sensor and the capacitor is a fixed value. Figure 4-15 also shows the interaction between the Propeller I/O pin and the RC circuit. On the left, the I/O pin charges up the capacitor, which behaves like a small battery, to 3.3 V by transmitting a high signal. Then, in the middle of Fig. 4-15, the code changes the I/O pin to input. From the circuit's point of view, its voltage source just disappeared, because when an I/O pin is an input, its high-input impedance makes it invisible to most circuits. Even though the I/O pin is invisible to the circuit, the circuit's voltage is visible to the I/O pin, so it can still monitor

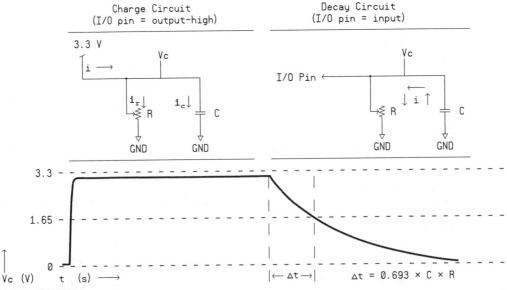

Figure 4-15   **Propeller RC decay measurement.** (*Excerpt from Propeller Education Kit Labs: Fundamentals*)

and detect when the voltage decays below the I/O pin's 1.65 V input threshold. Since a cog, or even a cog's counter module, can also measure the time between when the I/O pin was changed from output-high to input and when the voltage decayed below the I/O pin's 1.65 V threshold, it can capture that Δt needed to calculate the sensor's resistance or capacitance. More importantly, since the sensor's resistance or capacitance varies with some physical property, the microcontroller can determine the physical quantity the resistive or capacitive sensor measures by measuring the decay time.

RC decay is short for resistor capacitor decay, and it relies on the fact that the time it takes a capacitor to discharge through a resistor to half of its starting voltage is:

$$\Delta t = 0.693 \cdot C \cdot R$$

If C is a fixed value and R is a resistive sensor, the value of R is:

$$R = \Delta t/(0.693 \cdot C)$$

Likewise, if R is fixed and C is variable, $C = \Delta t/(0.963 \cdot R)$

In either case, so long as one value is fixed, a decay time measurement can be used to determine the other value.

RC DECAY EXAMPLE: MEASURE A POTENTIOMETER'S POSITION

The Test Simple RC Time.spin application interacts with the circuit in Fig. 4-16 to display a number that indicates the potentiometer's position based on the measurement of its variable resistance. In the previous section, the potentiometer was in a circuit that caused it to provide the Propeller I/O pin with a variable voltage source. In this circuit,

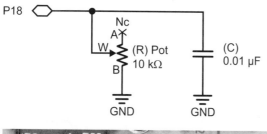

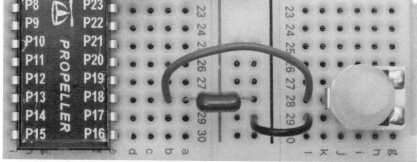

Figure 4-16   RC decay test circuit.

it is instead connected as a variable resistor with the A terminal disconnected, which is commonly referred to as *floating*. As the knob gets turned, the wiper (W) terminal's contact with the resistive element inside the potentiometer moves along its length. When the knob is turned in one direction, the contact point will get closer to the A terminal and the resistance between B and W will get larger. In the other direction, the contact point will get closer to the B terminal, and the resistance between W and B decreases.

The test application for measuring this circuit utilizes an object named RC Time, which comes from the PE Kit Tools: Measure Resistance and Capacitance post in the Propeller forum. The article in this post demonstrates many uses and applications of the RC Time object, which can be configured to take sequential measurements in the same cog or parallel measurements in one or more other cogs. It can also update the parent object's variable for storing the measurement results from another cog according to a sampling rate for control systems and data logging applications.

The simplest application of the RC Time object is demonstrated here with Test Simple RCTIME.spin, which repeatedly calls the RC Time object's Time method. The Time method expects three arguments: I/O pin, starting state (1 for decay or 0 for growth), and the address of a variable that the Time method should store the result in. In the example program, the Time method takes care of all the signaling in Fig. 4-15 and places the measurement result in the tDecay variable. In this particular example, all the measurements are happening in the same cog.

```
'' Test Simple RCTIME.spin

CON

    _clkmode = xtal1 + pll16x       ' Crystal and PLL settings.
    _xinfreq = 5_000_000            ' 5 MHz crystal x 16 = 80 MHz

OBJ

    rc  : "RC Time"                 ' RC decay/growth object
    pst : "Parallax Serial Terminal" ' Serial communication object

PUB Go | tDecay

    pst.Start(115200)               ' Start Parallax Serial Terminal

    repeat                          ' Repeat loop
      waitcnt(clkfreq/10 + cnt)     ' Refresh display at 10 Hz

      rc.time(18, 1, @tDecay)       ' Measure P17 decay circuit

      ' Display measurement
      pst.Str(String(pst#CE, pst#HM, "tDecay = "))
      pst.Dec(tDecay)
```

Tip: The reason the `Time` method requires the address of a result variable instead of just returning its value is to make simultaneous measurements in other cogs possible with the same method call that works for sequential decay measurements.

The potentiometer example in the on/off sensors section only returned a 1 or a 0. In this example, the potentiometer's analog range of motion is digitized into approximately 6000 different values, from 0 to just over 6000. A 0 measurement indicates that the dial is turned all the way counterclockwise, and 6000+ indicates all the way clockwise. The Parallax Serial Terminal in Fig. 4-17 indicates that the knob has been turned about 4000/6000, or two-thirds of the way clockwise. This digitized measurement represents the decay time in terms of 12.5 ns clock ticks. The RC Time object utilizes one of the cog's counter modules to take the measurement. Code in the object configures the counter module to count the number of clock ticks during which it detects a high signal (above 1.65 V) applied to the I/O pin by the RC circuit. While the counter module counts clock ticks, the cog waits until it detects the RC circuit's transition from high to low. Then it copies the accumulated clock ticks from the counter module to the memory address that the parent object passed to the `Time` method. For example, the variable address was `@tDecay` in Test Simple RCTIME.spin.

✓ Try replacing the potentiometer with the photoresistor. Simply remove the potentiometer and plug the photoresistor into rows 28 and 29 in Fig. 4-16. When you run the test application, the Parallax Serial Terminal will display smaller values corresponding to brighter light or larger values corresponding to dimmer light.

Since the decay time is $\Delta t = 0.693 \cdot C \cdot R$, a capacitor that's 10 times the size will result in decay times (and measurements) that are 10 times as large. Likewise, with a capacitor that's one-tenth the size, the measurements take one-tenth the time.

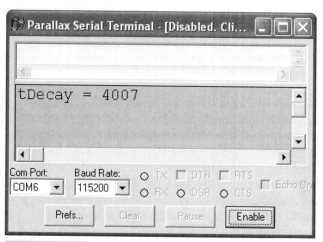

Figure 4-17   RC decay measurements in the Parallax Serial Terminal.

Try different size capacitors. The RC Time object has a `TimeOut` method that defaults to 10 ms (`clkfreq/100`). For larger capacitors, it may be necessary to call the RC Time object's `TimeOut` method and pass it a larger timeout value, such as 100 ms, which is `clkfreq/10`.

OTHER RC DECAY AND GROWTH CIRCUIT VARIETIES

Photodiodes generate current flow in proportion to light intensity, and devices such as the AD592 temperature probe allow current to pass proportional to temperature. Phototransistors also allow current through based on light intensity, but the relationship between current and light is not necessarily linear. In all these cases, instead of allowing the capacitor to discharge through the element, these devices conduct current either into or out of the capacitor. Depending on the current direction, voltage across the capacitor will either increase or decrease, and the resulting growth or decay measurement can be used to quantify the current and the physical property it represents.

Information: For more information on the AD592 temperature probe and blue-enhanced photodiode featured in this section, see the Applied Sensors Kit and PDF documentation available from www.parallax.com. This is also a good source of information for fluid level and salinity sensing with RC decay.

The AD592 temperature probe shown in Fig. 4-18 is designed to measure the temperature of liquids. For this circuit, the I/O pin has to apply a low signal to discharge the capacitor. When the I/O pin direction is changed to input, the AD592 allows current to pass in proportion to temperature, and the voltage at the upper capacitor plate increases.

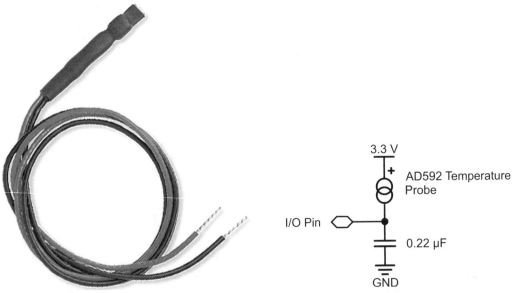

Figure 4-18 AD592 temperature probe.

Then, the Propeller measures the time from when the I/O pin changed to input to the time the voltage at the capacitor's top plate passes the 1.65 V logic threshold. Since this sensor circuit measures growth instead of decay, the starting state in the `rc.time` method call should be changed from 1 to 0: `rc.time(18, 0, @tDecay)`. This will cause the RC Time object to set the I/O pin low to discharge the capacitor and then measure the amount of time it takes the capacitor to charge from 0 to 1.65 V. The measurement will be in terms of clock ticks. It also would be a good idea in this case to rename the variable from `tDecay` to `tGrowth`.

> **Tip:** Converting clock ticks to other time increments is a simple matter of dividing the number of clock ticks in a given unit into the clock ticks in the measurement. First, define the unit. Microseconds can be defined by `us := clkfreq/1_000_000`. That's the number of ticks in 1/1_000_000th of a second. Next, `usResult := tDecay/us` converts the number of clock ticks into a number of microseconds.

A photodiode generates small currents that are proportional to light intensity. Figure 4-19 shows a photodiode and a schematic for an RC Decay circuit that incorporates the photodiode. The I/O pin has to be set high, but this time to discharge the capacitor and set the voltage difference across its plates to 0 V. When the I/O pin changes to input, the current from the photodiode charges the capacitor. As the voltage across the capacitor increases, the voltage at the lower plate drops since the upper plate is connected to the 3.3 V supply and its voltage cannot move. So the microcontroller measures the time it takes for the voltage at the lower plate to drop from 3.3 to 1.65 V. The code for this behaves identically to the RC decay measurements discussed earlier since it starts by applying a high signal. Then the I/O pin changes to input and waits for the voltage to drop from 3.3 to 1.65 V. RC measurements with a photodiode can be slow. The RC Time object's default measurement timeout should be adjusted to a larger value with a call to its `TimeOut` method.

Figure 4-20 shows a schematic of the QTI module, which measures infrared reflectivity of surfaces. It's popular with robotics hobbyists for determining whether the robot

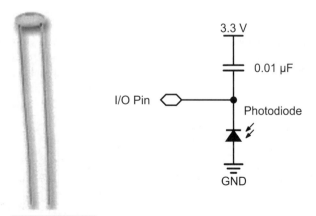

Figure 4-19   **Photodiode and circuit.**

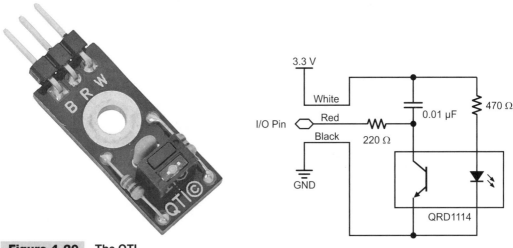

Figure 4-20   **The QTI.**

is above a black surface, which absorbs infrared, or a white surface, which reflects it. The letter q denotes charge, and the product name QTI was chosen as an abbreviation of *charge transfer infrared*. The module has a built-in infrared LED circuit that emits infrared light. If the infrared light reflects off a nearby (about one-eighth of an inch away) white surface and strikes the infrared transistor's light collecting surface, it will conduct much more current than if only a small amount gets reflected back from an infrared-absorbing black surface. The name "charge transfer" refers to the fact that the infrared transistor is controlling the current, or charge transfer, into the capacitor. The QTI module is another circuit that has to start with a high signal that discharges the capacitor by pushing the voltage at the capacitor's lower plate up to 3.3 V.

The QTI sensor needs 3.3 V connected to the White (W) terminal and 0 V connected to the Black (B) terminal. Then, the Red (R) terminal can be connected to an I/O pin for RC measurements. For example, if you connect the R terminal to P17, Test Simple RC Time can be used to measure this sensor.

> **Tip:** For best results, use black vinyl electrical tape on white paper or poster board. Not all printers make black marks that absorb infrared. Most, but not all, black/white laser printers do a good job, but photo printers are definitely hit-and-miss.

Pulse and Duty Cycle Outputs

Figure 4-21 shows examples of two sensors in the pulse and duty cycle output category: the Ping))) Ultrasonic Distance Sensor and the Memsic 2125 Accelerometer. Another example of a sensor that transmits pulses is the infrared receiver introduced in the On/Off Sensors That Depend on Signaling section, which sends pulses that relay a TV remote's infrared signals. The infrared LED inside the remote turns on/off rapidly,

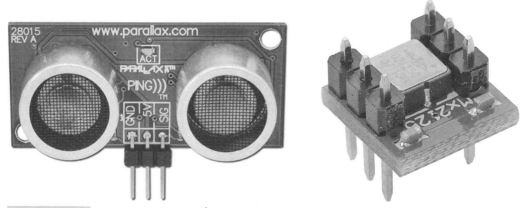

Figure 4-21 Ultrasonic Distance Sensor and dual-axis accelerometer.

typically in the 38 kHz range. Whenever the infrared detector inside the TV senses infrared flashing on and off at 38 kHz, it sends a low signal; otherwise, it sends a high signal. The infrared remote controls the amounts of time it transmits the 38 kHz signals to control the pulses that the receiver inside the TV sends to the TV's microcontroller. The microcontroller's firmware knows how to interpret these pulses and controls the TV accordingly. Radio control (also abbreviated RC) signals for RC cars, boats, and planes can also be monitored for the pulses they send with a Propeller microcontroller. This makes it possible to give programmed intelligence to remote-controlled hobby applications.

The Memsic 2125 Accelerometer and Ping))) Ultrasonic Distance Sensor provide two examples of sensors that communicate with pulse durations. Both of these devices send pulses to the microcontroller, and the pulse durations (the amount of time the pulses they send last) provide measurements of physical quantities. In the case of the Ping))), it provides an indication of the distance to an object; in the case of the accelerometer, it provides acceleration information, which can, in turn, be used to indicate velocity and position, or just simple tilt by measuring the acceleration due to gravity.

PING))) ULTRASONIC DISTANCE SENSOR—OBJECT EXCHANGE EXAMPLE

Figure 4-22 shows how a microcontroller gets a distance measurement from the Ping))) Ultrasonic Distance sensor. The microcontroller has to send the Ping))) sensor a pulse to make it emit an ultrasonic chirp. The duration of the pulse the Ping))) sensor sends in reply represents the time it takes for the ultrasonic chirp to make a round trip between the sensor and a nearby object. Knowing the speed of sound in air, the application code can use this time measurement to calculate the distance. The Ping))) sensor requires a pulse that's at least 2 μs, and returns a pulse that can last anywhere from 115 μs to 18.5 ms. Immediately after the microcontroller sends the pulse to initiate the chirp, it has to change its I/O pin direction from output to input and then measure the Ping))) sensor's reply pulse that indicates the chirp's round trip time.

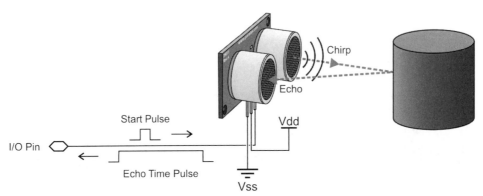

Figure 4-22   Ping))) sensor start and echo time signals.

One of the best things to do when presented with a new sensor is check the Propeller Object Exchange (http://obex.parallax.com) to find out if an object already exists for measuring the sensor. In the case of the Ping))) sensor, there's an object that takes care of all the signaling and time measurement tasks, and all you have to do is call its methods.

✓ Go to http://obex.parallax.com and enter <u>Ping</u> into the Search field. Then, click <u>Go</u>.
✓ Locate and download <u>Ping))) Demo by Chris Savage</u> from Parallax.
✓ Unzip the package.
✓ Open the Ping object, and click the <u>Documentation</u> button to view this object in Documentation view.

Figure 4-23 shows the Ping object in Documentation view. Note that it has `Inches`, `Centimeters`, and `Millimeters` methods, so no additional math is required to convert echo time to distance. Each of those methods takes care of the math and returns a distance value. Also note the 1 kΩ resistor in series between the Ping))) sensor's SIG terminal and the I/O Pin. In addition to series resistance built into the Ping))), the 1 kΩ resistor is sufficient to protect the Propeller I/O pin's 3.3 V input from the Ping))) sensor's 5 V output. If you decide to try out the Ping))) sensor with a Propeller chip, *make sure to include this resistor.*

Judging from the Ping object's documentation, no `Init` or `Start` method call is required, so a simple call to the `Centimeters` method should return the distance measurement in centimeters. This makes test code easy to write, but keep in mind that no `Start` method also means that the code is executed in the same cog. The code will be measuring pulse signals from the Ping))) sensor that last up to 18.5 ms, which can cause unwanted delays in a cog's code execution. In the event that 18.5 ms is too long for code in a given cog, the application can launch another cog for Ping))) measurements with the `cognew` command. Code executed by the new cog can then call the Ping object's methods and wait for results without slowing down code in the other cog. This is one of the beauties of multicore programming.

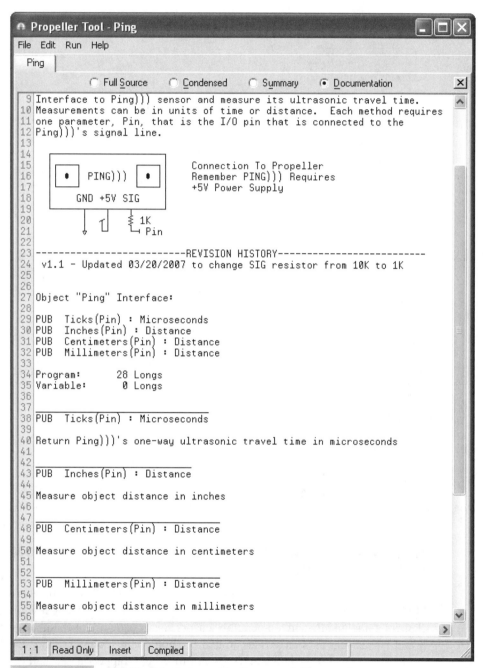

```
 9 Interface to Ping))) sensor and measure its ultrasonic travel time.
10 Measurements can be in units of time or distance.  Each method requires
11 one parameter, Pin, that is the I/O pin that is connected to the
12 Ping)))'s signal line.
13
14
15         ┌──────────────────┐        Connection To Propeller
16         │ •   PING)))   •  │        Remember PING))) Requires
17         │                  │        +5V Power Supply
18         │  GND +5V SIG     │
19         └──────────────────┘
20              ↓    ⌐ ⌐    ⋛ 1K
21                  │  │   └ Pin
22
23 -------------------------REVISION HISTORY-------------------------
24  v1.1 - Updated 03/20/2007 to change SIG resistor from 10K to 1K
25
26
27 Object "Ping" Interface:
28
29 PUB   Ticks(Pin) : Microseconds
30 PUB   Inches(Pin) : Distance
31 PUB   Centimeters(Pin) : Distance
32 PUB   Millimeters(Pin) : Distance
33
34 Program:      28 Longs
35 Variable:      0 Longs
36
37 _____
38 PUB   Ticks(Pin) : Microseconds
39
40 Return Ping)))'s one-way ultrasonic travel time in microseconds
41
42 _____
43 PUB   Inches(Pin) : Distance
44
45 Measure object distance in inches
46
47 _____
48 PUB   Centimeters(Pin) : Distance
49
50 Measure object distance in centimeters
51
52 _____
53 PUB   Millimeters(Pin) : Distance
54
55 Measure object distance in millimeters
56
```

Figure 4-23   Ping))) object viewed in Documentation mode.

```
{{ Test Ping Sensor and Object.spin
```

```
 ┌─────────────────────────────┐
 │ ┌───┐           ┌───┐ │
 │ │ ▪ │  PING)))  │ ▪ │ │
 │ └───┘           └───┘ │
 │                             │
 │     GND +5V SIG             │
 └─────────────────────────────┘
          │  ⊤│    ⧢ 1 kΩ
          ↓  └┘   └─ I/O Pin P15
}}
CON

  _clkmode = xtal1 + pll16x        ' Crystal and PLL settings.
  _xinfreq = 5_000_000             ' 5 MHz crystal x 16 = 80 MHz

OBJ

  ping : "Ping"                    ' Ping Ultrasonic Sensor object
  pst : "Parallax Serial Terminal" ' Serial communication object

PUB go | cmDist

  pst.Start(115200)                ' Start Parallax Serial Terminal

  repeat
    cmDist := ping.Centimeters(15) ' Get cm distance.

    ' Display measurement
    pst.Str(String(pst#CE, pst#HM, "cmDist = "))
    pst.Dec(cmDist)
    waitcnt(clkfreq/10 + cnt)      ' Update display at 10 Hz
```

MEMSIC 2125 ACCELEROMETER MODULE—CODE FROM OBJECT EXCHANGE

Figure 4-24 shows what happens inside the Memsic 2125 Accelerometer Module's MXD2125 chip when it gets tilted. This chip is actually a small chamber with a heating element in the middle and temperature sensors at each side of the chamber on both the X and Y axes. Just as soda sloshes around in a bottle in response to either acceleration or tilt, the distribution of hotter and cooler nitrogen gas trapped in the chamber behaves similarly, which is how the temperature sensors provide acceleration measurements. This is an example of microelectromechanical systems (MEMS) technology.

Circuits built into the MXD2125 chip convert the temperature values to pulse trains that indicate the distribution of hot and cold gasses in the chamber. These pulse trains, shown in Fig. 4-25, in turn indicate the acceleration sensed by each axis. Pulses that

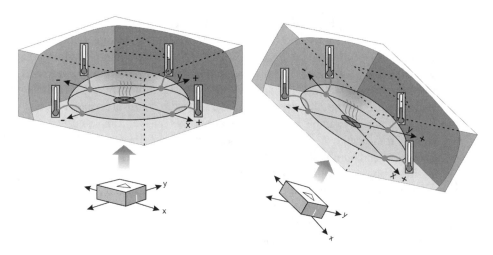

Figure 4-24   **Inside the Memsic Accelerometer** (*Excerpt from Smart Sensors and Applications, Parallax Inc.*).

last 5 ms indicate zero acceleration along an axis, and pulse durations can deviate from 5 ms by +/− 1.25 ms/g, where g is the acceleration due to gravity. If −1 g is applied to an axis, the pulse durations transmitted will be 3.75 ms. Likewise, accelerations from 0 to +1 would range from 5 to 6.25 ms.

> **Tip:** According to the device's datasheet, the MXD2125 reports acceleration in terms of duty cycle, which is the ratio of high time to signal cycle time. In practice, just the high pulses are equally accurate for this particular sensor.

Aside from examining object documentation, example programs that accompany objects from the Propeller Library and Propeller Object Exchange can provide clues on

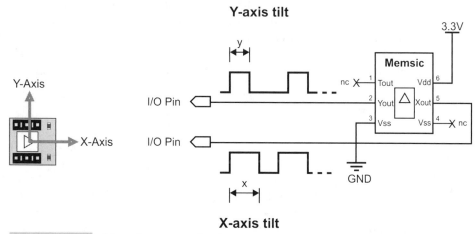

Figure 4-25   **Memsic communication.**

how to use them. There are several objects for the Memsic 2125 Accelerometer Module in the Propeller Library and on the Propeller Object Exchange, including MXD2125, MXC2125_Simple, and Memsic2125. They all also have demonstration objects in the Propeller Tool Software's Examples\Library folder. All of them are designed for a TV display, which is well and good if you are in the habit of using a TV for debugging, but they may need some adjustment if you don't have the hardware handy for connecting to a TV. Fortunately, example programs that utilize the TV_Terminal object tend to be easy to modify for use with the Parallax Serial Terminal object since the method calls that display characters and numbers are similar. Also, the main clues to look for in demonstration and example objects are how method calls to the building block object are implemented.

Of the three demonstration programs for the Memsic 2125 module, Paul Baker's MXD2125 Simple Demo.spin looks the most approachable in terms of modifying for use with the Parallax Serial Terminal. The first task is to examine the example program along with the MXD2125 Simple object to figure out what methods MXD2125 Simple Demo uses to get the accelerometer measurements. So check the nickname given to the MXD2125 Simple object in the OBJ block, and then look for *nickname.method* calls. In the case of MXD2125 Simple Demo.spin, the nickname is accel, and the object demonstrates two different sets of method calls: the first for using another cog to take the measurements, and the second for taking measurements with the same cog.

After declaring the MXD2125 Simple object with the nickname accel, the first example in the Setup method calls accel.Start to launch the accelerometer process into a new cog. The Start method call is embedded in a text.dec method call, so the TV_Terminal object displays the result value that the Start method returns. By convention, the Start method returns nonzero if it succeeded in launching a cog, or zero if it failed because all the cogs are already in use. After the Start method call, the example code uses accel.x to get the x-axis pulse measurement and accel.y to get the y-axis pulse measurement. Note that both the accel.x and accel.y method calls also are embedded in text.Dec method calls, so the TV_Terminal object displays the result returned by each method call. After 20 measurements, the application calls accel.stop to stop the cog. Next, it demonstrates measurements in the same cog by first calling the Init method, then accel.get_XY. Notice that the accel.get_XY(@XVal, @YVal) call passes variable addresses to the method, which indicates that the get_XY method stores the measurement results in those variables. After the method returns, both the XVal and YVal variables should store the results of the accelerometer x- and y-axis pulse measurements.

```
'MXD2125 Simple Demo.spin

CON
  _CLKMODE = XTAL1 + PLL16X
  _XINFREQ = 5_000_000

  'Constants used by the Accelerometer Object
  Xout_pin  = 0                         'Propeller P0 to MX2125 Xout
  Yout_pin  = 1                         'Propeller P1 to MX2125 Yout
VAR
  long XVal, YVal
```

```
OBJ

   text  :     "tv_text"              'Located in default Library
   accel :     "MXD2125 Simple"

PUB Setup
  text.start(12)

  'Separate cog example

  'load a cog with accelerometer driver
  text.dec(accel.start(Xout_pin, Yout_pin))
  text.str(string("Separate cog example",$0D))
  repeat 20
    text.dec(accel.x)                 'Retrieve X axis value
    text.out(" ")
    text.dec(accel.y)                 'Retrieve Y axis value
    text.out($0D)
    waitcnt(clkfreq>>1 + cnt)
  accel.stop                          'Stop the accelerometer driver cog

  'Now show in same cog example
  accel.init(Xout_pin, Yout_pin)
  text.str(string("Same cog example",$0D))
  repeat 20
    'Get X and Y values by passing pointers to variables
    accel.Get_XY(@XVal, @YVal)
    text.dec(XVal)
    text.out(" ")
    text.dec(YVal)
    text.out($0D)
    waitcnt(clkfreq>>1 + cnt)
  text.str(string("Demo complete"))
```

Let's try developing some simple test code with the Parallax Serial Terminal that relies on the portion of MXD2125 Simple Demo.spin that loads a cog with the accelerometer driver using `accel.Start` and then calls `accel.x` and `accel.y`. Figure 4-26 shows a schematic of the test circuit.

A call to `accel.Start(26, 27)` launches the accelerometer monitoring code and tells it that P26 is connected to the accelerometer's x-axis output and P27 is connected to the y-axis output. After examining the MXD2125 Simple object, it looks like the `x` and `y` methods both return the pulse measurements in terms of clock ticks. Since the accelerations are measured in terms of pulses that last milliseconds and fractions of milliseconds, displaying the results in terms of microseconds might make more sense. Since the pulses range from 3.75 ms to 6.25 ms, for +/−1 g with 5.0 ms level, this would translate to 3750 to 6250 µs, with 5000 µs level. A simple way to do this is by defining the number of ticks in a microsecond. Test MXD2125 Simple Object uses the expression `us := clkfreq/1_000_000` to assign the number of clock ticks in a microsecond to the local variable `us`. Then, the commands `ax := accel.x/us` and `ay := accel.y/us`

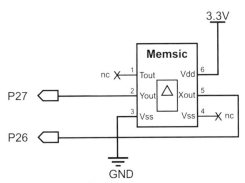

Figure 4-26 Memsic 2125
Accelerometer test schematic.

convert the clock tick pulse measurements into microsecond pulse measurements before
storing them in the ax and ay variables.

```
" Test MXD2125 Simple Object

CON

    _clkmode = xtal1 + pll16x          ' Crystal and PLL settings.
    _xinfreq = 5_000_000               ' 5 MHz crystal x 16 = 80 MHz

OBJ

    accel : "MXD2125 Simple"           ' Memsic 2125 object
    pst   : "Parallax Serial Terminal" ' Serial communication object

PUB go | ax, ay, us

    accel.Start(26, 27)                ' x-axis to P26, y-axis to P27
    pst.Start(115200)                  ' Start Parallax Serial Terminal

    us := clkfreq/1_000_000            ' Clock ticks in a microsecond

    repeat

      ' Get microsecond measurements for the x and y axes.
      ax := accel.x / us
      ay := accel.y / us

      ' Display measurement at 10 Hz
      pst.Str(String(pst#CE, pst#HM, "ax = "))
      pst.Dec(ax)
      pst.Str(String(pst#CE, pst#NL, "ay = "))
      pst.dec(ay)
      waitcnt(clkfreq/10 + cnt)
```

Frequency Output

Frequency output circuits are a common approach for determining the values of resistive and capacitive sensors. A common frequency output circuit that utilizes either a resistive or capacitive sensor employs a 555 timer IC to generate frequency signals that are determined by the values of resistors and capacitors connected to the chip. The circuit involves two resistors and a capacitor, and is called an *astable multivibrator* in electronics-speak. This circuit causes one of the 555 timer's pins to send a series of pulses, and resistor and capacitor values set both the frequency and the pulse width. Changes in either a capacitive or resistive sensor in this circuit result in changes in the multivibrator circuit's output frequency, making it a way to determine the sensor's value. Since the resistors and capacitors can also control the pulse durations, perhaps the temperature sensors inside the Memsic 2125 accelerometer from the previous section uses a similar design to send those pulses at certain durations.

Figure 4-27 shows examples of Parallax sensors and sensor modules that use frequency to indicate the values they measure; left to right they are the TSL230 Light to Frequency Converter, Piezo Film Vibra Tab, and Xband Motion Sensor. Optical interrupters such as the circuits that read black/white stripes on the inside of robot wheels are also common frequency output devices that indicate motor output speed.

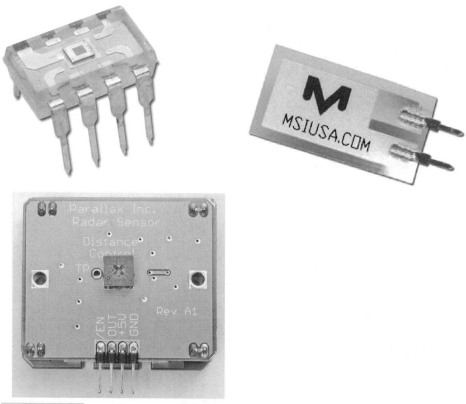

Figure 4-27 Frequency output sensor examples.

An example object that is prewritten for the TSL230 Light to Frequency Converter is the tsl230.spin object in the Propeller Library. Following is an excerpt from the tsl230 object in documentation view, including a connection diagram for the TSL230 chip. This object assumes that the user has studied the other documents on the TSL230 product page at www.parallax.com that explain how it works. The short version of these documents is as follows: The TSL230 transmits frequency signals that indicate light levels. When the TSL230 transmits higher frequencies, it indicates brighter light levels, and lower frequencies indicate dimmer light levels. The TSL230 has two different types of scaling, input sensitivity, and output frequency. Input sensitivity can be scaled for different lighting conditions, and output frequency can be scaled for the microcontroller application. Referencing the schematic with pin map in the tsl230 object documentation excerpt, Pins 1 and 2 are named S0 and S1, and signals applied to those pins scale the device's input sensitivity for a variety of lighting conditions. For example, if (S1, S0) = %01, meaning 0 V is applied to S1 and 3.3 V is applied to S0, the scaling is 1×, suitable for bright lighting conditions, including full sun. (S1, S0) = %10 is 10× sensitivity, which is useful for indoor lighting, and %11 is 100× sensitivity for low lighting conditions. Pins 8 and 7 are (S3, S2), and signals applied to them can optionally slow down the output frequency for slower microcontrollers.

```
*********************************************
* TSL230 Light to Frequency Driver v1.0   *
* Author: Paul Baker                       *
* Copyright (c) 2007 Parallax, Inc.        *
* See end of file for terms of use.        *
*********************************************

***************************************************************
    Taos TSL230 light to frequency sensor v1.0 driver
    with manual and auto scaling capabilities

                        ┌───┐
    ctrlpinbase  >──1│S0  o    S3│8──┐
                     │              │  │  │
                     │              │  │  ↓
    ctrlpinbase+1 >──2│S1       S2│7──┘
                     │      []      │
                  ┌──3│/OE    Out│6────< inpin
                  │  │              │  │ ↑
                  ├──4│GND    Vdd│5──┘  3.3V
                  ↓  └──────────────┘
***************************************************************
Object "tsl230" Interface:

PUB Stop
PUB Start(inpin, ctrlpinbase, samplefreq, autoscale) : okay
PUB GetSample : val
PUB SetScale(range)

Program:   50 Longs
Variable:   6 Longs
```

With a maximum output frequency in the neighborhood of 1.6 MHz, there's no need for the TSL230 to slow its output frequencies down for the Propeller. That's why the object documentation shows both pins 8 and 7 grounded, to disable any output frequency down scaling. The object documentation also explains that there are two input sensitivity scaling modes: manual and automatic. With manual scaling, the parent object can expect a value from 0 to 1.6 MHz for a given level of input sensitivity. If the measurements approach either of those values, the application object has to detect it and adjust the TSL230's input sensitivity scale. With automatic input sensitivity scaling mode, the Start method's autoscale parameter is set to true, and the object takes care of all this. So instead of results from 0 to 1.6 MHz with three different scale options (1×, 10×, and 100×), the object simply reports results from 0 to 160 M, which is a measurement range inclusive of all three scale settings.

When connecting the TSL230 to the Propeller chip, make sure to observe the ctrlpinbase and ctrlpinbase+1 convention in the object's documentation. It's shorthand for saying that the Propeller I/O pin number connected to TSL230 Pin 2 has to be one larger than the I/O pin number connected to TSL230 pin 1. For example, if you connect TSL230 pin 2 to Propeller I/O pin P25, TLS230 Pin 1 MUST be connected to Propeller I/O pin 24.

Test code that utilizes the tsl230 object also utilizes a TV display, but the code is simple enough that porting it for use with the Parallax Serial Terminal is not a problem. The tsl230 DEMO.spin object is in the Propeller Tool Software's Propeller Library Examples folder. The tv_text object's Term.out($0D) method has the same effect on a TV monitor that the Parallax Serial Terminal object's pst.NewLine method has on the Parallax Serial Terminal. According to the example code, the TSL230 chip's Out pin is connected to P0, and the TSL230's S0 (pin 1) and S1 (pin 2) are connected to Propeller I/O pins P1 and P2. Note that connecting TSL230 pin 1 to Propeller I/O pin P1 and TSL230 pin 2 to Propeller I/O pin P2 follows the tsl230 object's the ctrlpinbase and ctrlpinbase+1 convention. Also, note that since autoscale is set to true, the tsl230 object will automatically adjust its input sensitivity to the ambient lighting conditions.

```
' tsl230 DEMO.spin
CON
    _clkmode = xtal1 + pll16x
    _XinFREQ = 5_000_000
OBJ
 term  : "tv_text"
 lfs   : "tsl230"

PUB Go
 term.Start(12)
 lfs.Start(0,1,10,true)

 repeat
  waitcnt(80_000_000 / 10 + cnt)
  term.dec(lfs.GetSample)
  term.out($0D)
```

Voltage Output

Voltage output sensors are common, and many circuits transform resistive or capacitive sensors into voltage output sensors. Two examples of voltage output sensor circuits are in the "Converting Analog Sensors to On/Off Sensors" section. The potentiometer circuit in Fig. 4-9 and the photoresistor circuit in Fig. 4-10 both transform resistive sensors into voltage output sensors. These sensors transmit continuous ranges of output voltages (analog voltages) that indicate the quantities the sensors measure. Their outputs are commonly referred to as DC because they fluctuate gradually over time. An example of a capacitive voltage output sensor is the condenser microphone. A condenser mic has built-in capacitors that translate air pressure fluctuations across a diaphragm into voltage fluctuations that can be measured by a microcontroller. Since the voltage fluctuations are low amplitude and change rapidly with time, they are referred to as small signal and AC. Microcontrollers use analog-to-digital converters to measure and digitize sensor voltage outputs. Both A/D converter and ADC are common shorthand names for analog to digital converter. Propeller applications that require voltage measurements can make use of peripheral ADC chips or a technique called Sigma-Delta analog-to-digital conversion.

PERIPHERAL ADCs

Peripheral ADCs typically use two means of transmitting digitized voltage measurements to a microcontroller: parallel and serial. An ADC that transmits its measurements with parallel communication has a number of output pins connected to a number of Propeller I/O pins. For the sake of example, let's say that there are eight ADC output pins connected to eight Propeller I/O pins. With this scheme, the ADC can transmit eight binary digits at once to the Propeller chip. This arrangement makes it possible to report measurements quickly, but it takes a lot of I/O pins. In contrast, serial ADCs typically use two to four lines for communication. The most common variety of serial communication for ADCs is *synchronous serial,* where a clock signal is used to transfer each binary digit in messages to and from the microcontroller. For example, to get a measurement from a serial ADC with synchronous serial communication, the Propeller sends pulses to the ADC on a clock line; with each pulse, the ADC sends the next binary digit (a 1 or 0) in the measurement. Serial ADCs are slower than their parallel counterparts, but require fewer I/O pins. Serial ADCs are not typically fast enough to sample high-speed signals like video and radio intermediate frequency; however, some parallel ADCs are fast enough to keep up. Some serial ADCs, on the other hand, are fast enough to sample audio signals, while others are better suited to slower signals, such as sensors that measure lower-speed processes—ambient light, temperature, etc.

The measurement capabilities of peripheral ADCs are described in terms of resolution, which indicates the number of values the ADC can use to describe a range of input voltages. Resolution is typically given in bits, with values like 8-bit, 12-bit, 16-bit, and so on. If an ADC has 8-bit resolution, it means a measurement has 8 binary digits (bits) in the number that describes the voltage. An 8-bit value can store a number from 0 to 255, which is $2^8 = 256$ different values. A 12-bit value can store $2^{12} = 4096$ different values, ranging from 0 to 4095.

The digitized voltage an ADC returns to indicate a voltage measurement is typically:

$$\text{digitized voltage} = 2^{\text{bits}} \cdot \text{input voltage/input voltage range} \qquad (4.1)$$

For example, with a 12-bit ADC, a 0 to 5 V input range, and a 2.5 V input voltage, the digitized voltage measurement would be:

$$\text{digitized voltage} = 4096 \cdot 2.5/5.0 = 2048 \qquad (4.2)$$

Some ADCs are *single channel,* meaning they have one voltage input. Other ADCs have 2, 4, 8, or even more channels. ADC inputs typically have two flavors: *single-ended* or *differential.* Differential measures the voltage difference between one input and another, typically Vin+ and Vin−; single-ended inputs measure the voltage difference between the input voltage and ground.

Peripheral 4-Channel ADC Example with the MCP3204   One highly useful ADC object on the Propeller Object Exchange is the MCP3208 object, written by Chip Gracey and optimized and enhanced by Jim Kuhlman. This object uses a cog to communicate with the MCP3208 and is optimized for high speed. It supports the eight-channel Microchip MCP3208 synchronous serial ADC and the four-channel MCP3204. Figure 4-28 shows a schematic of a four-channel MCP3204 connected to some voltage output test circuits and to the Propeller chip.

Information: Other circuit examples in this chapter were built on the PE Kit 40-Pin DIP platform. This one was built on the PE Kit PropStick USB platform. The Propeller chip is on the module in the lower-right side of the photo. The 40-Pin DIP platform is recommended, especially for students. Aside from the worthwhile experience of wiring up the entire Propeller system, including voltage regulators, EEPROM program memory, crystal oscillator, reset circuit, and programming connection, the parts in the 40-Pin DIP kit are inexpensive and easy to replace. In contrast, the PropStick USB version is quick and easy to wire, but more expensive to replace because if a project mistake damages one part, the entire module has to be replaced.

Test MCP3208_fast.spin utilizes the MCP3208_fast building block object to acquire four channels of A/D measurements from the MCP3204 ADC, and it displays them in the Parallax Serial Terminal. The test code declares the MCP3208_fast object and gives it the nickname `adc`. The MCP3208_fast object's documentation comments specify the parameters for the `adc.Start` call as `Start(dpin, cpin, spin, mode)`, where `dpin` is the data pins tied together with a resistor, as shown in Fig. 4-28; `cpin` is the clock pin, `spin` is the enable pin, and `mode` is a parameter that was used in previous revisions of

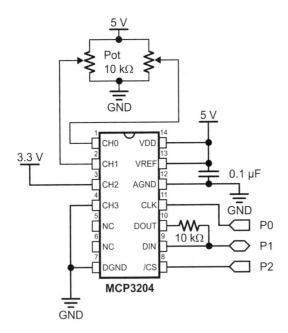

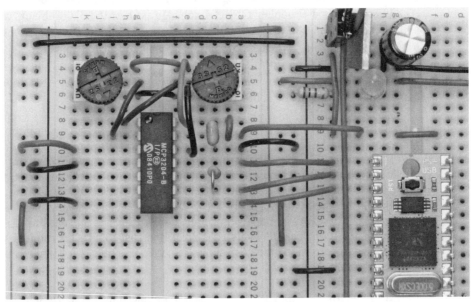

Figure 4-28   MCP3204 test circuit schematic.

the object. Inside the Main loop, the `repeat channel from 0 to 3` loop contains the expression `adcVal[channel] := adc.In(channel)`. When channel is 0, the expression stores the channel 0 ADC measurement in `adcVal[0]`. When channel is 1, the expression stores the channel 1 ADC measurement in `adcVal[1]`, and so on.

```
'' Test MCP3208_fast.spin

CON

  _clkmode = xtal1 + pll16x          ' Crystal and PLL settings.
  _xinfreq = 5_000_000               ' 5 MHz crystal x 16 = 80 MHz

OBJ

  pst : "Parallax Serial Terminal"   ' Serial communication object
  adc : "MCP3208_fast"               ' ADC object

PUB go | channel, adcVal[4]

  pst.Start(115200)                  ' Start Parallax Serial Terminal
  adc.Start(1, 0, 2, -1)             ' Start MCP3208, mode is legacy

  repeat channel from 0 to 3         ' Initialize display
    pst.Str(String("adcVal["))
    pst.Dec(channel)
    pst.Str(String("] = ", pst#NL))

  repeat                             ' Main loop
    waitcnt(clkfreq/10 + cnt)        ' Refresh display at about 10 Hz
    pst.Home                         ' Home position on PST
    repeat channel from 0 to 3       ' Get measurements.
      adcVal[channel] := adc.In(channel)
      pst.Position(12, channel)      ' Display measurements
      pst.Dec(adcVal[channel])
      pst.ClearEnd
```

Figure 4-29 shows the values Test MCP3208_fast.spin displays in the Parallax Serial Terminal. Each of the two potentiometers connected to MCP3204 channels 0 and 1 in Fig. 4-28 has 4096 possible values (0 to 4095), which represent voltages from 0 to (4095/4096) · 5 V. Another way of looking at 4095 is that it's about 1.22 mV below 5 V. Twist each potentiometer knob to adjust its voltage and monitor the results in the Parallax Serial Terminal. In Fig. 4-28, the MCP3204's channel 2 input is tied to 3.3 V, and the Parallax Serial Terminal in Fig. 4-29 shows a measurement of 2719. Since 2719/4096 is approximately 3.3/5, this measurement is correct and indicates that the system is functioning properly. The input voltage for channel 3 is tied to ground (0 V) in Fig. 4-28, and the result of `adcVal[3]` in Fig. 4-29 is 0, so that measurement provides a second indication that the system is functioning properly.

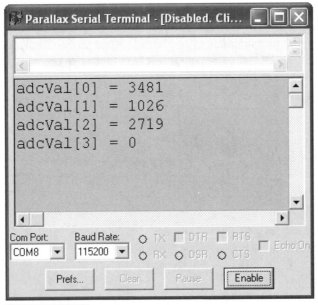

Figure 4-29 MCP3204 test measurements.

SIGMA-DELTA ADC

As mentioned earlier, a counter module is a configurable state machine that's capable of performing a variety of tasks independently for its cog. The On/Off Sensors that Depend on Signaling section demonstrated an object that used a counter module to transmit 38 kHz square waves. Another application of the counter modules built into each cog is Sigma-Delta analog to digital conversion. Figure 4-30 shows how a cog's counter module interacts with a Sigma-Delta ADC circuit. When in "POS detector with feedback" mode, the counter module opposes the state it detects at the input pin with

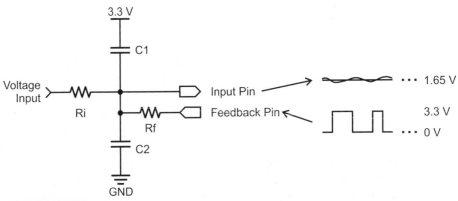

Figure 4-30 Sigma-Delta ADC circuit and signals.

the feedback pin, and keeps a running count of each clock tick at which the input pin detects a high signal. In this way, the counter module works to maintain 1.65 V at the input pin. The capacitors slow down the voltage response at the input pin to the highs and lows the feedback pin sends, so the voltage at the input pin only deviates slightly from 1.65 V, as the feedback pin rapidly transitions between 3.3 and 0 V every time it detects a slight deviation below or above the 1.65 V input threshold.

The interplay between a cog's counter module and the Sigma-Delta circuit makes it possible to digitize the analog voltage input. The cog configures the counter module to count the number of clock ticks when the voltage at the input pin is above 1.65 V, which turns out to be proportional to the voltage input. If it so happens that the voltage input is already at 1.65 V, the input pin detects high signals about half the time, adding 1 to the counter module's phase accumulation register (commonly referred to as the phase register) for every clock tick at which a high signal is detected. If the voltage input is instead 2.2 V, the input pin will detect and count high signals two-thirds of the time, and the feedback pin will send low signals two-thirds of the time to keep the voltage at the input pin at 1.65 V. Likewise, if the voltage input is 1.1 V, the input pin will detect a high signal only one-third of the time, and the feedback pin will only send low signals one-third of the time, and, of course, high signals the other two-thirds of the time. In each case, *the value stored by the counter module's phase register accumulates at a rate that's proportional to the voltage input.*

A cog typically works cooperatively with its counter module to perform Sigma-Delta A/D conversion. The cog managing the process has to wait a precise number of clock ticks, then copy the value stored in the counter module's phase register, and then clear it for the next sample interval. The resolution of the converter is defined by how many clock ticks the loop takes before it repeats. For example, if the converter takes 2000 clock ticks to repeat, the number of times the input pin detected a signal above 1.65 V could range from 0 to 2000. The typical Sigma-Delta ADC does not use all the clock ticks, so for a range of 0 to 2000 clock ticks, a voltage input of 0 V might correspond to 250 and a voltage input of 3.3 V might correspond to 1750. The actual resolution of the ADC in this case would be 1500, which is between 10 and 11 bits and would be used to describe the peripheral ADCs in the previous section. The range of measurements also depends on the output impedance of the sensor. If it has a high-output impedance, the entire measurement range might be reduced from 250–1750 to 600–1400, for example. The relationship between sensor output and ADC measurement should still be linear, just on a different scale that will take a few tests to determine. However, the testing can be worthwhile, since a couple of capacitors and resistors are inexpensive and don't require a lot of circuit board real estate in a project.

Test Sigma-Delta DC Signal Measurements Figure 4-31 shows a schematic of the DC signal version of the Sigma-Delta ADC along with an example circuit and Parallax Serial Terminal output for the Test Sigma-Delta ADC.spin object. Try running the program and then twisting the potentiometer. Plotting the potentiometer voltage versus ADC value should result in a linear graph. Test Sigma-Delta ADC.spin displays the ADC measurements in the Parallax Serial Terminal. The Sigma-Delta ADC object

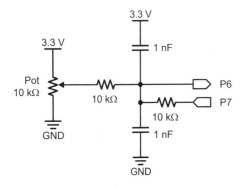

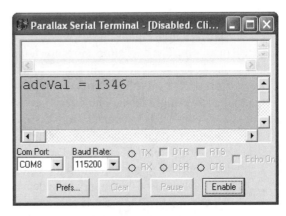

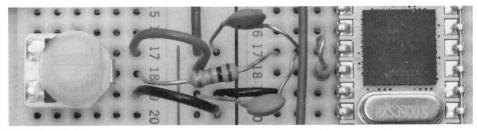

Figure 4-31   **DC Sigma-Delta ADC example circuit.**

takes care of all the interactions with the circuit in another cog and constantly updates the top file's adcVal variable.

Information: The Test Sigma-Delta ADC and Sigma-Delta ADC objects are included in the PE Kit Tools Sigma-Delta A/D Conversion article. It's available through the PE Kit Labs, Tools, and Applications sticky-thread at www.forums. parallax.com. Also, if you are using the Propeller Education Kit to try these programs and circuits, use the 100 pF capacitors included in the kit instead of the 1 nF capacitors shown in the schematic.

```
'' Test Sigma-DeltaADC.spin

CON

    _clkmode = xtal1 + pll16x        ' Crystal and PLL settings.
    _xinfreq = 5_000_000             ' 5 MHz crystal x 16 = 80 MHz

    FB_PIN = 7                       ' Sigma-delta ADC feedback pin
    IN_PIN = 6                       ' Sigma-delta ADC input pin

    SAMPLE_TICKS = 2000              ' Clock ticks per measurement.
```

```
OBJ

    adc : "SigmaDeltaADC(Spin)"        ' Declare Sigma-Delta ADC object
    pst : "Parallax Serial Terminal"   ' Serial communication object


PUB go | adcVal, t, dt                 ' Go method

    pst.Start(115200)                   ' Start Parallax Serial Terminal

    ' Start adc object, pass pins, the number of clock ticks in a sample,
    ' and the address of adcVal. The SigmaDeltaADC object stores the
    ' latest measurement in adcVal once every SAMPLE_TICKS.
    adc.start(FB_PIN, IN_PIN, SAMPLE_TICKS, @adcVal)

    t := cnt
    dt := clkfreq/10

    repeat
      waitcnt(t+=dt)
      pst.dec(adcVAl)                    ' Display current measurement
      pst.Str(String(pst#CE, pst#HM, "adcVal = "))
```

The SigmaDeltaADC(Spin) object, nicknamed adc in the test object, has a Start method with four parameters: feedback pin number, input pin number, clock ticks per Sigma-Delta sample, and the address of a variable in which to store the latest A/D conversion. Check the CON block for the IN_PIN, FB_PIN and SAMPLE_TICKs declarations. The IN_PIN and FB_PIN declarations match the I/O pins in the Fig. 4-31 schematic. After these constants and the address of the adcVal variable get passed to the SigmaDeltaADC(Spin) object's Start method with adc.Start(FB_PIN, IN_PIN, SAMPLE_TICKS, @adcVal), the SigmaDeltaADC(Spin) object updates the adcVal variable with the latest measurement every 2000 clock ticks. This results in a 40 kHz sampling rate because 2000 clock ticks at 80 MHz is 1/40,000th of a second. The test program only checks the adcVal variable at a rate of about 10 Hz, but applications that need to check at a faster sampling rate can do so. If a sampling rate needs to be faster than 40 kHz, the assembly version SigmaDeltaADC(ASM) should be used because it can accept sample intervals less than 2000 clock ticks.

Gain can be applied to A/D measurements by adjusting the resistor values in the Sigma-Delta ADC circuit shown in Fig. 4-32. By choosing a larger feedback resistor (Rf) and a smaller input resistor (Ri), it takes more cycles of low signals from the feedback pin to cause the voltage to swing down below 1.65 V if it's above, and vice versa. With the slower response, the input pin causes the counter module to accumulate high signals for more clock cycles than when the resistors were matched. The result is a larger ADC measurement for the same voltage input, which is essentially the same result that could be obtained by passing the signal through an amplifier. The advantage to setting gain with the resistors is it eliminates the need for an amplifier in some situations, which

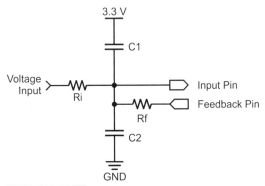

Figure 4-32 Sigma-Delta component values.

in turn reduces the number of components and system complexity. If the ratio of input to feedback resistors is reversed, it results in attenuated A/D measurements, which allows the ADC to work with larger voltage input ranges if needed.

Test Sigma-Delta AC Signal Measurements Figure 4-33 shows a Sigma-Delta microphone monitoring circuit built into the Propeller Demo Board. The 0.1 μF capacitor in the figure is called a *coupling capacitor,* and it allows the mic's AC signals to pass through stripped of any DC component. Regardless of the DC offset the microphone signals were riding to the left of the coupling capacitor, the voltage fluctuations (AC signals) transfer to the 1.65 V offset that the counter module maintains on the right side of the coupling capacitor.

Tip: This circuit can also be built with the Propeller Education Kit and a condenser mic (available at hobby electronics stores). For the PE Kit, make sure to use the 100 pF capacitors marked 101 that come with the kit instead of the 1 nF capacitors the Propeller Demo Board uses.

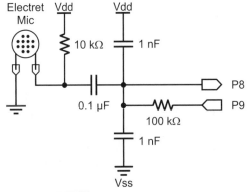

Figure 4-33 AC Sigma-Delta circuit connected to a microphone.

As mentioned earlier, condenser microphones (including the electret mic) convert fluctuations in air pressure to fluctuations in voltage. The Sigma-Delta ADC does a great job of measuring these voltages, and the ViewPort software introduced in the previous chapter can also do a great job of graphically displaying the time-varying voltage signal the Propeller chip measures with the Sigma-Delta ADC. Figure 4-34 shows the mic's output in response to a whistle in ViewPort's analog view. The Oscilloscope pane displays the ADC measurements versus time, and the Spectrum Analyzer pane displays signal strength versus frequency. The cursor on the Spectrum Analyzer pane can be positioned with the mouse cursor at the top of the spike in the graph to determine that the tone is 1.24 kHz at a glance.

Although the whistle is a single frequency, which could also be measured by dividing the period of the sine wave into 1, the Spectrum Analyzer pane is exceedingly useful for displaying the component sine waves and magnitudes in signals such as spoken

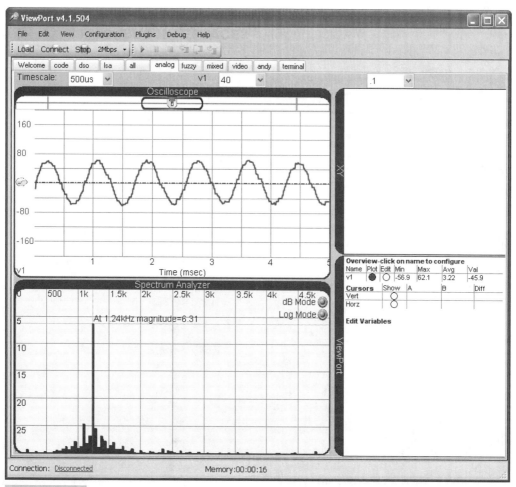

Figure 4-34   ViewPort analog view.

phonemes and telephone dial tones. Figure 4-35 shows the phoneme "aaaaahhhh" pronounced at a fairly high pitch. The component frequencies of the sound might be difficult to discern from the Oscilloscope pane, but the Spectrum Analyzer pane's cursor tool can be placed at the top of each spike to measure its frequency and amplitude. The sound is primarily composed of three sine waves, with the lowest frequency at 468 Hz.

Like the potentiometer version of the code, the SigmaDeltaADC(Spin) object takes care of all the interactions with the circuit in another cog and constantly updates the top file's adcVal variable. The Conduit object is running in a third cog and streaming the values of the adcVal variable to ViewPort for display in the analog view's Oscilloscope and Spectrum Analyzer panes. The program has some vp.config calls that were set first with ViewPort's dials and knobs. After that, they were copied from ViewPort using Configuration → Copy to Clipboard. When pasted into the top file, they appear as vp.config calls. These vp.config calls cause ViewPort to automatically set the dials and knobs accordingly after the ViewPort software's Connect button is clicked.

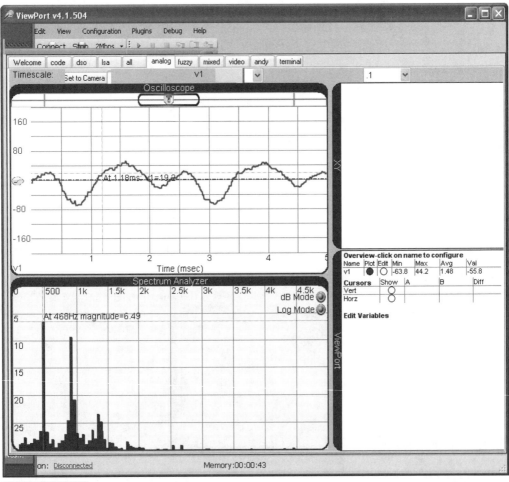

Figure 4-35   Say aaaaaahhhhh.

```
' Display Mic with ViewPort.spin

CON

    _clkmode = xtal1 + pll16x        ' Crystal and PLL settings.
    _xinfreq = 5_000_000             ' 5 MHz crystal x 16 = 80 MHz

    SAMPLES = 2000                   ' Clock ticks per measurement.

    FB_PIN = 9                       ' Sigma-delta ADC feedback pin
    IN_PIN = 8                       ' Sigma-delta ADC input pin

OBJ                                  ' Object declarations

    adc : "SigmaDeltaAdc(Spin)"      ' Declare Sigma-Delta ADC object
    vp  : "Conduit"

PUB go | adcVal                      ' Go method with local variables

  ' Configure ViewPort
  vp.config(string("var:v1(acdc=ac)"))
  vp.config(string("start:analog"))
  vp.config(string("dso:view=v1(offset=0.4037,scale=40), Ð
      trigger=v1>auto, timescale=500µs,ymode=manual"))
  vp.config(string("lsa:timescale=10ms,timeoffset=0"))
  vp.share(@adcVal,@adcVal)

  ' Start adc object, pass feedback and input pins, number of ticks
  ' per sample & address of adcVal.
  adc.Start(FB_PIN,IN_PIN,SAMPLES,@adcVal)

  repeat                             ' Repeat loop keeps this cog going
```

Note: New ViewPort functions and features are on the horizon at the time of this writing. To check for updated versions of code examples from this section that are compatible with the latest version of ViewPort, go to ftp.propeller-chip. com/PCMProp/Chapter_04. As mentioned earlier, the latest version of ViewPort is available for a 30-day free trial from www.parallax.com.

Multicore Signal Acquisition, Analysis, and Display   Since some applications require signal analysis independent of a PC, one of the Propeller microcontroller's cogs can be devoted to this task while another cog is devoted to signal acquisition. This multicore approach greatly simplifies signal acquisition and analysis, which can be especially challenging to implement with a single-core microcontroller. The next example application demonstrates the multicore approach, and it uses other cogs to graphically display a spectrum analysis on a TV display for good measure. TV display is another difficult task for single-core microcontrollers. In contrast, the Propeller Library's TV

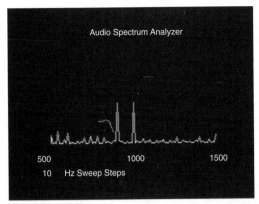
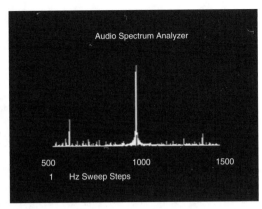

Figure 4-36   Propeller Audio Spectrum Analyzer application.

and Graphics objects distill these multicore tasks down to a few method calls made by the top-level application object.

Figure 4-36 shows TV display examples from Beau Schwabe's <u>Propeller Application DEMO: Spectrum Analyzer (for Audio)</u> posted at http://forums.parallax.com. The microphone and Sigma-Delta ADC circuit the application uses is similar to Fig. 4-33, and the TV connection was introduced in the previous chapter. The code utilizes the Propeller Library's TV and Graphics objects to draw the display and an object named Fast IO Grab to buffer digitized microphone voltage measurements over time. The top-level object then applies a noise cancellation and frequency detection algorithm to determine the signal strengths at frequency intervals specified in the code by a variable named FrequencyStep. The application uses four cogs: one for the TV display, one for the graphics engine, one for digitizing the microphone voltages, and one for the top-level application, which does the frequency detection and then sends the data to the Graphics object. This example still has four cogs available, and they could be devoted to incorporating this application's features into a larger application, perhaps robotic, mechatronic, biomedical, or consumer.

Synchronous Serial

Synchronous serial communication protocols are common for peripheral IC and module sensors. The MCP3204 ADC featured in the "Peripheral 4-Channel ADC Example with the MCP3204" section used synchronous serial communication, and examples of sensors with built-in synchronous serial interfaces include the DS1620 Digital Thermostat, Parallax Digital Compass, Tri-Axis Accelerometer, and Solid State Gyro, all shown in Fig. 4-37.

Synchronous serial communication protocols have numerous variations; examples include Serial Peripheral Interface (SPI), three-wire, four-wire, and Inter-Integrated Circuit (I^2C). Each integrated circuit that uses a synchronous serial protocol has a datasheet with timing diagrams and explanations of the signaling the microcontroller

Figure 4-37   Sensors with synchronous serial communication.

has to use to communicate with the IC. Since sensors with synchronous serial interfaces tend to be popular, objects that support them also tend to be published on the Parallax Object Exchange.

PARALLAX 1-AXIS GYRO MODULE

A gyro sensor is a useful ingredient for self-balancing robots, as well as inertial guidance and autopilot systems. The Parallax Gyro Module measures the rate of angular rotation velocity about the axis perpendicular to the module's PCB shown in Fig. 4-38. In other words, you can think about the axis sticking up from the top of the module as the handle of a top that you can twist to spin, and the gyro provides information about how fast and in which direction it's spinning (clockwise or counterclockwise). This angular velocity is called the *yaw rate*.

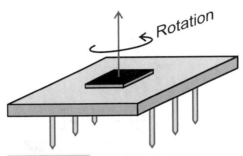

Figure 4-38   Rotational velocity.

Gyroscope Demo.spin is demonstration code for this gyro that integrates rotational velocity measurements over time to calculate rotational position. It displays the rotational position with the Parallax Serial Terminal as you rotate the board, as shown in Fig. 4-39. The first time you run the code, the angle will probably drift. Adjust the TRIM constant at the beginning of the program until the angle stays still when the board is not rotating. Then, rotate the board, and the angle and needle gauge in the display will faithfully follow the board's rotational position.

Information: At the time of this writing, the Parallax Gyro has not yet been released, and the photo and example code here are all in the prototype phase. For the latest test, demonstration, and object code, check the LISY300 Gyroscope Module product page at www.parallax.com.

The LISY300AL yaw rate sensor on the Gyro Module transmits the module's rate of rotation as an analog voltage, which gets measured by an ADC101 10-bit ADC

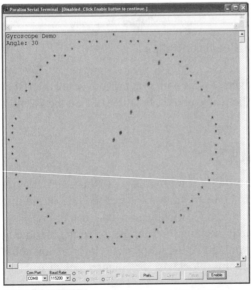

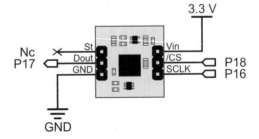

Figure 4-39   Rotational position display.

that's also on the module. This ADC digitizes the voltage value for the microcontroller. Remember from the "Peripheral ADCs" section that the digitized voltage value from an ADC is

$$\text{digitized voltage} = 2^{\text{bits}} \cdot \text{input voltage/input voltage range}$$

The analog voltage indicating no rotation should be 1.65 V. With a 3.3 V supply and reference voltage, the 10-bit ADC measurement should be

$$\text{digitized voltage} = 1024 \cdot 1.65/3.3 = 512 \qquad (4.3)$$

The output range of 1.65 V +/− 1.65 V corresponds to +/− 300 degrees per second rotational velocity. Another way to look at it is that ADC measurements of 512 +/− 512 correspond to +/− 300 degrees per second. So the microcontroller can calculate the yaw rate with this equation:

$$\text{Yaw rate} = (\text{adcVal} - 512) \cdot 300 \text{ degrees/second} \div 512 \qquad (4.4)$$

Figure 4-40 shows the synchronous serial communication timing diagram for the Gyro Module's ADC101S021 A/D converter. This timing diagram is an excerpt from the ADC101S021 datasheet. The steps for this communication are

1. Initialize the nCS and SCLK high.
2. Set the I/O pin connected to the nCS pin low.
3. Apply 16 clock pulses, and record the SDATA after each rising edge.
4. Set nCS high.

FIGURE 1. Timing Test Circuit

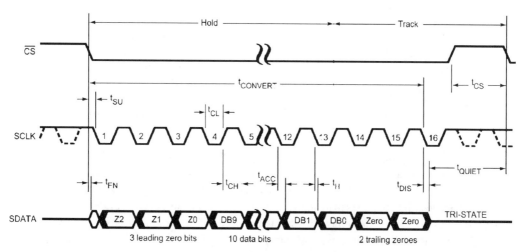

20145306

FIGURE 2. ADC101S021 Serial Timing Diagram

Figure 4-40   **ADC101S021 timing diagram** (*Reprinted with permission of National Semiconductor Corporation*).

In Test Gyro.spin, a call to Init followed by a call to the Measure method implements the timing in the diagram. The call to Init covers step 1, making nCS and SCLK output-high. Measure takes care of the rest of the steps. It sets the nCS pin low, and then takes the SCLK pin low, then high. After each transition from low to high, the command adcVal := adcVal << 1 | ina[SDATA] does two things. First, it shifts whatever is in the adcVal variable left by one, leaving a 0 in the rightmost binary digit. Then, it copies the value (1 or 0) that the ADC101's SDATA pin is sending to the Propeller chip's I/O pin (SDATA = 17 in this case) to the empty rightmost bit in the variable. After repeating this 16 times, the entire measurement has been shifted right three extra digits because of those two zeros and one tristate at the end of the timing diagram. So the command adcVal >>= 3 shifts the entire value right by three binary digits to correct this before returning the ADC measurement.

```
'' Test Gyro.spin
' ...
'

  SCLK    = 16              ' I/O pin Assignments
  SDATA   = 17              ' "DOUT" in the schematic
  nCS     = 18              ' "/CS" in the schematic
' ...
'

PUB Init

  outa[SCLK]~~             ' SCLK & /CS → output-high
  outa[nCS]~~
  dira[SCLK]~~
  dira[nCS]~~

PUB Measure : adcVal        ' Get gyro ADC value

  outa[nCS]~                ' /CS → low
  repeat 16                 ' 16 clock pulses
    outa[SCLK]~             ' Clock pulse falling edge
    outa[SCLK]~~            ' Clock pulse rising edge
    adcVal := adcVal << 1 | ina[SDATA] ' Get data bit from DOUT
    outa[nCS]~~             ' /CS → high
    adcVal >>= 3            ' Shift right to get rid of
                            ' last three digits
```

ViewPort's lsa tab makes it possible to examine the synchronous serial signaling to verify that it's correct. Figure 4-41 shows the communication signaling on lines 16 through 18. Line 18 is nCS, and it does indeed go low for the entire exchange. Line 16 is SCLK, and the Propeller sends 16 pulses to it; line 17 is SDATA, the binary values the ADC transmits in response to each clock pulse. Remember that the last three SDATA that accompany the last three rising edges are discarded. With that in mind, the binary value is %0111111101 = decimal 509.

The ViewPort package's QuickSample object uses another cog to monitor the I/O pins at up to 20 MHz. Here are the minimal steps for adding LSA ViewPort functionality to your code:

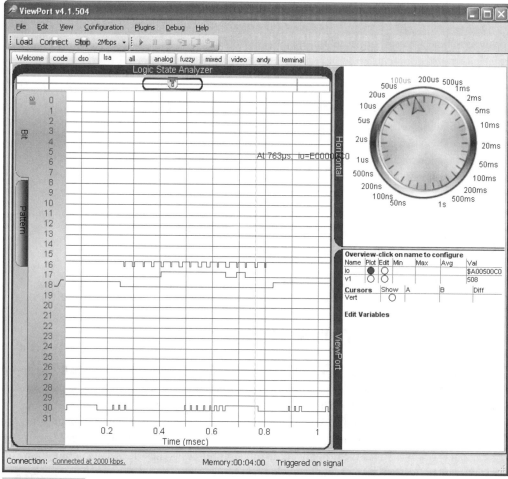

Figure 4-41   ViewPort synchronous serial communication.

✓ Add a 400 element frame array.
✓ Declare the QuickSample object along with the Conduit object.
✓ Add the following vp.register call before the vp.share call.

```
" Test Gyro with ViewPort.spin
'
' ...
'


VAR

    long frame[400]                    ' Stores measurements of INA port
```

```
OBJ

  vp : "Conduit"                  ' Transfers data to/from PC
  qs : "QuickSample"              ' Samples INA continuously in 1
                                  ' Cog- up to 20Msps
PUB Go | yawAdc

  vp.register(qs.sampleINA(@frame,1))' Sample INA into <frame> array
  vp.share(@yawAdc,@yawADC)           ' Share the <freq> variable



' ...
'
```

To display the data as shown in Fig. 4-41

✓ Load Test Gyro with ViewPort.spin into the Propeller chip.
✓ Click ViewPort's <u>Connect</u> button.
✓ Click the <u>lsa</u> tab.
✓ Click the <u>Plot</u> button next to <u>io</u> in the ViewPort overview table.

Asynchronous Serial

Sensors that communicate with asynchronous serial protocols are typically equipped with microcontroller coprocessors. Examples of such asynchronous serial sensors include the Parallax GPS module and RFID reader shown in Fig. 4-42.

The Propeller chip uses asynchronous serial signaling to communicate with the Parallax Serial Terminal. The serial communication activity is visible on I/O pin P30 in Fig. 4-41. That's the Propeller chip transmitting serial messages that contain ViewPort data to the USB adapter on its way to the PC. For most Propeller boards, there is a USB-to-serial converter in the Propeller's programming tool that converts incoming USB

Figure 4-42   GPS and RFID modules.

signals to serial messages, and it also converts outgoing messages from serial to USB signals. On the PC side, a driver that supports the USB/serial chip on the Propeller's programming tool converts incoming USB messages back to serial messages, and it also converts outgoing messages from serial to USB signals.

Tip: The Propeller chip can also be programmed directly through a COM port using RS232 serial communication instead of USB; see the Propeller Tool Help.

EXAMINING SERIAL COMMUNICATION SIGNALS

For a closer look at serial communication, the Parallax Serial Terminal object can be configured to transmit 4800 bps serial messages to an I/O pin, and the QuickSample object can be configured to monitor and stream the I/O pin activity to the ViewPort software for display in its logic state analyzer. Figure 4-43 shows an "A" character

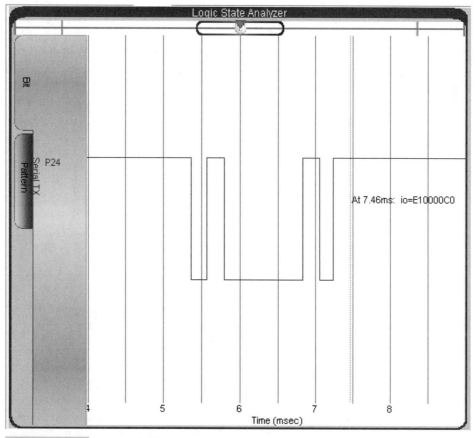

Figure 4-43 Letter "A" at 4800 bps 8N1.

transmitted by the P24 I/O pin at 4800 bps, 8 data bits, no parity, 1 stop bit. These serial port settings are often listed in the abbreviated form: 4800 bps 8N1.

The ASCII numeric code that represents "A" is 65, or in binary %01000001. The asynchronous serial message shown in the figure starts high, or in resting state. Since the baud rate is 4800 bps, every binary bit is given a 1/4800 s time window. The first (leftmost) low signal, which is called the start bit, lasts for 1/4800 s. The start bit signals that eight more data bits are on their way. The first data bit is a high, or binary 1. Since asynchronous serial messages transmit the least significant bit first, this 1 is the value in bit-0. In the case of the serial "A" byte, the least significant bit, or bit-0, is the rightmost 1 in the value %01000001. The next five 1/4800 s time windows are all low, transmitting five binary 0s for bits 1 through 5, then a high signal for bit-6 and a low signal for bit-7. After the last data bit, the I/O pin returns to resting state for at least 1/4800 s, assuming the protocol requires one stop bit.

View Serial Character with ViewPort.spin uses the Parallax Serial Terminal object to repeatedly send an "A" character out P24 with the 4800 bps 8N1 serial communication settings. The example program also uses ViewPort's QuickSample object to monitor the I/O pin activity from another cog. The repeat loop in the example program uses P24 to transmit an "A" character approximately once every 1/200th of a second. P23 sends alternating high/low signals with each successive "A" character, which makes it convenient to set the ViewPort Logic State Analyzer's trigger to make the serial messages stay still in the display. The reason the application transmits serial messages with P24 instead of with the Parallax Serial Terminal object's default P30 I/O pin is because the Conduit object needs P30 to stream I/O pin data to the ViewPort software running on the PC. P30 is connected to the Propeller programming tool's USB-to-serial converter along with P31 for programming and bidirectional serial communication with the PC. These two pins are also used for communication with the debugging software packages introduced in the previous chapter.

```
'' View Serial Character with ViewPort.spin

CON

  _clkmode = xtal1 + pll16x        ' Crystal and PLL settings.
  _xinfreq = 5_000_000             ' 5 MHz crystal x 16 = 80 MHz

VAR

  long frame[400]                  ' Stores measurements of INA port

OBJ

  vp  : "Conduit"                  ' Transfers data to/from PC
  qs  : "QuickSample"              ' Samples INA continuously in 1
                                   ' cog- up to 20Msps
  pst : "Parallax Serial Terminal" ' Serial communication object
```

```
PUB Go | yawAdc

  vp.register(qs.sampleINA(@frame,1))'sample INA into <frame> array
  vp.config(string("var:io(bits=[Serial TX[24P24]]),v1"))
  vp.config(string("start:lsa"))
  vp.config(string("dso:timescale=10ms"))
  vp.config(string("lsa:view=io,trigger=io[23]r,timescale=500µs"))
  vp.share(@yawAdc,@yawADC)          'share the <freq> variable

  ' Start Parallax Serial Terminal on I/O pins RX = P25 and TX = P24
  pst.StartRxTx(25, 24, 0, 4800)     ' Baud rate = 4800 bps

  dira[23]~~                         ' Set P23 to output

  repeat                             ' Main loop
    pst.Char("A")
    waitcnt(clkfreq/200 + cnt)
    !outa[23]                        ' Toggle P23 with each "A"
```

After the display was configured as shown in Fig. 4-43, ViewPort's <u>Configure</u> → <u>Copy to Clipboard</u> menu item was selected, which copied a number of `vp.config` calls to the Clipboard. The resulting `vp.config` calls were then pasted into View Serial Character with ViewPort.spin between the `vp.register` and `vp.share` calls. The `vp.config` calls make the Propeller chip send the configuration information to ViewPort so that it can duplicate the settings that were in effect at the time you selected <u>Configure</u> → <u>Copy to Clipboard</u>.

VIEW MORE VIEWPORT

For more information on topics like configuring the Logic State Analyzer to display the activity of particular I/O pins, see the Propeller Education Kit Lab: *Propeller + PC Applications with ViewPort*, available from the Downloads & Articles link at www.parallax.com/Propeller. Also, www.mydancebot.com has a 60-page PDF manual, and a Videos link with some great video tutorials that demonstrate how to configure ViewPort for a variety of applications.

ASCII stands for American Standard Code for Information Exchange. Each time you press one of your computer keyboard's printable character keys, it transmits the information to your PC as an ASCII code. The ASCII codes 0...33 are nonprintable characters that are used for terminal control and configuration. The Parallax Serial Terminal uses 16 of these for commands such as ClearScreen and NewLine. The ASCII code for space is 32, and 33...126 are displayed by the Printable Ascii Table.spin program and listed in Fig. 4-44.

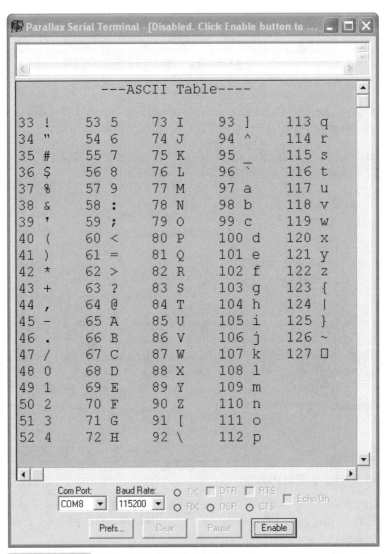

Figure 4-44   Printable ASCII characters.

```
" Printable Ascii Table.spin

CON

    _clkmode = xtal1 + pll16x        ' Crystal and PLL settings.
    _xinfreq = 5_000_000             ' 5 MHz crystal x 16 = 80 MHz

OBJ

    pst : "Parallax Serial Terminal" ' Serial communication object
```

```
PUB go | c

  pst.Start(115200)                      ' Start Parallax Serial Terminal

  pst.str(String("    ---ASCII Table----"))

  repeat c from 33 to 126

    ' Place cursor for each character in table.
    case c
      33..52   : pst.Position(0, c-31)
      53..72   : pst.Position(8, c-51)
      73..92   : pst.Position(16, c-71)
      93..112  : pst.Position(24, c-91)
      113..127 : pst.Position(32, c-111)

    ' Display ASCII code next to ASCII character.
    pst.dec(c)                    ' ASCII code (decimal)
    pst.Char(" ")                 ' Print a space
    pst.Char(c)                   ' ASCII character
```

GPS MODULE EXAMPLE

An example of a synchronous serial sensor is a global positioning system (GPS) receiver. These receivers calculate the time difference between the arrival of a number of satellite signals to determine geographic position, and they typically report the position information using one of the National Marine Electronics Association (NMEA) protocols. Although NMEA has a new protocol named NMEA 2000, the older NMEA 0183 protocol is still widely used, and the receivers are inexpensive. NMEA 0832 receivers communicate at 4800 bps, 8 bits, no parity, and 1 stop bit with no handshake. This is the same serial configuration the Parallax Serial Terminal software and object would use if they were both configured to communicate at 4800 bps, and it's also the protocol that was used to transmit the letter "A" with P24 that was just displayed by ViewPort in Fig. 4-43.

Parallax carries two different GPS receivers for prototyping, shown in Fig. 4-45. The Parallax GPS Receiver Module on the left has a coprocessor that allows a microcontroller to request predigested information. The raw GPS receiver without the coprocessor is shown on the right. Since the Propeller chip has multiple processors, an application can incorporate an object that launches a cog as a coprocessor to serially communicate with the GPS receiver, digest the NMEA sentences, and return requested information.

Figure 4-46 shows a circuit with the raw GPS receiver connected to the Propeller chip. The wire harness that accompanies the receiver has six lines. If you only plan on using the device with microcontrollers, consider removing the RS232 TX and RX lines; if the RS232 TX line is inadvertently connected to a microcontroller I/O pin, the +/− 3 to +/− 15 V output could damage the I/O pin or even the microcontroller.

Figure 4-45   GPS receivers with and without coprocessors.

While numerous objects on the Propeller Object Exchange support GPS, two that really stand out for simplifying application development and reducing prototyping times are I.Kövesdi's GPS Float packages: GPS Float Demo and GPS Float Lite Demo. The demonstration programs use the Parallax Serial Terminal to display GPS info and take care of the floating-point calculations required for a variety of navigation tasks. In addition to the well-documented GPS_Float and GPS_Float_Lite objects, the demonstration programs in each package feature method calls to the GPS object methods that show how to use their many features.

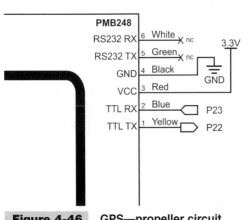

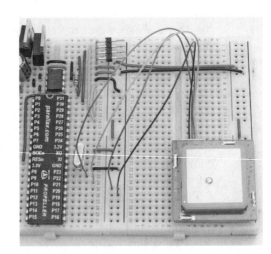

Figure 4-46   GPS—propeller circuit.

The one "gotcha" with the GPS Float packages at the time of this writing is that the I/O pins are declared in building block objects. Follow these steps to make the objects compatible with the circuit in Fig. 4-46:

✓ Download GPS Float Demo and GPS Float Lite Demo from ftp.propeller-chip.com/PCMProp/Chapter_04.
✓ Unzip each package into a folder.
✓ Open the GPS_Float and GPS_Float_Lite objects with the Propeller Tool software.
✓ Update the _RX_FM_GPS and _TX_TO_GPS constant declarations to 22 and 23, respectively.

```
_RX_FM_GPS    = 22
_TX_TO_GPS    = 23
```

The GPS_Float_Lite_Demo and GPS_Float_Demo objects also communicate with the Parallax Serial Terminal at 57,600 bps instead of 115, 200.

✓ Set the Parallax Serial Terminal's Baud Rate drop-down menu to 57600.

Figure 4-47 shows the GPS_Float_Lite_Demo object's output, taken at Parallax Incorporated in Rocklin, CA. The latitude and longitude shown can be entered into

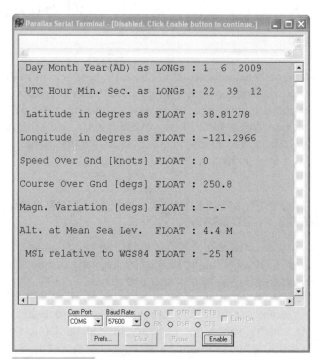

Figure 4-47   GPS_Float_Light_Demo–Where in the world is Parallax?

http://maps.google.com. After clicking the <u>Satellite</u> button, you'll see an aerial view of Parallax. To get a similar indication of your location, follow these steps:

✓ Load GPS_Float_Light_Demo into the Propeller chip.
✓ As soon as the Propeller Tool software's Communication window reports "Loading…", click the Propeller Serial Terminal's <u>Enable</u> button.

The GPS receiver needs enough open sky in view to receive at least three, but preferably four or more, satellite signals to calculate its position. On power-up in a new location, the receiver may take several minutes to detect enough satellites. During that time, a red indicator LED on the module will blink. When the receiver receives signals from enough satellites, its indicator LED will stop blinking and remain on.

✓ Take the setup to an area with plenty of sky exposure and wait for the GPS receiver's indicator light to stop blinking and remain on.

When the GPS receiver's indicator light stays on, the Propeller will relay the date/ time, latitude, longitude, and the rest of the information, similar to Parallax' location shown Fig. 4-47.

Questions about Processing and Storing Sensor Data

There are several frequently asked questions about processing and storage of Propeller-acquired sensor data. The most common question regarding processing is how to synchronize an application object running at a lower speed with a higher-speed building block object. Questions about storage typically involve long-term media that cannot be erased if the application restarts, and another common question that comes up is how to transfer information to a PC application.

HOW DO I SYNCHRONIZE MY SLOWER APPLICATION CODE WITH A HIGH-SPEED PROCESS?

Let's say that the ADC object samples the signal at 40 kHz, but an application needs the information at a rate of 1 kHz. This might happen because an example algorithm is available that is designed for 1 kHz, or maybe other peripherals in the system can only respond at 1 kHz. Suffice it to say that one object needs the information at 1 kHz, but the building block object samples at 40 kHz. A common mistake is to try to slow down the building block object, which is simply not necessary. Instead, the application should have a loop that repeats at 1 kHz and grabs the most up-to-date measurement from the building block object. Remember the convention for a loop with `waitcnt(t+=dt)` to synchronize the loop to repeat in `dt` clock ticks? Immediately after the `waitcnt` command, store the latest sensor measurement from the building block object. A low-speed (10 Hz) example of this was demonstrated by the Test Sigma-DeltaADC.spin object in the section on Sigma-Delta ADC.

HOW DO I STORE DATA FOR LATER RETRIEVAL AND KEEP IT FROM GETTING ERASED?

The portion of the Propeller's external EEPROM not used for storing the program can be useful for storing small amounts of data. The *PE Kit Tools: EEPROM Datalogging* article features a Propeller EEPROM object that allows the application to make backup copies of segments from the Propeller chip's Main RAM to the external EEPROM and other methods for retrieving the stored data. This article and its objects are available from http://forums.parallax.com.

Be careful with how you apply the Propeller EEPROM object; the Propeller Tool software overwrites the lowest 32 KB in EEPROM when the Load EEPROM feature is used for programming. There are three ways to prevent overwriting your data:

- If data is stored in the lower 32 KB and a separate program is intended to be loaded for retrieving the data, the Load RAM feature can be used to program the Propeller, and the data on the EEPROM will remain intact.
- A feature for reporting collected data can be incorporated into the application so that a separate program does not need to be loaded into the Propeller chip.
- Replace the 32 KB 24LC256 EEPROM, which is just enough for storing Propeller applications, with a larger one, such as the 64 KB Microchip 24LC512. Values stored at any address above 32 K in these larger EEPROMs will be safe from being overwritten by EEPROM programming. This approach is probably the best for prototyping.

These techniques are discussed in greater detail in the *PE Kit Tools: EEPROM Datalogging* article.

For larger volumes of data, the Propeller Object Exchange has an entire section devoted to data storage. One object that stands out in terms of usefulness is Tomas Rokicki's "FAT16 routines with secure digital card layer." The beauty of this system is that both the Propeller and a PC can read from and write to an SD card that has been formatted with the FAT16 file system, and the object supports SD cards up to 4 GB. John Twomey also expanded on this object with methods that can store strings and numerical values.

An SD card adapter from www.ucontroller.com, shown in Fig. 4-48, makes prototyping Propeller applications with an SD card simple and convenient. Note in Fig. 4-49 that it only takes six wires to connect.

Tip: Aside from being used to store large volumes of sensor data, SD cards have been used for numerous synchronized MP3 displays, such as fireworks, stage lighting, and even fountains. The Propeller chip's multicore architecture lends itself to these projects because separate cogs can be assigned to playback

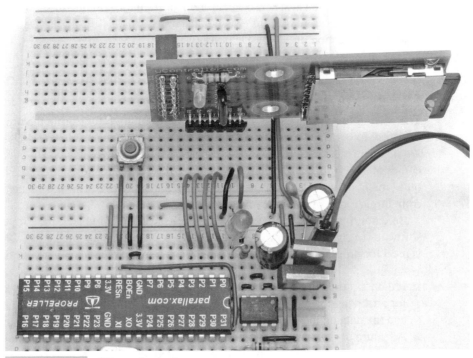

Figure 4-48   SD card adapter.

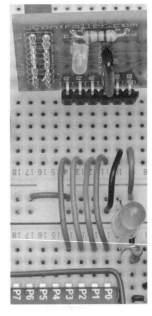

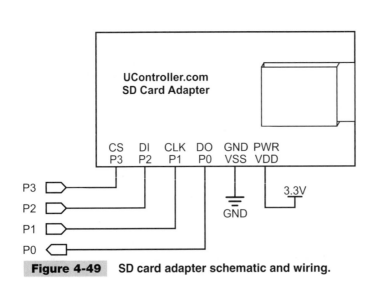

Figure 4-49   SD card adapter schematic and wiring.

and synchronized events. Likewise with sensor applications: One or more cogs can be acquiring sensor data, and another cog can be storing the most recent measurements and timestamps to the SD card.

Figure 4-50 shows a test from the fsrw-and-friends-1.6 package that was adapted for the Parallax Serial Terminal. It was originally designed for a television display. The test opens the SD card, reads the root directory, creates a file with some "R" characters, and then closes the SD card. The program can be modified to examine the contents of the SD card as well, or you can examine it with your PC, provided you have an SD card reader or SD card-to-USB adapter. If you don't, SD card-to-USB adapters are inexpensive and available at many office supply, computer retail, and electronics outlets.

Although sdrw_test.spin was written for a TV display, porting it to sdrw_test for PST.spin for use with the Parallax Serial Terminal was simple because the TV Terminal and Parallax Serial Terminal objects' text and number display method calls are similar. It was mainly a matter of replacing the TV_Terminal object declaration with one for the Parallax Serial Terminal. Then, the `term.Start` method was updated to

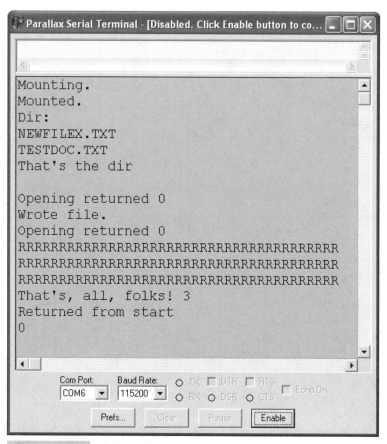

Figure 4-50   SD card test.

pst.Start(115200), and all term.out calls were replaced with term.Char. The method calls that you should examine in the top-level object start with sdfat: sdfat.mount, sdfat.opendir, sdfat.popen, sdfat.pputc, sdfat.pcclose, and sdfat.pgetc. Although many of their functions can be guessed, make sure to read the documentation comments in the fsrw object to find out more.

```
' sdrw_test for PST.spin

CON

  _clkmode = xtal1 + pll16x          ' Crystal and PLL settings.
  _xinfreq = 5_000_000               ' 5 MHz crystal x 16 = 80 MHz

OBJ

  pst    : "Parallax Serial Terminal"  ' Serial communication object
  sdfat  : "fsrw"

VAR

  byte tbuf[20]

PUB go | x

  pst.Start(115200)                  ' Start Parallax Serial Terminal
  x := \start
  pst.Str(string("Returned from start", pst#NL))
  pst.Dec(x)
  pst.NewLine

PUB Start | r, sta, bytes

  pst.Str(string("Mounting.", pst#NL))
  sdfat.mount(0)
  pst.Str(string("Mounted.", pst#NL))
  pst.Str(string("Dir: ", pst#NL))
  sdfat.opendir
  repeat while 0 == sdfat.nextfile(@tbuf)
    pst.Str(@tbuf)
    pst.NewLine
  pst.Str(string("That's the dir", pst#NL))
  pst.NewLine
  r := sdfat.popen(string("newfilex.txt"), "w")
  pst.Str(string("Opening returned "))
  pst.Dec(r)
  pst.NewLine
  sta := cnt
  bytes := 0
```

```
repeat 3
  repeat 39
    sdfat.pputc ("R")
  sdfat.pputc (pst#NL)
sdfat.pclose
pst.Str (string ("Wrote file.", pst#NL))
r := sdfat.popen (string ("newfilexr.txt"), "r")
pst.Str (string ("Opening returned "))
pst.Dec (r)
pst.NewLine
repeat
  r := sdfat.pgetc
  if r < 0
    quit
  pst.Char (r)
pst.Str (string ("That's, all, folks! 3", pst#NL))
```

```
" Excerpt from sdrw_test, Copyright (c) 2008 Tomas Rokicki,
" obex.parallax.com
```

HOW DO I TRANSMIT DATA FROM THE PROPELLER TO THE PC?

One quick and simple way to bring data from the Propeller chip to the PC is by displaying the values in the Parallax Serial Terminal. The values can then be copied with CTRL+C and pasted into a text document. Provided the values the Propeller sends to the Parallax Serial Terminal are formatted so that each number on a given line is separated by a comma, most data analysis and number crunching software packages can import the text file as comma delimited. Tab delimited tends to be the default import setting for these software packages, but they all have comma delimited option as well, and it displays much better in the Parallax Serial Terminal that way.

Most programming environments also have COM port features that can be programmed to exchange data with the Propeller chip. This makes it possible to use an actual COM port, serial-over-USB, or even serial-over-Bluetooth to transfer the information. Just a few examples of programming environments with COM port features include Microsoft Visual BASIC, Microsoft Visual C#, MATLAB, and LabVIEW. Of course, the PC application has to be configured for the right COM port, and the PC code and Propeller code have to agree on a communication protocol. A LabVIEW example is available in the Propeller Education Kit Applications section at http://forums.parallax.com.

Summary

Instead of grouping sensors by the physical quantities they measure, this chapter grouped them into common microcontroller interfaces. Some sensors had more than one interface option because the circuits that connected them to microcontroller I/O pin(s) determined the types of outputs. Resistive sensors, for example, can be measured with RC decay time in an RC or resistor-capacitor circuit, or with an A/D converter if placed in series with another resistor. Each interface section represented a variety of sensors. For example, the RC decay interface can measure a potentiometer, resistive or capacitive humidity sensors, a color-enhanced photodiode, or an infrared transistor, to name just a few.

This chapter also introduced a number of different resources for writing code to get information from sensors in each interface group, including the Propeller Object Exchange, Propeller Library, and Propeller Education Kit resources. The Propeller chip's object-based Spin language lends itself to objects that solve tricky or complex sensor problems, and the Propeller Tool Software's Propeller Library and the Propeller Object Exchange are two fantastic resources for getting objects that can save many hours that might otherwise be spent on writing code from scratch. This chapter also included some examples of writing code from scratch, for on/off sensors as well as for an A/D converter, using the data sheet and timing diagrams.

Although this chapter featured brief explanations of each sensor, keep in mind that any given sensor typically has a wealth of published information to support it. Many companies that manufacture and sell sensors have a vested interest in making sure that people who want to incorporate their sensors into projects and products succeed. Any sensor Parallax manufactures and/or distributes has a product page, and on that product page there will almost always be a Downloads section with PDF documentation that explains how the sensor works and includes a test circuit and example code that shows how to get measurements from it.

Exercises

1 Use Timestamp.spin to record when an object is detected with infrared.
2 Write a program that displays the gyro's rate of rotation in degrees per second. Consider making use of the Propeller Library's FloatMath object.
3 Design a program that allows you to control the cursor's placement in the Parallax Serial Terminal with an accelerometer. Hint: Axis acceleration measurements vary with tilt.
4 Design an application that uses the SD Card Reader to datalog GPS coordinates.

5

WIRELESSLY NETWORKING PROPELLER CHIPS

Martin Hebel

Introduction

This chapter looks at how your Propeller can be part of a *wireless sensor network* (WSN) to share data through wireless communications. WSNs are not intended for large data transfers, such as files, but small amounts of data back and forth. The Propeller is an amazing controller, and its ability to perform parallel processing makes data communications fast and simple for use in a WSN. While the main task is being carried out, other cogs can be sending or receiving data on the network.

With a lot to discuss and learn along the way, the final completed project of this chapter, depicted in Fig. 5-1, will be a three-node network that has:

- A tilt-controller node transmitting drive and control data.
- A robot (bot) node that receives the data; has a compass and ultrasonic range finder; and is transmitting data on drive, range, and direction. It also has the ability to "map" what is in front of it for remote display.
- A node that accepts data from the bot and displays the information graphically on the TV.

This chapter highlights communications to, from, and between Propeller chips using XBee® transceivers from Digi International. Topics covered in this chapter include:

- Networking and XBee overview
- PC-to-XBee communications
- Configuring the XBee manually and with the Propeller

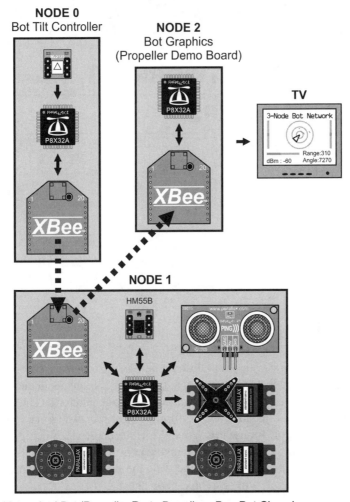

Networked Bot (Propeller Proto Board) on Boe-Bot Chassis

Figure 5-1   **Three-node network for monitoring and control.**

- PC-to-Propeller and Propeller-to-Propeller communications with the XBee
- Transparent and API data modes of the XBee
- Forming a multi-node Propeller network for robot control and monitoring

This chapter will work through several examples of communications, but really, the intent and focus is on *how* to perform the communications with the Propeller. It is left to you, the reader, to take the principles discussed, combine them with your imagination or needs, and develop a Propeller network of your own. Many other projects and information from this text can be combined with this chapter for truly amazing projects!

Resources: Demo code and other resources for this chapter are available for free download from ftp.propeller-chip.com/PCMProp/Chapter_05.

Overview of Networking and XBee Transceivers

The ability to communicate wirelessly has had such a significant impact on personal and data communications that many today cannot envision life without the use of cell phones, Wi-Fi networks and Bluetooth® features in personal devices. The ability of these devices to communicate on their respective networks (even your Bluetooth headset forms a network with the player) relies on key features:

- The use of addressing to send data to specific destination devices and to identify the source of the data
- The use of framing and packets to encompass the data itself in a "package" with necessary information (such as the destination address)
- The use of error checking to ensure the data arrives at the destination without errors
- The use of acknowledgements back to the source so that the sender knows the data arrived correctly at its destination

Simple two-device (or two-node) systems may not need all these features. It's really dependent on the needs of the network, but if ensuring data arrives correctly to an intended destination is vital, then these features are a must.

The XBee uses a fully implemented protocol and communicates on a *low-rate wireless personal area network* (LR-WPAN), sometimes referred to as a *wireless sensor network* (WSN) with RF data rates of 250 kbps between nodes. For the seasoned network readers, LR-PANs operate using IEEE 802.15.4, a standardized protocol similar to Wi-Fi (IEEE 802.11) and Bluetooth (IEEE 802.15.1). The XBee is currently available in the XBee 802.15.4 series and the XBee ZigBee/Mesh series. The 802.15.4 series (often referred to as Series 1) is the simplest and allows point-to-point communications on a network. The ZigBee/Mesh series (Series 2) uses the ZigBee® communications standard on top of 802.15.4 for WSNs to provide self-healing mesh networks with routing. This chapter will focus exclusively on the XBee 802.15.4 and its higher-power sibling the XBee-Pro 802.15.4. These will be referred to as simply the XBee.

Key benefits of using the XBee include the ability to perform addressing of individual nodes on the network, data is fully error-checked and delivery acknowledged, and data can be sent and received transparently—simply send and receive data as if the link between devices were directly wired. XBees operate in the 2.4 GHz frequency spectrum.

An image and a drawing of an XBee are shown in Fig. 5-2. The XBee is a 20-pin module with 2.0 mm pin spacing. This can cause some aggravation when working with breadboards and protoboards, which have 2.54 mm (0.1 in) pin spacing, but solutions to this will be addressed.

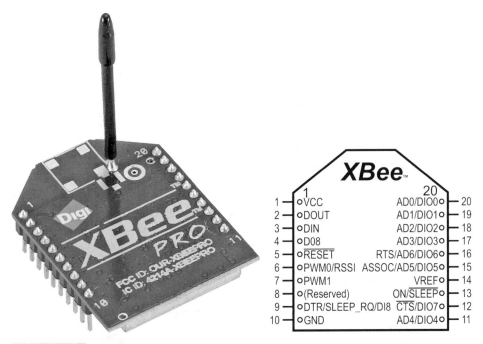

Figure 5-2   XBee module and pins.

Don't get scared! The XBee has a large number of pins, but for most of this chapter, we will use only four:

- Vcc, Pin 1: 2.8 V to 3.4 V (Propeller Vdd voltage)
- GND, Pin 10 (Propeller Vss)
- DOUT, Pin 2: Data out of the XBee (data received by Propeller)
- DIN, Pin 3: Data into the XBee (data to be transmitted by Propeller)

Other pins include a sleep pin (Sleep_RQ) for low power consumption, flow control pins (RTS/CTS), analog-to-digital (ADC) inputs, digital inputs and outputs (DIO), among others. This chapter will discuss some of these other pin functions, but the focus is on simply sending and receiving data between the Propeller and XBees using the DOUT and DIN pins.

Note: Please see the XBee manuals on Digi's web site for in-depth discussion and information: www.digi.com and included in the distribution files.

The XBee has a current draw of around 50 mA and a power output of 1 mW with a range of about 100 m (300 ft) outdoors. The XBee-Pro has a current draw of 55 mA when idle or receiving data and 250 mA when transmitting. With a power output of 100 mW, it has a range outdoors of 1600 m (1 mi) line sight. They both have sleep modes, with current draws of less than 10 μA, but can't send or receive data while sleeping. There are different antenna styles as well, though the whip antenna is probably the most popular.

Tip: Don't get too excited about the distances. Line-of-sight communications rely on height as well as distance. Due to ground reflections and deconstructive interference (Fresnel losses), the heights of the antennas need to be taken into account. For good communications at 100 m, a height of 1.4 m (4.6 ft) is recommended.

Information: For more insight on distance, height issues, and calculations, search the web for "Fresnel clearance calculation."

Though the XBee is ready to go right out of the box, it is feature-rich and can be configured for specific applications.

Hardware Used in This Chapter

The following is a list of hardware used in this chapter and their sources, but as you read through, you'll find it's not written in stone. We recommend you read through the chapter to understand how the hardware is used before making an expensive investment.

■ 2—Propeller Demo Boards (Parallax)
■ 1—Propeller Proto Board (Parallax)
■ 1—Prop Plug (Parallax)
■ 3—XBee 802.15.4 (Series 1) modem/transceivers (www.digikey.com)
■ 3—AppBee-SIP-LV XBee carrier boards (www.selmaware.com or other styles available on www.sparkfun.com)
■ 1—PING))) ultrasonic sensor (Parallax)
■ 1—HM55B compass module (Parallax)
■ 1—Memsic 2125 accelerometer/inclinometer (Parallax)
■ 1—Boe-Bot chassis (Parallax)
■ 1—Ping Servo Mounting Bracket Kit (Parallax)
■ 2—Additional Boe-Bot battery holders or other portable battery source
■ Miscellaneous resistors

Testing and Configuring the XBee

An important step in constructing a complex project is to make sure the individual devices work properly and their use is understood. In this section, the XBees will be tested, configuration settings explored, and means of configuring these devices discussed.

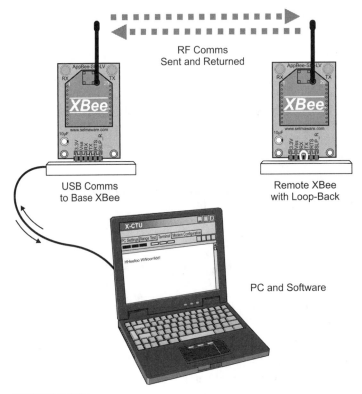

Figure 5-3   **Configuration and testing diagram.**

Figure 5-3 shows the diagram for this test. A PC will communicate directly to an XBee, and a remote XBee is set up with a loop-back jumper. In the loop-back, the DOUT line of the XBee is tied to its DIN so that any RF data it receives is looped back into the device to send it out again via RF.

The following is a list of the hardware and software used for this test, but there are many ways to achieve the same results. Essentially, a means is needed to communicate to an XBee serially from the PC and means to supply power to the base and remote XBees.

Equipment and other software:

- 2—Propeller Demo Boards (Parallax)
- 2—XBees (www.digikey.com)
- 1—Prop Plug (Parallax)
- 1—AppBee-SIP-LV from Selmaware Solutions (www.selmaware.com)
- X-CTU software from Digi International (www.digi.com)

The AppBee-SIP-LV is simply a carrier board for the XBee providing 3.3 V power from the Demo Board and access to I/O in a breadboard-compatible header. Figure 5-4 shows the AppBee-SIP-LV and a drawing of the physical connections to the XBee.

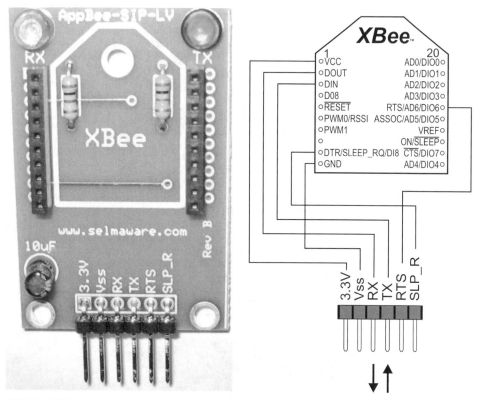

Figure 5-4   AppBee-SIP-LV carrier board and drawing with physical connections.

Tip: Another good source of carrier boards and other XBee accessories is www.sparkfun.com. Search their web site for XBee.

ESTABLISHING PC-TO-XBee COMMUNICATIONS

The first task is to communicate with the XBee directly from the PC for configuration changes and monitoring. Figure 5-5 shows two ways of establishing communications: using the Propeller as a serial pass-through device or communicating directly with the XBee using the Prop Plug as a serial interface. Either method allows the serial connection between the PC and the transceiver.

If you are using the Propeller to pass serial communications, the program Serial_Pass_Through.spin should be downloaded using F11. If the serial communications port is closed in the software, the Propeller may be cycled when the DTR is toggled, reloading the Propeller from EEPROM. Using F11 ensures a cycling of the Propeller will reload the correct program.

The program itself is simple but highlights the power of Propeller. Microcontrollers that provide multiserial communications are difficult to find. Two instances of the

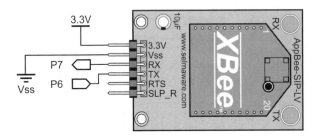

(a) Using Propeller Serial Pass Through

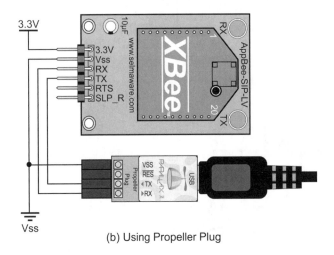

(b) Using Propeller Plug

Figure 5-5   **Two methods of PC communications with XBee.**

FullDuplexSerial object establish the transparent link. Data from the PC is sent to the XBee, and data from the XBee is sent to the PC; with each method in separate cogs, it allows transfer speeds tested up to 115,200 bps. But for now we need to stick to 9600 bps since that is the default configuration on the XBee.

```
OBJ
  PC   : "FullDuplexSerial"
  XB   : "FullDuplexSerial"

Pub  Start

  PC.start(PC_Rx, PC_Tx, 0, PC_Baud) ' Initialize comms for PC
  XB.start(XB_Rx, XB_Tx, 0, XB_Baud) ' Initialize comms for XBee
  cognew(PC_Comms,@stack)        ' Start cog for XBee--> PC comms
```

```
    PC.rxFlush                        ' Empty buffer for data from PC
    repeat
      XB.tx(PC.rx)                    ' Accept data from PC and send to XBee

Pub PC_Comms
    XB.rxFlush                        ' Empty buffer for data from XB
    repeat
      PC.tx(XB.rx)                    ' Accept data from XBee and send to PC
```

Caution: Watch the I/O numbers! If another configuration is used, modify the pin numbers in the CON section of the code.

If you are using the Propeller for passing serial data:

✓ Connect the hardware as shown in Fig. 5-5a.
✓ Download the Serial_Pass_Through.spin program to the Propeller using F11.

If you are using the Prop Plug to communicate directly, connect it as shown in Fig. 5-5b.

✓ If you haven't yet, download and install the X-CTU software available in the distributed files or from Digi's web site. There is no need to check for updates—this can take a long time and the basic installation has all that is needed for now.
✓ Open the X-CTU software. It should look similar to Fig. 5-6. Select the COM port that your Propeller is communicating through.
✓ At this point, use the Test/Query pushbutton to test communications with the XBee.

Caution: As always, only one software package can access the same COM port at any time. You'll get used to slapping your head when you can't communicate as you go between the Propeller tool software and X-CTU!

Tip: If communications fail, recheck your hardware and pin numbers, reload the Propeller program, and verify no other software is using the COM port. If you continue to have problems and it is not a brand-new XBee, the serial baud rate may have been changed or the XBee may be in API mode—test various baud rates and check the API box to test.

If all went well, you may have seen the RX and TX lights blink on the board and received a message informing you communications were okay, along with the firmware version on the XBee.

✓ Select the Modem Configuration tab on the X-CTU software.
✓ If your XBee was reconfigured, this would be a good time to click the Restore button to return it to the default configuration.
✓ Click the Read button.

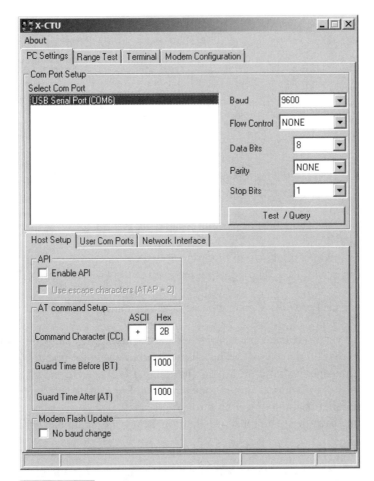

Figure 5-6   X-CTU software showing COM port selection.

The screen should have loaded with the configuration setting of the XBee as shown in Fig. 5-7. Many of them will be explained shortly—we're only going to use a handful of the settings available. But for now, let's test out some wireless communications.

TALKING XBee TO XBee USING LOOP-BACK

With a second XBee, supply power and connect a jumper between DOUT and DIN (or RX and TX on the carrier board), as illustrated in Fig. 5-8, using the AppBee-SIP-LV carrier board (or similar). Do not connect to any Propeller I/O at this time—we are simply using the board for power. We used a second Demo Board for this test.

✓ Power up the remote XBee with loop-back jumper in place.
✓ Click the X-CTU Terminal tab.
✓ Type "Hello World!"

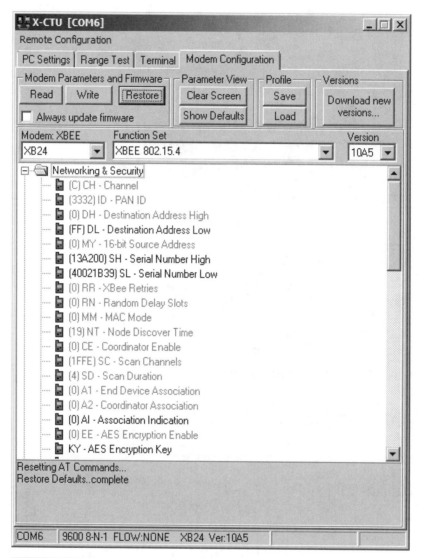

Figure 5-7   X-CTU software showing XBee configuration settings.

You should see the TX and RX lights flashing on both units (if using the AppBee carrier) and text in your Terminal window. You should see two of each character—what you typed in blue and what was echoed back and received in red—as shown in Fig. 5-9.

Tip: Having problems? If you don't see any data returning, be sure the remote XBee is connected properly. If it is not a new XBee, if may have been configured differently. Turn off both units and swap the XBees. After powering up, "Restore", the XBee to default configuration using the X-CTU button, read the second XBee using the X-CTU software, and test again.

Figure 5-8   Remote XBee connections for loop-back.

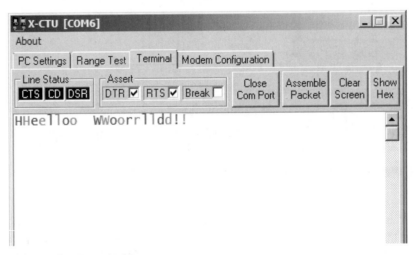

Figure 5-9   X-CTU Terminal window.

Tip: Beyond testing purposes, the X-CTU software is not essential, and any terminal program or other serial software package may be used, such as the PST Debug-LITE software used in previous chapters. Just ensure baud rates match between the software and the devices.

As noted, each character is transmitted as it is typed. The XBee can actually send a string of characters at once (up to 100), but it only waits so long before assembling a packet to be transmitted. We type too slowly to get multiple characters quickly enough with the default configuration, but we can assemble a packet of characters that will be kept together:

✓ On the X-CTU Terminal window, click <u>Clear screen</u>, and then click <u>Assemble Packet</u>.
✓ Type "Hello World!" in the packet box, and click <u>Send Data</u>.

You'll notice your text is returned as a single packet.

One last test is the range test. This allows you to monitor the signal strength from −40 dBm to the XBee's sensitivity limit of around −100 dBm by having the software repeatedly send out a packet to be echoed.

✓ Check the <u>check box</u> below the vertical RSSI (receiver signal strength indication).
✓ Click <u>Start</u>.
✓ Monitor the number of good packets received and signal strength.
✓ Block the area between the XBees or move the remote XBee to another room, and test the effect on RSSI level.

Note: In theory, you should never see a bad packet (malformed data) in the received data from the XBee, such as in the Terminal window. All data is error-checked and retried if there is no response or if the error check fails. You should receive either good data or no data at all. The serial-link issue with the XBee is a more probable cause than an RF issue with bad data.

Now that we have an RF link going, it's time to discuss and test some XBee configurations.

XBee CONFIGURATION SETTINGS

As seen, the XBee has numerous settings that can be configured. This configuration can be performed through the Configuration window, through the Terminal window, or through strings sent out from the Propeller. Let's first take a look at some of the more important settings shown in Table 5-1 for this chapter (we will use only a few) and others of interest should you delve deeper with your experiments. Click the <u>Modem Configuration</u> tab of the X-CTU software to view the settings. Clicking any setting will give a brief description and range of values at the bottom of the window.

OK, let's test out a few things:

✓ Test and verify your loop-back setup by sending a string.
✓ Under <u>Modem Configuration</u>, change DL to 1.
✓ Click <u>Write</u>.
✓ The XBee should be updated. Click <u>Read</u> and verify.
✓ Go to the Terminal window and type once again. You should get no response, and the remote RX light on the AppBee should not blink.

TABLE 5-1 SUMMARY OF PERTINENT XBee SETTINGS	
COMMAND CODE	**MEANING & USE**
Networking & Security	
CH	Channel: Sets the operating frequency channel within the 2.4 GHz band. This may be modified to find to a clearer channel or to separate XBee networks.
ID	PAN ID: Essentially, the network ID. Different groups of XBee networks can be separated by setting up different PANs (personal area networks).
DL	Destination Low Address: The destination address where the transmitted packet is to be sent. We will use this often to define which node receives data. A hexadecimal value of FFFF performs a broadcast and sends data to all nodes on the PAN. The default value is 0.
MY	Source Address: Sets the address of the node itself. This will be used often in all our configurations. The default value is 0.
Sleep Modes	
SM	Sleep Mode: Allows the sleep mode to be selected for low power consumption (<10 µA). While we won't use it, a good choice is 1—Pin Hibernate. This would allow an output of the Propeller to put the XBee to sleep (using the Sleep Request pin) when it is not sending or expecting data.
Serial Interfacing	
BD	Interface Data Rate: Sets baud rate of the serial data into and out of the XBee.
AP	API Enable: Switches the XBee from transparent mode (AT) to a framed data version where the data must be manually framed with other information, such as address and checksum. This is a powerful mode and will be explored in this chapter.
RO	Packetization Timeout: In building a packet to be transmitted, the XBee waits a set length of time for another character. If not received in the set time, the packet is sent. This is why as we typed characters, each was sent and echoed back. This can be important to change if you have multiple units sending data to one node to ensure that all data sent is received as a single transmission from one unit; otherwise, you may get data from various nodes intermixed.
I/O Settings	
D0 – D8	Sets the function of the I/O pins on the XBee, such as digital output, input, ADC, RTS, CTS, and others.
IR	Sample Rate: The XBee can be configured to automatically send data from digital I/O or ADCs. It requires the receiving node to be in API mode and the data parsed for the I/O values.
Diagnostics	
DB	Received Signal Strength: The XBee can be polled to send back the RSSI level of the last packet received.
EC	CCA Failures: The protocol performs clear channel assessment (CCA)—that is, it listens to the RF levels before it transmits. If it cannot get an opening, the packet will fail and the CCA counter will be incremented.

EA	ACK Failures: If a packet is transmitted but receives no acknowledgement that data reached the destination, EA is incremented. The XBee performs two retries before failure. Additional retries can be added by using the RR setting.
AT Command Options	
CT	AT Command Timeout: Once in command mode, this sets how long of a delay before returning to normal operation.
GT	Guard Time: When switching into AT command mode, this defines how long the guard times should be (absence of data before the command line) so that accidental mode change is not performed.

By changing DL to 1, data is intended for an XBee at address 1. The default settings on XBees are a DL of 0 and an MY of 0. Previously, we were sending data to a node at address 0 from a node at address 0 and vice versa. Be aware, the XBee actually does receive data, sees it is not the intended node, and then dumps it instead of passing it to the DOUT pin (to which the RX LED is connected).

Let's now try configuring using the Terminal window. Due to timeouts, you may have to type a little fast, so you may need a few attempts. Enter the following lines—*do not* type what is in parentheses. Press enter after each line except for +++.

✓ (Wait three seconds since you typed anything last—this is guard time.)
✓ **+++** (*Do not* press ENTER.)
✓ (Wait a few more seconds and you should see that it is now in command mode.)
✓ ATDL (Requests the current DL value; it should return 1)
✓ ATDL 0 (Sets the DL address to 0)
✓ ATDL (Again requests the DL address, which should be 0)
✓ ATCN (Exits AT command mode)
✓ Hello World?

If all went well, you should once again be getting echoes after changing the destination address back to 0. The waiting before and after the +++ is called the *guard time,* and it ensures that if a string containing +++ is sent, the unit won't flip into command mode inadvertently.

Tip: Permanent changes? Using the Modem Configuration feature of the X-CTU software, all changes are saved to nonvolatile memory and will still be in place after cycling power. Using the AT commands, the settings will revert to original values after cycling power, unless the ATWR (write) command is sent to write to nonvolatile memory.

The important aspect here is that just as we sent data strings to the Xbee for configuration changes, so can your Propeller configure the XBee through code. Multiple commands can be used in one line by separating them with commas. For example, the following sets DL to 0 and exits command mode: ATDL 0, CN.

TRY THESE!

✓ Try changing your MY address to 1 and sending data. You should see the remote unit receive and transmit, but you get nothing back. Why?

✓ Change your DL to FFFF. This is the broadcast address—any nodes on your network would receive it. Be sure to set MY back to 0 for the loop-back to work!

✓ Use the command ATND (Network Discovery). After a few seconds, you should see a list of other nodes in the network, including their MY address, two lines of the physical address (like a MAC address), and the RSSI level in hexadecimal.

✓ Use the command ATED (Energy Detect). You should see a list of about 11 hexadecimal values. This is the energy level seen on the various channels. Higher values are less noisy—a value such as 5A (hexadecimal), for example, converts to a level of −90 dBm.

✓ Use the Configuration tab to restore the XBee to its default values when done testing, or use the AT command ATRE, followed by ATWR, to save to memory.

UPDATING THE XBee VERSIONS

Just a note about the version of the XBees: In the Modem Configuration tab, you can see the version of firmware on your XBee, such as 1083, 10A5, or 10CD. Later versions are more capable. The majority of this chapter requires at least 1083. The firmware on the XBee can be updated by selecting a new version, checking Always update firmware, and clicking Write, but this requires more data lines than we have available with our configurations. A board such as the XBIB-U from Digi International or the WRL-08687, the XBee Explorer, from www.sparkfun.com (which can also double as a carrier board) is recommended. These boards can be used for direct USB access to the XBee as well as changing the firmware, and they supply power to the XBee.

Now that we can send and receive data and configure the XBee, we are ready to start using Spin and the Propeller to communicate via the XBee.

Sending Data from the Propeller to the PC

In this section we will equip a remote Propeller/XBee system with a couple of sensors and then transmit the data from the sensors back to the base XBee to send the data to the PC for monitoring. The base can be the Propeller using serial pass-through, using the Prop Plug to the XBee, or using a dedicated XBee-to-PC board, as previously mentioned. The sensors used for testing are Parallax's HM55B compass module and the PING))) ultrasonic range finder. These devices will eventually assist in our robot project, but you are free to modify the code to use any of the sensors previously explored in this text.

Additional equipment:

- HM55B Compass Module
- PING))) Ultrasonic Range Finder
- Or other sensors as desired, with appropriate code

Figure 5-10 is an image of the nodes. Even though we don't need to just yet, we will use this opportunity to set the DL address of the remote unit to 0 to ensure it is sending data to the base unit.

✓ Connect the PING))) sensor and HM55B compass on the remote unit as shown in Fig. 5-11. If a different I/O pin is used, update the pin numbers accordingly in the CON section of the code. Connect the LEDs as well; we will use them shortly.
✓ For the base unit XBee, open and clear the X-CTU Terminal window. Open the COM port if closed. Having that port in use will help ensure the correct Propeller is programmed.
✓ Download Simple_PC Monitoring_from_Remote.spin to the remote unit.
✓ Monitor the remote unit's LEDs—they should blink rapidly a few times after several seconds as the XBee is configured.
✓ Monitor the base unit's Terminal window. A "ready" message should be displayed, then the readings of the sensors should be reported every half-second.
✓ Test the compass bearing. It should read 0 to 8191 (roughly) as you rotate it, with 0 being approximately magnetic north.
✓ Test the range finder by placing an object in front and moving it in and out. The PING))) sensor will report distances from roughly 30 to 3000 mm (3 cm to 3 m).
✓ If either sensor fails to respond properly, check your connections and code.

Tip: The range finder has a fairly large angle of emission and detection. Test this by putting an object to the side of range finder and going in and out to determine how wide the angle is at different distances.

After initializing the XBee and compass, there is a three-second delay, +++ is sent followed by another three-second delay and the string of "ATDL 0, CN." Finally, a byte of 13 representing a CR or ENTER key is sent. The destination address is set to 0 and command mode is exited (CN) in exactly the same fashion as you did in the Terminal window.

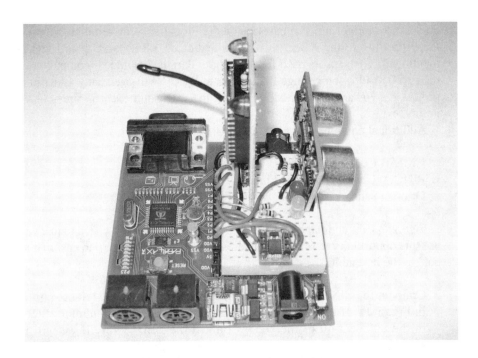

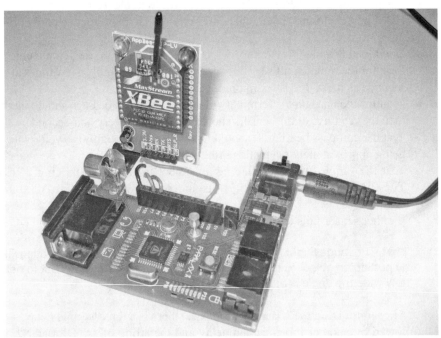

Figure 5-10   Base and remote nodes.

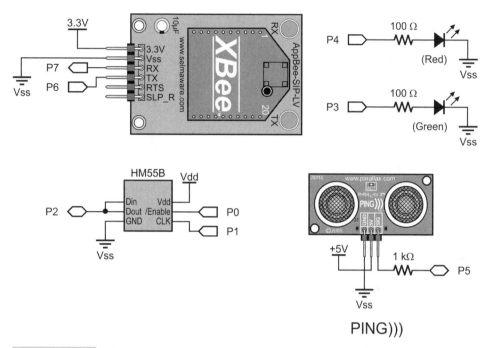

Figure 5-11   Remote unit with PING))) and compass.

```
delay (3000)                          ' Guard time for AT mode
XB.str (string ("+++"))               ' Send AT command request
delay (3000)                          ' Guard time
XB.str (string ("ATDL 0,CN"))         ' Send code to set DL = 0
XB.tx (13)                            ' Send carriage return
```

The command codes from the Propeller are passed to the XBee using the FullDuplexSerial object duplicating your serial terminal. Through this method any number of commands may be sent to the XBee for configuration changes on initialization or during operation. Those 3-second guard times can cause lag during operation, but we'll deal with that soon.

In the SendData method, you can see that the range and direction (theta) are read from the devices. Using a combination of text strings and XB.dec (decimal) methods, the data is sent to the base XBee, where it is passed through to the PC for monitoring in the Terminal window. Figure 5-12 is a sample output of received data.

```
repeat
    range := Ping.Millimeters (PING_Pin)      ' Get range in mm
    theta := HM55B.theta                      ' Get bearing (0-8191)
    XB.str (string (13,13,"Ping Range (mm) : "))' Send string to base
    XB.dec (range)                            ' Send range as decimal
    XB.str (string (13,"Direction (0-8192) : "))' Send string to base
    XB.dec (theta)                            ' Send bearing as decimal
    delay (500)                               ' Short delay before repeat
```

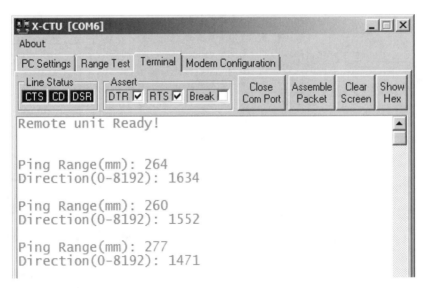

Figure 5-12   Sample output in Terminal window of range and bearing.

TRY IT!

✓ Try adding a simple device, such as a pushbutton, and reporting its state back to the PC. If you are out of I/O, you may remove the LEDs.

Polling Remote Nodes

In an LR-PAN, nodes typically come in one of three flavors:

- *Coordinators* help manage the network, from controlling communications to assigning information to devices.
- *End devices* are used to read and control devices on the network.
- *Routers* are used to pass data between nodes at distances too far to reach directly.

There is nothing prohibiting end devices from talking to one another, and once a network is established, the coordinator's job may come to an end. In this chapter we will refer to the base unit, the one at the PC, as a coordinator because it will help control communications and be a common collection point. Our remote nodes will be end devices that we will monitor and control.

Multinode communications can be tricky. Aspects to be dealt with include: Which node can send data when? When data arrives, who is it from? Do nodes need permission to talk or can they do so at any time? We need to ensure that nodes don't talk over one another (causing collisions on the network) and that the receiving units know who the data is from in order to respond appropriately or take some other action. XBee, using IEEE 802.15.4, works similar to Wi-Fi. A node listens before it transmits to help ensure that no other node is transmitting at the time (this is *Clear Channel Assessment,* or CCA). Delivery of data is verified through acknowledgements. If the sender does not get a response, it tries again. This method is known as CSMA/CA or *Carrier Sense, Multiple Access/Collision Avoidance.* Unlike Ethernet, which uses collision detection (CSMA/CD), a node cannot listen once it starts transmitting so it cannot detect collisions.

So the data link layer of communications helps ensure data gets passed properly, but it still doesn't assist in higher-level functions controlling the who and when of communications. In the next section we will look at a method of using a Propeller acting as a coordinator to poll end devices for their data. USB works in much the same way—each device is polled one at a time to see if they need access or have data to send.

COORDINATOR MANUALLY POLLING REMOTE END DEVICES

A hardware configuration similar to the one from the previous section will be used, but this time, the Propeller needs to be in the communications chain at the base instead of simply using a Prop Plug for XBee communications. Also, to demonstrate control action, the two LEDs on the remote end device provide control action. You are welcome to have as many end points as you desire (well, up to 65,000), or just use one and change the end point's address to test. Figure 5-13 is a diagram of our network and hardware.

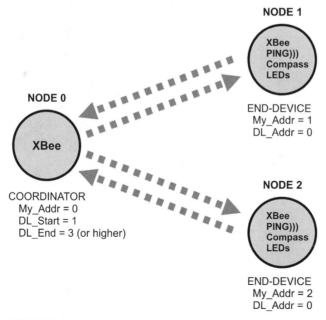

Figure 5-13   Hardware for coordinator polling.

In this example, the coordinator cycles through a range of end-point addresses by changing the DL value of the coordinator's XBee. It sends out codes and values to request data from each end point and to control the LEDs on each. Before allowing the coordinator to have control, we are going to manually test the control and responses.

✓ Add the LEDs to the remote end device.
✓ Open Acquisition_with_Control_End.spin.
✓ For each end device, number the constant MY_Addr in the CON section of the code sequentially from 1 up, skipping a few numbers to test "unresponsive nodes."
✓ Download Acquisition_with_Control_End.spin to each remote end device.
✓ Use the Propeller for serial pass-through or another PC-to-XBee configuration at the PC.
✓ Change the DL of the coordinator/base XBee to 1.
✓ In the Terminal window, type some p's and c's. If your end point at address 1 is awake, you should get values back for compass bearing and range finder distance.
✓ For this next test, use the "Assemble Packet" window. Type and send the following:
  o Type i3 and then hit Enter.
  o Type 1 and then hit Enter.
  o Click Send.
✓ Change the 1 to a 0 and send again.
✓ Test again by using 4 instead of 3.
✓ What you should see is LEDs on P3 and P4 turning on with 1 and off with 0.

Figure 5-14 is an image of our communications test.

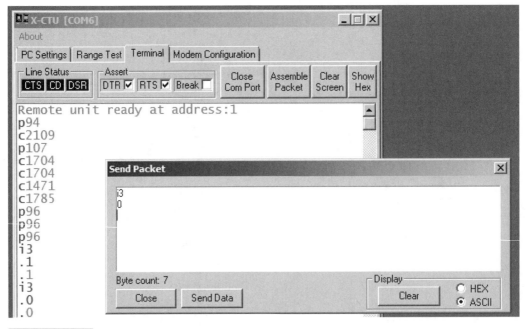

Figure 5-14   End-device responses to requests.

Looking at the end-device's code, data communications with the XBee is now through the XBee_Object. This is an object I wrote for easing some data communication and configuration issues. It uses FullDuplexSerial but greatly extends it.

Tip: The "XBee_Object" can be downloaded from Parallax's Object Exchange (http://obex.parallax.com). If you have previously downloaded it, be sure it is version 2 or higher. It is also included in the book's distributed files.

XB.AT_Init initialized the XBee to AT mode, allowing for short guard times (using ATGT), so instead of six seconds to modify a configuration, it can be done quickly in code. XB.AT_ConfigVal allows passing an AT command and a value to set configurations, such as the DL and MY addresses. The underlying code switches the XBee to command mode, sends data, and exits using the short guard times.

```
" Enable XBee for fast configuration changes
XB.AT_Init

" Set MY and DL (destination) address.
XB.AT_ConfigVal(string("ATMY"), MY_Addr)
XB.AT_ConfigVal(string("ATDL"), DL_Addr)
```

In the ProcessData method, XB.rx is used to tell the Propeller to wait for one character or byte of data. It then tests this character to determine what set of actions to take:

```
dataIn = XB.rx
Case dataIn
   "p":                                   ' p = PING distance
      range := Ping.Millimeters(PING_Pin) ' Read PING in mm
      XB.dec(range)                        ' Send range as ASCII decimal value
      XB.cr                                ' End decimal string with CR

   "c":                                   ' c = Compass
      theta := HM55B.theta                 ' Read Compass
      XB.dec(theta)                        ' Send theta of bearing as decimal
      XB.cr                                ' End with carriage return

   "i":                                    ' i = I/O control
      IO := XB.rxDecTime(timeout)          ' Accept IO number w/timeout
      state := XB.rxDecTime(timeout)       ' Accept state (1/0) w/timeout
      if state <> -1
         dira[IO]~~                         ' Set direction of pin
         outa[IO] := state                  ' Set state of pin
         XB.dec(outa[IO])                   ' Send state back for verification
         XB.cr                              ' End decimal string with CR
```

If p, send back the decimal value of the range finder.
If c, send back the decimal value of the compass bearing.

If i, accept the next two decimal values and use them for I/O and State, which sets the I/O direction to be an output and the state of the I/O. RxDecTime is used to accept the decimal values with a timeout. This allows the program to continue to run if incorrect data is received following a timeout period. Should a timeout occur, a −1 is returned to the value. In accepting the data, note that each decimal value must end in an ASCII 13 or CR (or comma, see Sec. Data Acquisition and Control Using API Mode). Finally, the actual value of the I/O is sent back.

✓ Change the coordinator/base to a nonexisting end-device address (DL) and try again. You should get no data back.

What we are designing here can be considered a *protocol*—rules of communication. If you don't follow the rules set forth, nothing, or even incorrect things, may happen. When coding protocols, we attempt to cover all contingencies regarding what could go wrong and how they will be dealt with, such as i3 and no further data. What happens if you enter something other than a 1 or 0 for state? That contingency is not covered!

The data between the units is kept simple—byte codes and decimal strings. This allows short packets between the units and eases using the data in the code.

Caution: Be aware that currently the code can control *any* of the Propeller chip's I/O pins, so be careful of what you send for your IO values!

AUTOMATIC POLLING WITH THE PROPELLER

In this next exercise the Propeller will operate as the coordinator, polling each of the end devices in succession.

✓ Ensure you have downloaded Acquisition_with_Control_End.spin to your end device(s) using F11, with sequential MY_addr values while skipping a few values.
✓ In Acquisition_with_Control_Coor.spin, modify the values of DL_Start and DL_End in the CON section to match the range of your end-device addresses.
✓ Download Acquisition_with_Control_Coor.spin to the coordinator Propeller.
✓ Once downloaded, open the Terminal window.
✓ Wait and watch… you should see results similar to Fig. 5-15. Note that in this test only an end device with a MY_addr of 2 is responsive.

In Pub Start, once configured, the code loops through the range of defined end-device values, passing the address to the Poll method. Pub Poll accepts the address, sets the DL address, and informs the user. It then goes through a series of steps for acquisition and control.

Calling Control_IO, the I/O number and state are passed to turn on the LEDs. This method will send the correct i-instruction to control the end-device IO. The returning value with a timeout is accepted, passed back, and displayed.

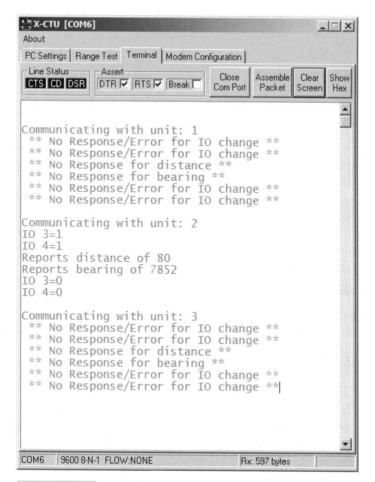

Figure 5-15   Coordinator responses from automated polling.

```
Pub Control_IO(pin, state) : Value
  XB.tx("i")                           ' Send i for IO control
  XB.dec(pin)                          ' Send pin as decimal value
  XB.cr                                ' Send CR
  XB.dec(state)                        ' Send state as decimal value
  XB.cr                                ' Send CR
  Value := XB.rxDecTime(200)           ' Accept value with timeout
```

Next the GetDistance method is called, which sends the p-instruction, accepts returning data, and passes it back for display. Then the GetAngle method is called; sends the c-instruction; and accepts, returns, and displays data. Finally, the LEDs are again turned off using `Control_IO` sending 0s.

```
Pub GetDistance : mm

    XB.tx("p")                        ' Send p to get range
    mm := XB.rxDecTime(50)            ' Accept data with timeout
```

The cycle repeats each end-device value, pauses longer, and starts over. In each step of the way, timeouts are used to ensure nonresponsive end devices do not lock up the system and that they are reported as being nonresponsive.

In this example we are simply collecting data and controlling LEDs for testing purposes while displaying information for the user. The returned data could be used by the coordinator for some logical decisions or to control a local output or send data to another end device for action.

TRY THESE!

✓ Add another sensor and the code to request and respond with data.
✓ Use a returned value in some way at the coordinator, such as lighting an LED if the distance is within 100 mm (10 cm).
✓ Rapidly collect a remote value and plot it using ViewPort.

Though not used in our code, reading configuration values from the XBee can be done by sending the AT command and accepting returning data. The XBee uses hexadecimal for all values. The receiver is flushed to ensure that no data exists in the Propeller's object buffer. In this example, the dB level of a recent XBee reception is read and displayed.

```
XB.rxFlush
XB.AT_Config(string("ATDB"))
dataIn := XB.RxHex
PC.DEC(-dataIn)
```

Using the XBee API Mode

API MODE AND DATA FRAMING

Continual polling can take a lot of time and resources to check for data that may change infrequently. It is good to have the coordinator control the communications, but this requires the remote units to be awake. Another mode for the XBee is called API mode, for *application programming interface*. Instead of sending or receiving the data alone, the entire frame is manually constructed for transmission and manually parsed on reception. The frame consists of sender's address, RSSI level, options, frame IDs, and the data or message itself. Depending on the frame type, different types of data are carried. Some benefits to using API mode include:

- Pull sender's address directly from received frame.
- Pull RSSI level from certain received frame types.
- Place the destination address for the packet directly in the frame.
- Use frames for local XBee configuration as opposed to AT mode.
- Use frames for REMOTE XBee configuration (firmware version 10CD required).
- Pass analog and digital data from the XBee's I/O pins *without* a controller on the remote (firmware version 10A3 or higher required).
- Use frames that provide delivery notification to the sender.
- Data is received in a single frame (up to 100 bytes), ensuring it is from a single source.

As you can see, using API mode opens many doors to fast and powerful communications, but it can be a little complex. The XBee Object supports the means for constructing and retrieving data for many of the API frame types. Let's look at how a packet must be framed to be accepted for transmission, as shown in Fig. 5-16, taken from Digi International's XBee manual. This frame type is for sending strings between units, such as our data.

Note: Data is *always* sent in frames between XBees, but when in AT mode (transparent mode), the only thing we deal with is the data, or message, itself.

First, all frames start with a *start delimiter* so the receiving unit can locate where the start of a frame is as data pours in. Next is a 16-bit length (MSB and LSB), which has to match the number of bytes from after the length through to the checksum, but not including it. This is followed by the *API identifier,* a unique value telling the receiving unit what type of message it is.

Next is the *identifier-specific data,* consisting of the *frame ID* (if set to 0, it will suppress acknowledgement packets back to the controller; we will ignore these packets) and then the 16-bit destination address as 2 bytes. To send data to a unit at address 1, these would be values of 00 01. Options are set to disable acknowledgements or to send the data as a broadcast. Next is the actual data—up to 100 bytes. And finally, all the byte

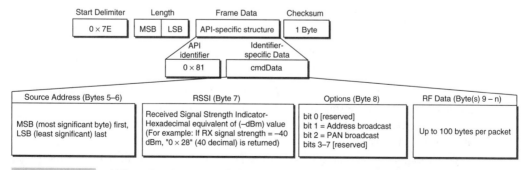

Figure 5-16   **API packet for transmitting string using 16-bit address.** (*Reprinted by permission of Digi International.*)

values up to that point are summed together to create a checksum value. The receiving unit will perform a summation itself, verifying against this value before using the data. If the packet is well formed, the XBee will accept this frame and transmit it. If not, it will be discarded.

Simple huh? Actually, it's not all that bad, but much more complex than just sending a string to be transmitted. Let's look at the Spin code that forms a packet when we send a string, such as XB.API_Str(String("Hello!")).

From the XBee Object:

```
Pub API_Str (addy16,stringptr) | Length, chars, csum,ptr
{{
  Transmit a string to a unit using API mode - 16 bit addressing
  XB.API_Str(2,string("Hello number 2"))      ' Send data to address 16
  TX response of acknowledgement will be returned if FrameID not 0
  XB.API_RX
  If XB.Status == 0 '0 = Acc, 1 = No Ack


}}
  ptr := 0
  dataSet[ptr++] := $7E
  Length := strsize(stringptr) + 5   ' API Ident + FrameID + API TX cmd +
                                     ' AddrHigh + AddrLow + Options
  dataSet[ptr++] := Length >> 8      ' Length MSB
  dataSet[ptr++] := Length           ' Length LSB
  dataSet[ptr++] := $01              ' API Ident for 16-bit TX
  dataSet[ptr++] := _FrameID         ' Frame ID
  dataSet[ptr++] := addy16 >>8       ' Dest Address MSB
  dataSet[ptr++] := addy16           ' Dest Address LSB
  dataSet[ptr++] := $00              ' Options '$01 = disable ack,
                                     ' $04 = Broadcast PAN ID
  Repeat strsize(stringptr)          ' Add string to packet
     dataSet[ptr++] := byte[stringptr++]
  csum := $FF                        ' Calculate checksum
  Repeat chars from 3 to ptr-1
    csum := csum - dataSet[chars]
  dataSet[ptr] := csum

  Repeat chars from 0 to ptr
    tx(dataSet[chars])               ' Send bytes to XBee
```

As you look through the code, you can see how all the individual bytes that make up a well-formed frame for transmission are combined into an array of bytes, the bytes are summed (actually subtracted from $FF one at a time) for the checksum, and the array of bytes is transmitted.

When the data is received by the XBee, the frame is checked. If in API mode, the frame shown in Fig. 5-17 is sent to the Propeller for processing. Based on the API

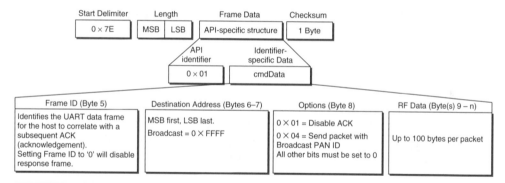

Figure 5-17   **API packet for received string using 16-bit address.** (*Reprinted by permission of Digi International.*)

identifier, the XBee Object can decide how to handle the frame data, and the top-level code can determine what to do with that type of frame data.

We won't go into the details, but again, specific bytes have specific meanings. In API mode, all this data is sent out to the Propeller. The XBee Object accepts the data and processes it accordingly using the RxPacketNow method. This method is actually private (PRI). The method called is API_RX or API_RxTime(ms), which looks for the start delimiter ($7E). Once found, execution is passed to RxDataNow to accept remaining data. Once accepted, the identifier is checked to determine the type of packet, which in the case of our received string, would be $81. Next the packet is parsed, pulling out the data and placing it into variables that can be accessed from the top-level code, such as XB.RxRSSI to find out the value of RSSI for the packet, or XB.srcAddr to get the sources address. Note that the data is actually accessible through XB.rxData—it is not the actual data, but a pointer to where the data string resides in memory.

Other methods can help us pull decimal data out. After receiving data, calling XB.ParseDEC(XB.rxData,2) would pass the location of the string and pull out the second decimal value in the string (values can be separated by ASCII 13—CRs—or by commas).

In sending decimal values, an API_DEC does not exist. Numbers, unless sent as raw byte values, must be converted to a string and sent that way. The Numbers.spin object can aid in the conversion, such as sending the range in API mode:

```
XB.API_str(num.ToStr(range,num#DEC))
```

But that's the only thing that could be sent, since once called, the string is transmitted in a frame. To keep our data together, another method is used to assemble a string (packet) manually before sending it. This will be demonstrated in the example coming up. In API mode, all data to be sent in one transmission must be assembled first.

Note: Both transmitter and receiver *do not* need to be in API mode. One side can be using transparent transmission and the receiver using API reception and vice versa. This makes our job a little easier.

DATA ACQUISITION AND CONTROL USING API MODE

In this example, we will continue using the coordinator and end-device(s) hardware, but use API mode instead for data reception and transmission on the coordinator. The end devices have the ability to transmit at any time; while we have it on a delay, another option may be to use sleep mode for low power consumption and have it wake to transmit, or have it transmit only when some event takes place, such as a range being too close (someone is near!).

The end device will send a string for the values `range` and `theta` without being prompted. The receiver will accept the string in API mode and pull out the source address, RSSI level, and data. It will then send back strings to blink the LED on the end device.

✓ Open and modify the DL value in API_Mode_End.spin for each of your end devices. The value doesn't matter, as long as it is not more than 255 ($FF). We are sending only one byte to hold the address in our example. Note that the X-CTU software uses hexadecimal values when configuring as opposed to decimal.

✓ Download API_Mode_End.spin to your end device(s).

✓ Download API_Mode_Coor.spin to your coordinator.

✓ Open and monitor the coordinator's Terminal window.

The resulting data should be similar to that shown in Fig. 5-18.

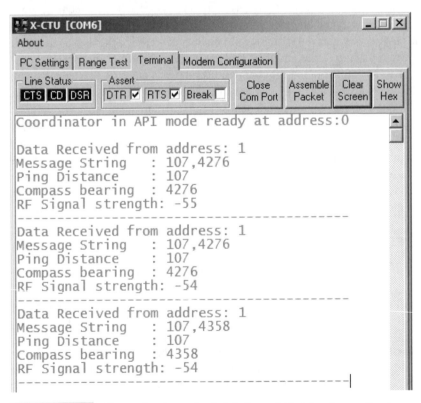

Figure 5-18   Example of terminal data from API data at coordinator.

In the end-device's code, the method SendUpdates is running in a separate cog to allow GetData to monitor for incoming data continuously. This allows data to be sent or received independent of the timing. The XBee is *not* in API mode, and SendUpdates sends the string values for range, theta every two seconds as decimal strings, such as the characters "1," "0," and "5" for the value of 105—three bytes' worth of data for one value. The unit does this endlessly.

```
Pub SendUpdates | range, theta
    HM55B.start (Enable,Clock,Data)
    XB.Delay (1000)
    repeat
      range := Ping.Millimeters (PING_Pin) ' Read range
      theta := HM55B.theta        ' Read Compass
      XB.dec (range)              ' Send range as decimal value
      XB.tx (",")                 ' Send a comma to separate
      XB.dec (theta)              ' Send bearing
      XB.Delay (2000)             ' Wait 2 seconds and send again
```

Caution: The XBee only waits so long in assembling a packet for transmission. If the delay between data sent is too long (send some data, read a sensor and process new data, then send the new data), it may send it as two different frames. To increase the time it waits for more data, the RO (Packetization Timeout) value can be increased.

In the GetData method, the Propeller endlessly awaits data in one cog. Once received, if the byte is "i," it accepts the next two bytes and uses them for IO number and state to control an output pin. This is different from prior examples where we collected a decimal value.

```
    dataIn := XB.rx             ' Wait for incoming byte
    If dataIn == "i"            ' i = I/0 control
      IO := XB.rx               ' Accept IO number as byte value
      value := XB.rx            ' Accept state (1/0) as byte value
      dira[IO]~~                ' Set direction of pin
      outa[IO] := value         ' Set state of pin
```

Using bytes instead, the packet is always three bytes long. To control P20 to turn on, the structure would be

| i | byte value 20 | byte value 1 |

Instead of

| i | string "20" (2 bytes or characters) | string "1" |

By using bytes as values instead of decimal string, the packet size can be compressed. For values greater than 255 (maximum byte value), two bytes can be used and combined:

Value = byte1 << 8 + byte2

... where byte1 (MSB) is shifted over by eight bits and then added to byte2 (LSB). We used a similar technique in the RxPacketNow methods in the XBee Object to assemble the 16-bit address from two received bytes.

In the coordinator's code, XB.AT_Config(string("ATAP 1")) shifts the XBee in API mode and the Propeller waits for an API packet to be received in ProcessFrame. If the Identifier (RxIdent) is $81, the packet is of the message variety, as opposed to a status or other type. The source address is accessed and displayed.

```
XB.API_Rx                             ' Wait for API data
if XB.RxIdent == $81                  ' If data identifier is a msg string
                                      ' Display source address
    PC.Str(string(13,"Data Received from address: "))
    PC.DEC(XB.srcAddr)
```

Since the actual message contained values separated by commas, the ParseDEC method is used to pull out and display the range and bearing. The signal strength, RSSI, is accessed and displayed.

```
    PC.str(string(13,"Ping Distance    : "))
    Range := XB.ParseDEC(XB.RxData,1)
    PC.DEC(Range)
    PC.str(string(13,"Compass bearing  : "))
    theta := XB.ParseDEC(XB.RxData,2)
    PC.DEC(theta)

    PC.str(string(13,"RF Signal strength: "))
    PC.DEC(-XB.rxRSSI)               ' Display RSSI level
```

The ControlPin method is used to send data back to the end device. It is passed the address to send the packet to (the source address of the incoming packet), the IO pin number, and the state (0 or 1). In order to packetize the data, a new packet is constructed and then passed to be transmitted.

```
Pub ControlPin(destAddr, pin, state)
    XB.API_NewPacket                 ' Clean out packet of old data
    XB.API_AddStr(string("i"))       ' Add an i to packet
    XB.API_AddByte(pin)              ' Add a byte of pin number
    XB.API_AddByte(state)            ' Add a byte of pin state
                                     ' Send the packet
    XB.API_txPacket(destAddr,XB.API_Packet,3)
```

In ControlPin, the packet string in which the data will be sent is cleared out (API_NewPacket). All bytes in the packet are set to 0 when cleared. The string and byte values of i, pin number, and state are added to the packet (API_AddStr or AddByte). API_txPacket is used to send the data to the correct address, the pointer for the packet is given, and the number of bytes to be sent is provided.

The difference between the XBee Object's `API_str` and `API_txPacket` is that strings cannot have byte values of 0—a string ends with a byte value of 0. Our packet has a byte value of 0 for possibly either pin or state, so we needed to specify that it would be sent as a packet and then provide the number of bytes in it. Here are some examples of transmitting API data to address 5:

Sending a simple string:

```
XB.API_str (address, string)
XB.API_str (5, string ("Hello!"))
```

To send a string with a value, such as "Range = range value" (in objects, declare `num`: `"numbers"`):

```
XB.API_NewPacket
XB.API_addStr (string ("Range = "))
XB.API_addStr (num.ToStr (range,num#dec))
XB.API_str (5, XB.API_Packet)
```

To send just a byte in the packet:

```
XB.API_tx (5, 13)
```

Working in API mode can be intimidating, but its benefits are many. The XBee Object has multiple methods for interfacing with the XBee in both modes with example code. It would be of benefit to read through the object documentation as well as the XBee manual.

TRY THESE!

✓ Modify the end-device code so that it sends data only if range < 100 mm.
✓ Add a pushbutton to the coordinator. Have it control an LED on the end device (it's not a good idea to have two different cogs trying to send data; comment out the code to blink the LED on reception of data).

A Three-Node, Tilt-Controlled Robot with Graphical Display

OVERVIEW AND CONSTRUCTION

A three-node network for controlling and monitoring a robot will be explored for the last project in this chapter. The system shown in Fig. 5-19 has:

■ A Propeller Demo Board network node (address 0) with an accelerometer to measure angle of inclination on two axes for the tilt controller

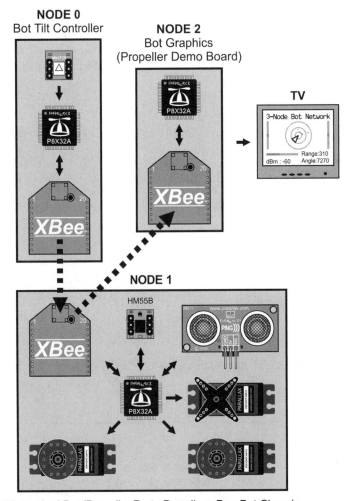

Networked Bot (Propeller Proto Board) on Boe-Bot Chassis

Figure 5-19   3-Node bot network diagram.

- A robot on the network (address 1) using a Propeller Proto Board on a Boe-Bot robot chassis with HM55B compass, PING))) range finder on a servo bracket, and LEDs
- A Propeller Demo Board on the network (address 2) driving a TV for video display of the graphical display

Figure 5-20 shows the wiring connection diagram for each of the nodes. Note that in switching to the Proto Board we will change the I/O pins used for the XBee.
Hardware construction tips for bot:

✓ If you are not familiar with the Boe-Bot robot, you may want to look through "Robotics with the Boe-Bot" by Andy Lindsay, available for download at www.parallax.com/education to familiarize yourself with the basic hardware and servo operation.

NODE 0
Bot Tilt Controller (Propeller Demo Board)

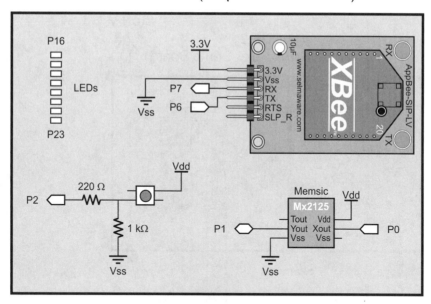

NODE 1
Bot (Propeller Proto Board on Boe-Bot Chassis)

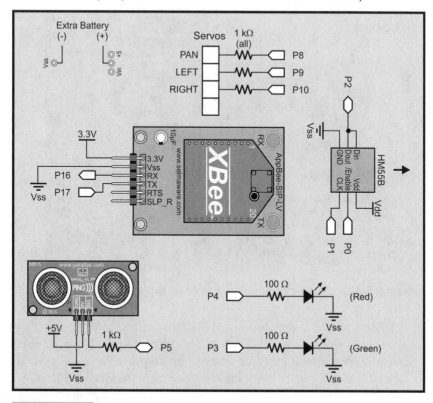

Figure 5-20 Hardware wiring diagram for bot system.

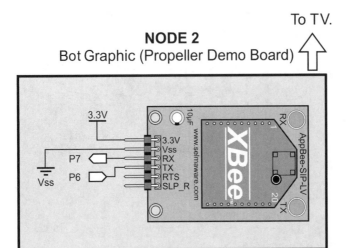

Figure 5-20   (*Continued*)

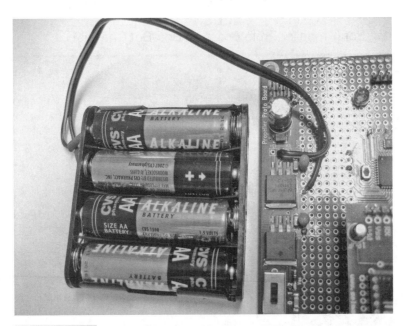

Figure 5-21   Supplying separate power for bot's servo drive.

✓ For the bot, two 4-AA packs were used with the spare tied under the normal battery pack. The second battery pack is used only for servo power. Attempts at using a single supply caused voltage and current spikes affecting the Propeller chip's operation. The connector of the battery pack was cut off and soldered to the servo header power and Vss (see Fig. 5-21). Other sources may be used, but supply voltage should not exceed 7.5 V or the servos can be damaged. Use coated wire to strap the batteries under the bot.

- Ensure the HM55B compass is mounted facing forward and that it is away from large current loads, such as batteries and servos. The magnetic fields will cause problems with proper compass bearing.
- The PING))) sensor is mounted using the PING))) Mounting Bracket Kit. Manually rotate the servo to find the center prior to mounting the bracket. Mount the cable header so that servo rotation does not hit it (the servo turns further manually than with the code—about 45 degrees each way).
- A battery supply was used on the tilt controller as well for unfettered operation.

OVERVIEW OF SYSTEM OPERATION

<u>Tilt Controller:</u> This is used to read the Memsic 2125 accelerometer module, calculate right and left motor drives based on inclination, and then send drive values to the bot. The tilt controller, shown in Fig. 5-22, also receives data from the bot and uses the PING))) range measurements to light the eight Demo Board LEDs, showing distances of 100 mm for local range indication. Pressing the pushbutton will send a panning map instruction to the bot to map the area in front of it for display by the TV graphics node. This node has a MY address of 0 and a DL address of 1 (the bot).

<u>Bot:</u> This controls all function of the networked bot shown in Fig. 5-23, including:

- Accepting drive values for motor drive. Should no data be received for 1.5 s, the bot will stop and blink the red LED.

Figure 5-22   Tilt-controller board with Memsic 2125 Accelerometer.

Figure 5-23 Networked bot with range finder, compass, and XBee.

- Accepting panning map instruction (p) to perform mapping operation. When this instruction is received, the bot will stop and pan the PING))) sensor from right to left, measuring distances and sending map data (m) to the TV graphics node and the tilt controller. When mapping is complete, the bot will remain steady and blink the green LED until the tilt controller releases it from mapping mode (user presses button again). The bot will send a clear (c) code to the video to clear the display when map mode exits.
- While driving, the bot will transmit updates (u) of right and left drive values, PING))) range, and direction of travel from the HM55B compass (0-8191).
- This node has a MY address of 1 and sends data to $ffff—all nodes on the network for a broadcast.

Bot Graphics: Shown in Fig. 5-24, drives the graphics TV display showing:

- The bot bearing as a rotating triangle and text
- Distance to object as a red point in front of the bot and text
- Yellow range marker circles at 0.25 m, 0.5 m, and 1 m
- Left and right drives as bar indicators
- Signal strength for RSSI (dBm) as bar and text

The update packets contain information for much of the display, and the RSSI is pulled from the received frame through API mode. When the bot is in mapping mode, mapping

Figure 5-24   Bot TV graphics controller.

packets contain data to plot the range map. It also accepts clear codes from the bot to clear the mapped display. This node has a MY address of 2 and sends out no data.

The output display of the graphics controller is shown in Fig. 5-25 in both normal driving and with the bot performing a panning map operation.

Note: The PING))) range finder has a wide angle of emission and reception. Do not expect pinpoint accuracy when mapping.

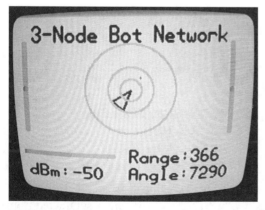
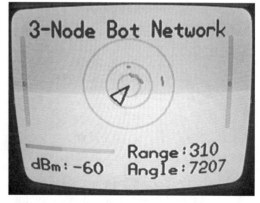

Figure 5-25   TV displays for normal and panning map data.

BOT NETWORK CODE

Bot Tilt Controller   For the tilt controller, in the SendControl method, if the button is pressed, a series of p's is transmitted. With the amount of data flying, some missed bytes on reception are normal, and this helps ensure the bot gets a p-instruction for a panning map operation. If not pressed, the accelerometer is read for the X- and Y-axis (−90 to 90 degrees, 0 level) and the drive for each servo is calculated by mixing the two axes of tilt for a final servo value of 1000 to 2000 (the range of servo control) for each. The data is sent as a "d" packet for drive.

```
Forward := (accel.x*90-offset)/scale * -1

' Read and calculate -90 to 90 degree for turn
Turn := (accel.y*90-offset)/scale

' Scale and mix channels for drive, 1500 = stopped
Left_Dr := 1500 + Forward * 3 + Turn * 3
Right_Dr := 1500 - Forward * 3 + Turn * 3
```

In the AcceptData method (which is running in a separate cog), incoming packets are analyzed if update data (u) or map data (m) and the local LEDs are updated. Based on the range, eight 1s are shifted to the left eight positions, then shifted right again based on the range/100. This allows one LED to light for every 100 mm or 0.1 m, out to 800 mm or 0.8 m.

```
outa[16..23] := %11111111 << 8  >> (8 - range/100)
```

Bot Code   For the bot controller, in Start, received bytes are analyzed with a timeout. If any data is not received for 1500 ms, the red LED will begin to blink. Received data is analyzed for either "d" for drive data or "p" to begin a mapping scan.

```
case DataIn                        ' Test accepted data

    "d":                           ' If drive data
        Right_dr    := XB.RxDEC    ' Get right and left drive
        Left_dr     := XB.RxDEC
        SERVO.Set(Right, Right_dr) ' Drive servos based on data
        SERVO.Set(Left, Left_Dr)

    "p":                           ' p = pan and map command
        mapping := true            ' Set flag for mapping
        outa[grnLED]~~             ' Turn on green LED
        Map                        ' Go map
```

The SendUpdate method is run in a separate cog to continually send out the status of the range, direction, and drive values led by "u." The value of theta is subtracted from

8191 to allow the direction of rotation to be correct in the graphics display. If a panning map is in progress, updates are suspended due to mapping being true.

```
Repeat
  if mapping == false                 ' If not mapping
    XB.Delay(250)
    Range := Ping.Millimeters(PING_Pin)  ' Read range
    theta := HM55B.theta                 ' Read Compass
    XB.TX("u")                           ' Send "update" command
    XB.DEC(Range)            ' Send range as decimal string
    XB.CR
    XB.DEC(8191-theta)       ' Send theta of bearing (0-8191)
    XB.CR
    xb.DEC(Right_Dr)         ' Send right drive
    XB.CR
    XB.DEC(Left_Dr)          ' Send left drive
    XB.CR
```

When mapping, the value of pan is looped from 1000 to 2000, the range of allowable servo values. The range is measured, and the PanOffset is calculated. The value of the "pan" has 1500 subtracted (recall that 1500 is a centered servo). The result is multiplied by 2047 (90 degrees, with 8191 being a full 360 degrees) and divided by the full range of pan. Finally, an "m" is sent followed by range and angle of the servo plus the pan offset. This repeats for each value of pan, from 1000 to 2000, in increments of 15 steps or 1.35 degrees (15 · 90 degrees/1000 steps = 1.35 degrees). Once mapping is complete, the system will wait until another "p" is received to exit pan mapping mode while sending a "c" to clear the video display. The variable "mapping" is used as a flag to prevent the SendUpdates code running in a separate cog from sending updates while mapping.

```
Pub Map | panValue
" Method turns servo from -45 to + 45 degrees from center in increments
" and gets ping range and returns m value at each increment

    SERVO.Set(Right, 1500)      ' Stop servos
    SERVO.Set(Left, 1500)

    SERVO.Set(Pan, 1000)        ' Pan full right
    XB.Delay(1000)
                                ' Pan right to left
  repeat pan from 1000 to 2000 step 15
    SERVO.Set(Pan,panValue)
    Range := Ping.Millimeters(PING_Pin)    ' Get range calculated
                                           ' based on compass
                                           ' and pan
    PanOffset := ((panValue-1500) * 2047/1000)
```

```
        XB.TX ("m")                          ' Send map data command
        XB.DEC (Range)                       ' Send range as decimal
        XB.CR
        XB.DEC ((8191-Theta) + PanOffset)    ' Send theta of bearing
        XB.CR
        XB.delay (50)
    XB.delay (1000)

    SERVO.SET (Pan, 1500)                    ' Re-center pan servo
```

TRY IT!

✓ Add another device, such as speaker, to your bot. Add a button on the tilt controller and modify code to control the device from the tilt controller.

Bot Graphics   The bot graphics code is responsible for accepting the data and displaying it graphically on a TV screen. Note that this XBee is in API mode so that the RSSI level may be pulled out of the received frame. The code looks for one of three incoming byte instructions: "u," "m," and "c." Updates, "u," are messages with update data as the data moves, with range, bearing, and drive values (limited between 1000 and 2000), and it retrieves RSSI level for display creation.

```
Repeat
  XB.API_rx                              ' Accept data
  If XB.RxIdent == $81                   ' If msg packet...
    if byte[XB.RxData] == "u"            ' If updates, pull out data
        ' Get DEC data skipping 1st byte (u)
      range    := XB.ParseDEC (XB.RxData+1,1)
      bearing  := XB.ParseDEC (XB.RxData+1,2)
      rDrive   := XB.ParseDEC (XB.RxData+1,3) <#2000 #>1000
      lDrive   := XB.ParseDEC (XB.RxData+1,4) <#2000 #>1000
      RSSI     := XB.RxRSSI
      Update
```

Mapping (m) strings are used to map what the bot "sees" without clearing off old data while a mapping pan is in progress. Clear, "c," is received once the bot switches back into drive mode after mapping.

We aren't going to delve too deeply into the graphics creation here, as it's not a major subject for this chapter. One point of interest is in that many graphic programs the video data is written into one part of memory (such as `bitmap_base`), and when the complete display change is ready, it is copied into the section of memory that the graphics driver uses to display the actual display. It is effectively double-buffered to prevent flicker on the screen. We do not have the luxury of the memory needed for that operation. Instead, to reduce flicker, values of the old data are saved. When updating, the graphics are redrawn in the background color to "erase" them, then the new data is used to draw the graphics in the correct color, such as in this code:

```
' Draw bot vector image
gr.width(2)
gr.color(0)                             ' White
gr.vec(120,120, 100, (bearing_l), @bot) ' Erase last image
gr.color(1)                             ' Black
gr.vec(120,120, 100, (bearing), @bot)   ' Draw new image
```

Many features of graphics.spin are used, including text, lines, arcs, and vector-based graphics. The code is fairly well commented for adaptation.

TRY IT!

✓ Add another sensor to your bot. Modify both the bot and graphics code to send and display the value.

Summary

In this chapter we looked at what the XBee is and how it can be configured and used in a wireless sensor network. Using AT codes sent from the controller, the XBee can be configured for specific applications, such as unique addresses used in polling operations. In API mode, frames are sent and received with specific data. Using the networking capabilities, a three-node bot system was developed for control and monitoring using the graphics ability of the Propeller chip.

The ability to configure the device and send data between Propeller chips efficiently and with addressing leads to a wide array of projects that can be implemented. Allowing different Propeller chips to perform their own processing and easily communicating with each other brings the excitement of parallel processing to a whole new level.

In my research with institutions, such as Southern Illinois University, University of Florida, USDA in Texas, and the University of Sassari, Italy, I have been involved in many XBee/Propeller (and some other controller) projects. These projects include monitoring corn irrigation needs, biological monitoring, and monitoring the vibration of citrus fruit as it's shaken from the tree.

Wireless sensor networks are a powerful and quickly expanding field for remote monitoring and control. They are finding use in research and in building, plant, and home automation. The Propeller, with its ability to perform parallel processing, is an outstanding choice for monitoring and control. As mentioned at the outset of this chapter, whether you build the projects in this chapter or simply gain an understanding of the material, I hope you can use the base code and principles in projects of your own invention.

Exercise

A FINAL PROJECT FOR YOU—DIRECT XBee ADC/DIGITAL DATA

For our final exploration into the Propeller/XBee combination, let's exercise the XBee's ability to measure and transmit analog and digital data without a controller. The received data has a packet identifier of $83 (the XBee needs to have firmware version 10A3 or higher to be able to this).

✓ Apply an analog voltage of up to 3.3 V (using a potentiometer or other device) to ADC 0 (pin 20) and ADC 2 (pin 19) and a pushbutton to DIO2 (pin 18).
✓ Using X-CTU software, starting from the default settings, configure for a MY of 6 and for I/O settings, such as:
D0 = mode 2: ADC
D1 = mode 2: ADC
D3 = mode 3: DIN (Digital Input)
IR = 3E8 (sample rate of 1 second. $3E8 = 1000 decimal or 1000 ms of time).
✓ Connect the Vcc (pin 1) and the Vref pin (pin 14) of the XBee to 3.3 V. Connect Vss to GND (pin 10). Do not connect anything else, including the Propeller.
✓ Download ADC-Dig Output Sample.spin to your coordinator board.
✓ Open the Terminal window and monitor. You should see something similar to Fig. 5-26.

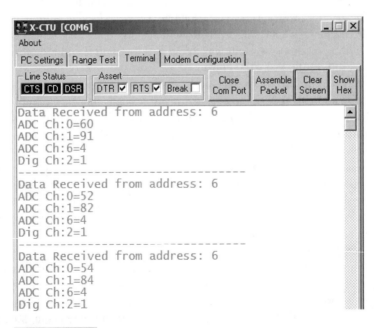

Figure 5-26   Displaying data from an XBee sending raw ADC/digital data.

The remote XBee is reading the ADC and digital channels specified and sending a packet containing the data. You will not see the LED blink on the sending XBee because no communication enters or exits it through the serial port.

On the coordinator, when a frame with an identifier of $83 (ADC/Digital data) arrives, valid data is pulled out and displayed (nonenabled channels are −1).

```
PUB Start | channel
  ' Configure XBee & PC Comms
  XB.start(XB_Rx, XB_Tx, 0,9600)
  PC.start(PC_Rx, PC_Tx, 0, 9600)


  XB.AT_Init                  ' Fast config
  XB.AT_ConfigVal(string("ATMY"), MY_Addr)
  XB.AT_Config(string("ATAP 1"))' Switch to API mode


  PC.str(string("Coordinator in API mode ready at address:"))
  PC.dec(MY_Addr)
  PC.Tx(13)


  Repeat
    XB.API_Rx                 ' Wait for API data
    if XB.RxIdent == $83      ' If data identifier is a ADC/Dig data
      PC.Str(string(13,"Data Received from address: "))
      PC.DEC(XB.srcAddr)
      repeat channel from 0 to 6      ' Cycle through ADC channels
        if XB.rxADC(Channel) <> -1    ' Display if not -1
          PC.str(string(13,"ADC Ch:"))
          PC.dec(channel)
          PC.tx("=")
          PC.DEC(XB.rxADC(Channel))


      repeat channel from 0 to 7      ' Cycle through Digital channels
        if XB.rxBit(Channel) <> -1    ' Display if not -1
          PC.str(string(13,"Dig Ch:"))
          PC.dec(channel)
          PC.tx("=")
          PC.DEC(XB.rxBit(Channel))


      PC.str(string(13,"--------------------------------"))
```

If you were to use an analog accelerometer, you could read the accelerometer and a digital pushbutton on the tilt controller and have the XBee send those values automatically. Then the Propeller could accept data at the bot and process it for control action. Or you may have an array of sensors in the area and collect data from them as they wake, sample, send, and go back to sleep.

6

DANCEBOT, A BALANCING ROBOT

Hanno Sander

Introduction

The Propeller's multicog architecture allows you to tackle complex projects one step at a time. A project that's keeping me busy is my balancing robot that uses vision to interact with people. In the next two chapters, I'll describe how I built my robotic dance partner. In this first chapter, I'll cover the basics of building a balancing robot:

- The parts: Sensors, motors, and more
- The code: Which way is up?
- Achieving balance: How to avoid falling over
- Steering a balanced robot

Resources: Demo code and other resources for this chapter are available for free download from ftp.propeller-chip.com/PCMProp/Chapter_06.

The Challenge

Balancing robots make a great platform for mobile robots. They are highly maneuverable, have great traction, and move more smoothly and naturally than other designs. Balancing robots are mechanically very simple. They only have two wheels with which they move through their environment. Each wheel is driven by its own motor, which makes the robots highly maneuverable. Since the robots maintain their own balance, they can be quite tall while maintaining a small footprint. They can turn on a dime, navigate precisely, and are a pleasure to watch while they keep their balance. See Fig. 6-1 for a picture of my DanceBot and this link for videos of the DanceBot in action: http://mydancebot.com/dancebot/videos.

Figure 6-1   DanceBot balancing on two wheels.

Building a robot that balances and maintains position robustly in any environment is not easy. Balancing robots are complex in that they require input from multiple sensors and must react quickly to changing conditions. A related problem typically taught in physics classes is the classic *inverted pendulum problem*. Unlike a normal pendulum, which hangs downward and is stable, the inverted pendulum has its mass above the pivot point and is, therefore, quite unstable. To keep the mass from falling, it must be actively balanced by moving the pivot point horizontally as part of a feedback system. For our balancing robot, we need a high-performance microcontroller to process the sensor input and calculations. The Propeller's multiple cogs allow you to split your complex project into simple parts that can be built and tested one at a time.

We're now blessed with wonderful hardware that makes it much easier and cheaper to build a balancing robot. In the past, the necessary sensors, motors, and processing power were more difficult to locate and expensive to buy. In particular, nanotechnology and parallel processing have opened the doors to today's balancing robots. Low-cost accelerometers using nano-scale cantilevers, and small, solid-state ceramic gyroscopes are readily available and easy to use. Being able to run multiple programs in their own processor while sharing a global memory makes it easier to build and test the software required for the robot.

The fun really starts when a balancing robot is combined with computer vision (see the next chapter). Computer vision allows the robot to see the world and interact with it. By tracking a target object, like a human, the balancing robot can literally dance with its partner; it becomes a "DanceBot"! By being tall and maneuverable, the DanceBot more closely resembles a dance partner than a four-wheeled car or stationary mechanism. It can face its partner by turning on its axis and moving forwards and backwards to maintain a desired distance. Interacting with a robot by dancing is more intuitive and pleasing than using remote controls or, even worse, programming it. Even young children can get in the action; see Fig. 6-2, in which my kids are interacting with my DanceBot.

Like people, balancing robots continually move to stay balanced; they don't stand still. They continually move slightly forward and slightly back to keep from falling over and to maintain to their desired position. They only stay still for the short time that their center of gravity is perfectly positioned over their wheels' contact point on a level ground. This tendency to keep moving makes them great attention-getters. The human vision system is particularly sensitive to things that move. This helped our ancestors to survive by finding food and keeping us safe from predators. Now, advertisers use this trait to their advantage by using blinking lights and images that move or change over time. Builders of balancing robots can use it to their advantage, too, by drawing onlookers wherever they go!

Figure 6-2   Anja and Kyle with my DanceBot.

Building the DanceBot

Okay, let's build our DanceBot. What will we need? I've experimented with many designs and I'm happy with my setup (see Table 6-1). At some point, I'm planning to build a kit that will make it easier for others to build their own robots; if you're interested, look at my web site. The following sections will describe each subsystem of the robot. I'll start with an overview of the system and then get into the nitty-gritty details of which part I used, how to interface it, and how the Propeller will use it.

MECHANICS—MAKE THE ROBOT MOVE

The mechanics of a balancing bot are simple: Two powered wheels control a rigid body. For the body, we could use everything from erector set parts (like VEX) to a small section of large-diameter transparent acrylic pipe (see Fig. 6-3 for inspiration). The pipe solution looks clean and lets you keep all the electronics inside while still letting people see what you've done. It's also easy to drill mounting holes for the motors.

For the motors, we'll use the Lynxmotion Geared Motor (GHM-16). It runs on 12 VDC and has a 30:1 metal gear for an impressive stall torque of 4.6 kg-cm. At 12 V, with no load, it turns the axle at 200 RPM. This combination of high torque and good speed enables your robots to recover from hard pushes or uneven terrain. Lynxmotion also sells rubber wheels for this motor, so let's use the 3.6 in diameter Dirt Hawg tires.

Before we dive into the details of the DanceBot, let's look at an overview of how to allocate the Propeller chip's 32 I/O pins to the various sensors and actuators (see Table 6-2). Luckily, the Propeller's input and output pins are all general purpose. We can configure the behavior of each pin through software.

TABLE 6-1 DANCEBOT PARTS	
PART	**DESCRIPTION**
Parallax Proto Board	Prototyping board with Propeller, power supply circuitry, and crystal
GWS PG-03 Gyro	Piezoelectric ceramic gyroscope
2* GHM-16 Motors	Geared robot motor
LIS3LV02DQ	Accelerometer
Acrylic tube	Frame for the DanceBot
2*LMD18200	H-bridge motor controller
3.6 in diameter tires	Robot tires
Motor Encoder QME-01	Quadrature encoder to sense position
12 V Battery	Rechargeable battery to power motors and circuit
ViewPort	Debugging environment

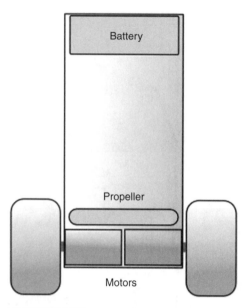

Figure 6-3   DanceBot mechanics showing frame, motors, and circuit board.

TABLE 6-2 PROPELLER PINOUTS	
PROPELLER I/O PIN	**PURPOSE**
3, 4, 26, 27	Motor control: Drive 2 H-bridge chips with two pins to control the motor's direction and speed.
12,13	Position: Read pulses from quadrature encoder to calculate position and speed of one motor.
6, 7	Accelerometer: Communicate with Accelerometer chip using I2C to measure acceleration.
5, 8	Gyroscope: Send and receive timed pulses to the gyroscope to measure the robot's change in tilt.
30, 31	Serial communication: Allow ViewPort, running on the PC, to monitor and configure the robot.
28, 29	EEPROM: Read the robot's program from the EEPROM.
19, 23, 24	Vision (see next chapter): Clock the ADC with a 10 MHz signal and read the digital representation of the camera's signal.

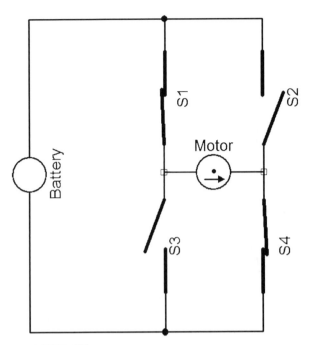

Figure 6-4   H-bridge schematic, showing four switches controlling motor's direction.

To control the direction of each motor, we'll use an electronic circuit called an H-bridge! The H-bridge applies voltage across the motor in either direction to drive forwards, backwards, and even turn. Looking at Fig. 6-4, you'll see why the circuit is called an H-bridge: The four switches surrounding the motor resemble an "H."! Let's look at what happens when we set the switches to different positions. When the switches S1 and S4 are closed and S2 and S3 are open, a positive voltage will be applied across the motor. By opening the S1 and S4 switches and closing the S2 and S3 switches, this voltage is reversed, allowing reverse operation of the motor. Switches S1 and S2 should never be closed at the same time, as this would cause a short circuit on the input voltage source. The same applies to switches S3 and S4.

To control the speed of the motor, we'll use *pulse width modulation* (PWM). We can't directly tell the H-bridge to supply only half power to the motor because the switches can only be turned on or off. However, we can tell the H-bridge when to turn the switches on and for how long. So if we want to drive at half power, we can turn the switch on for 50 percent of the time and later turn the switches off for 50 percent of the time. The motors will average this out and drive the wheels at a medium speed.

The LMD18200 implements an H-bridge that can deliver up to 3 A of continuous output at up to 55 V. It has a low on resistance of $0.3\ \Omega$ per switch and can be driven directly from the Propeller's outputs. Besides implementing an H-bridge, it also protects our robot with an undervoltage lockout, an overcurrent detector, and a thermal shutdown. These features

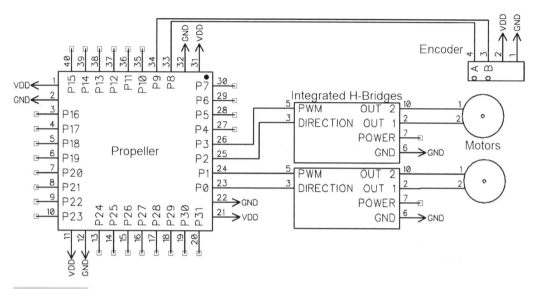

Figure 6-5 Schematic showing the motor controller's interface to the Propeller.

ensure that the robot won't hurt itself, even if we make mistakes in our programming!! We power the chip through a fuse from the batteries and connect the outputs to the motor we wish to drive. To control it, we connect the "direction" and "pwm" input pins to the Propeller pins designated as output pins for this motor; see Fig. 6-5 for a schematic. These inputs determine the state of the switches in the H-bridge.

To continually drive the H-bridge, we'll use the following assembly code. It runs in its own cog and continually outputs configurable patterns to the output pins, which are connected to the Propeller. Changing the output patterns and timing allows us to change the direction and speed of the two motors.

```
        org
doPWM   or dira,dirav
 loop   mov ptr,par
        mov :i,#3
:duty   rdlong orV,ptr     'Get value to or
        add ptr,#4
        mov :t,outa
        rdlong timeV,ptr 'Get wait
        add ptr,#4
        and :t,andV
        or :t,orV
        mov outa,:t
        add timeV,cnt
        waitcnt timeV,#0
        djnz :i,#:duty
        jmp #loop
```

```
:t        long 1
:i        long 1
diraV     long 1
andV      long 1
ptr       long 1
orV       long 1
timeV     long 1
```

To initialize the code that will drive our motors, we'll use the following Spin code. It initializes the output patterns, configures the DIRA register, and starts the previous assembly code in a cog.

```
pub setupMotor
  'pwm is structure with bits to turn on, then duty for that pattern
  pwm[1]:=5
  pwm[3]:=5
  pwm[5]:=5
  diraV:=m1Dir|m1Pwm|m2Dir|m2Pwm
  andV:=!diraV
  cognew (@doPWM, @pwm)
```

To update the speed and direction of the two motors, we'll use the following Spin code. It limits the speed for each motor, sets the output patterns to turn the motors in the specified direction, and sets the time values to modulate the motors at the correct speeds.

```
pub Update(speed,speed2)|p0,p2,t
  p0:=m1pwm+m2pwm
  if speed>0
    p0+=m1dir
  if speed2>0
    p0+=m2dir
  pwm[0]:=p0
  speed:= (||speed) <#MAXSPEED
  speed2:= (||speed2) <#MAXSPEED
  if speed > speed2
    p0-=m2pwm
    pwm[1]:=(speed2)+5    '2505
    pwm[3]:=(speed-speed2)+5 '2505
    pwm[5]:=5+(5000-speed)    '15
  else
    p0-=m1pwm
    pwm[1]:=(speed)+5
    pwm[3]:=(speed2-speed)+5
    pwm[5]:=5+(5000-speed2)
  pwm[2]:=p0
  pwm[4]:=0
```

MEASURING POSITION—WHERE AM I?

In order to measure how far your robot has traveled, you'll need a quadrature encoder. A quadrature encoder is an electromechanical sensor that outputs two pulse signals, which can be analyzed to determine how quickly and in which direction a wheel is moving. The sensor typically consists of a disk covered in black and white stripes and two photo detectors that read the optical pattern coming from the disk. Counting how many black/white transitions are made at a sensor tells us how fast the wheel is turning. For example, if there are 100 black/white lines on the disk, turning the disk by one rotation will produce 100 black/white cycles. The photo detectors are offset by 90 degrees relative to each other; when one sensor is on the middle of a black line, the other will be on the transition from white to black (see Fig. 6-6 for an illustration). By looking at the relative phase of the signals coming from the detectors, we're able to determine if the wheel is moving forwards or backwards. To balance my DanceBot, I use just one quadrature encoder to measure the position of one of the wheels; this simplifies the control logic. To track where the robot has traveled after it has made turns, you'll need to use an encoder on each wheel.

Lynxmotion sells the Motor Encoder QME-01, which registers 100 cycles per revolution. Since the encoder is mounted to the motor, you'll get:

$$100 \text{ cycles/motor rev} \cdot 30 \text{ motor rev/wheel rev} = 3000 \text{ cycles/wheel rev}$$

On a 3.6 in diameter wheel, this means you get:

$$3000 \text{ cycles/}\pi \cdot 3.6 \text{ in} = 265 \text{ cycles/in}$$

This lets us measure both the speed and position of the wheel quite accurately. In the real world, our wheel will slip and suffer from accumulated error as the robot drives over different surfaces like linoleum and carpets, so we can't actually measure our actual position that accurately. When the wheels are turning at their top speed of 200 revolutions/min, the processor will need to count and process

$$200 \text{ RPM} * \text{min/60 s} \cdot 3000 \text{ cycles/revolution} = 10,000 \text{ cps}$$

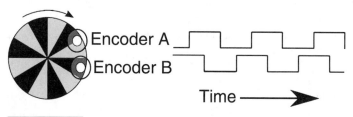

Figure 6-6   Quadrature encoder with disk and offset detectors.

To measure these pulses, we'll use the following assembly code to count pulses and determine the relative phase of the signals from the photodetectors. The code uses bit manipulation to determine which way each motor is turning.

```
Dat           '
        org
count
        mov m2w,par
        add m2w,#4
        mov :t,ina                    'Get ina- 1011
:loop  ' 21 instr= 80 cnts=1Ms/sec
        and :t,mAEnc                  't shows before flank, what A bits are
        waitpne :t,mAEnc             'Wait for encoder A bits to change
        mov :t1,ina
        and :t1,mABEnc                't1 shows after flank, what AB bits are
        mov :t2,:t1
        and :t2,mAEnc
        xor :t2,:t                   ' t2 shows which A bits changed
        mov :t,:t1                   ' Get t ready for next measurement
        shr :t2,#12
        rcr :t2,#1    wc             ' Check if A on motor 1 changed
if_nc shr :t1,#14                    ' Leave both t2, t1 shifted if take jump
if_nc jmp #:doM2                     '
        shr :t1,#12
        rcr :t1,#1    wc             ' Check if A=1
if_nc shr :t1,#1
if_nc jmp #:doM2
        rcr :t1,#1    wc             ' Check fwd/bkw on motor 1
        rdlong m1Cnt,par
if_c  add m1Cnt,#1
if_nc  sub m1Cnt,#1
        wrlong m1Cnt,par
        'Ready to do motor 2, t1 shifted 14, t2 shifted 13
:doM2
        shr :t2,#9
        rcr :t2,#1    wc             ' Check if A on motor 2 changed
if_nc jmp #:loop
        shr :t1,#8
        rcr :t1,#1    wc             ' Check if a=1
if_nc jmp #:loop
        rcr :t1,#1    wc             ' Check if fwd/bkw on motor 2
        rdlong m2cnt,m2w
if_nc add m2Cnt,#1
if_c  sub m2Cnt,#1
        wrlong m2Cnt,m2w
        jmp #:loop
:t2     long 1
:t1     long 1
```

```
:t       long 1
m2W      long 1
mAEnc    long %0100_0000_0001_0000_0000_0000 '12
mABEnc   long %1100_0000_0011_0000_0000_0000' 12,13 '00011
m1Cnt    long 0
m2Cnt    long 0
```

The code object that we will use to both drive and measure the motor is motor.spin. It lets us control the motor's speed and direction, and keeps track of where the wheel is using the pulses from the quadrature encoder.

MEASURING TILT—WHICH WAY AM I FALLING?

We humans have an inner ear that tells us which way we're tilting. This mechanism relies on fluid moving in a channel, where it is sensed by small hairs. When we drink alcohol, the alcohol changes the properties of the ear's fluid and our sense of balance is affected. Apart from this, our sense of balance is good and helps us to balance quite well. A good sense of balance, or rather, a sensor that will tell us which way we're tilting, is critical to building a robust balancing robot. How do we best teach our robot to measure its tilt?

One technique is to measure the distance to the ground from a sensor offset from the wheels. Measuring the distance can be done either using ultrasound or measuring the amount of light reflected from a beam. This can work in some environments, but the surface has to be absolutely flat; otherwise, the robot will fall because it's not able to accurately measure which way is down.

The approach that most resembles that of the human ear are various inclinometers. These range from mercury-filled glass tubes to pendulums, which all use the earth's gravity to determine which way is down. All these devices work by measuring the acceleration of gravity. Recently, small *micro electromechanical systems* (MEMS) accelerometers have become available to the hobbyist. MEMS accelerometers use a small cantilever beam with a known mass to measure the direction of a force by sensing how the cantilever bends (see Fig. 6-7). They're a single-chip solution, which means they're small and inexpensive. The one problem with all these "inclinometers" is that on a moving robot, gravity is not the only acceleration. When the robot speeds up, is bumped, or drives over a rough surface, the acceleration due to gravity might be small

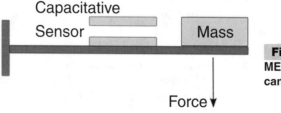

Figure 6-7 Mechanics of a MEMS accelerometer showing cantilever sensor.

relative to the other forces. Averaged out over a long time, these devices may be accurate, but for short periods in a moving vehicle, their output isn't accurate enough for a balancing robot.

Gyroscopes don't directly measure tilt, but are great for accurately measuring the change in tilt of an object. Traditionally, gyros had a spinning disk gimbaled inside an enclosure. The disk remains in the same orientation even when the sensor is rotated about an axis. By monitoring the position of the disk, the sensor can determine changes in orientation. This technique is accurate, but the implementation is typically large and complex. Recently, piezoelectric technology has dramatically simplified these sensors. Modern devices rely on the same gyroscopic physics, but instead of depending on a spinning disk, they vibrate a piezoelectric material. Similarly to rotating gyroscopes, the vibrating object tends to keep vibrating in the same plane as it is rotated. This is measured electronically and yields an inexpensive but accurate sensor for measuring change in tilt.

We can try to calculate tilt by integrating the change in tilt measured by the gyroscope sensor. This will work for a while, but our calculation will drift, as the sensor's value isn't perfect. When we integrate the sensor's value, we also integrate any error, and this will add up quickly and make our result meaningless. So, at this point we have two sensors that are great for measuring *tilt* and *change in tilt* under special conditions. The MEMS accelerometer is great for measuring tilt over long periods, and the piezoelectric gyroscope is great for measuring rate of tilt over short periods. What to do?

Luckily, a mathematical function exists that's appropriate especially for this case. The *Kalman filter* can be used to "fuse" the raw sensor readings from the gyroscope and the accelerometer to produce clean tilt measurements. The complete theory behind the Kalman filter is beyond the scope of this book, but let's look at the basics. According to Wikipedia: "The Kalman filter is an efficient recursive filter that estimates the state of a linear dynamic system from a series of noisy measurements." (http://en.wikipedia.org/wiki/Kalman_filter)

What does this mean? Being an "efficient recursive filter" means that the algorithm processes signals to accentuate the good while reducing the bad. Recursive filters have an internal state that is updated every timestep. "Estimating the state of a linear dynamic system from a series of noisy measurements" means that the Kalman filter's internal state is an estimate of a system that is linear and dynamic. In our case, the two inputs are from a linear and dynamic system—one (the change in tilt as measured by the gyroscope) being the derivative of the other (tilt as measured by the accelerometer). A good analogy is that the Kalman filter automatically selects the filter values so that the low-pass filtered accelerometer measurement corresponds with the high-pass filtered gyroscope measurement. So, we can use the Kalman filter to filter and fuse our noisy signals to estimate the state of our system—we can accurately calculate tilt from our signals!

For our code, we have to understand that we need to run the Kalman filter continually and that it runs in two phases: predict and update. During the predict phase, the state estimate from the previous timestep is used to produce an estimate of the state at the current timestep. So it's predicting the current state from the last state. In the update phase, measurement information at the current timestep is used to refine this prediction to arrive at a new, more accurate state estimate, again for the current timestep. For the math wizards, here's the formula:

$$x_k = F_k x_{k-1} + B_k u_k + w_k$$

where F_k = state transition model, which is applied to the previous state $x_k - 1$
B_k = control-input model, which is applied to the control vector u_k
w_k = the process noise, which is assumed to be drawn from a zero-mean multivariate normal distribution with covariance Q_k

Now that we've understood the theory, let's look at the actual devices and code that we'll be using. The GWS PG-03 is an inexpensive hobby gyroscope that's typically used to stabilize small remote control helicopters. It has two servo connectors. The first is typically plugged into an RC receiver, from which it gets a pulsed signal that indicates the desired position for the servo. The second connects to the servo and provides a pulsed signal whose length is the sum of the input signal and the measured signal. For our application, we'll feed it with a constant input signal and measure the signal on the output. This is straightforward for the Propeller. The following code outputs a configurable signal on the gyro's input pin and uses CTRA to measure the length of the gyro's output signal.

```
Con
  gyroOutP =5
  gyroInP  =8
pub doGyro(gyroTime)
  dira[gyroOutP]~~
  ctra := (%10101 << 26 ) | (%001 << 23) | GyroInP
  'set up counter for gyroin, measured lag till pulseout
  frqa:=2
  repeat
    outa[GyroOutP]~~
    phsa~
    GyroDone:=cnt+gyroTime
    waitcnt(GyroDone)
    outa[GyroOutP]~
    gyro:= (phsa-gyroOffset)
```

For our accelerometer, let's look at the LIS3LV02DQ, a three-axis +/−2 g digital MEMS Linear Accelerometer. It's a low-cost sensor that uses multiple miniature cantilevers to measure acceleration in three axes. It runs on 2.16 to 3.6 V and uses the common I²C or SPI serial protocols for communication. Since it is digital, we just need to provide it with power and ground and two lines to the Propeller for communication. To use it, we need to first initialize several control registers that set various filters and ranges, and then we query it periodically to read the accelerometer's values. Here's the Spin code, which uses James Burrows' i2cObject from the Propeller Object Exchange:

```
'init accelerometer
  a:=i2c.Init(accelSDAP, accelSCLP, true)
  i2c.writeLocation(ACC3D_Addr,$20,%1101_0111 ,8,8)
  'ctrl_reg1 :power on, 160hz, no self test, all axis on
  i2c.writeLocation(ACC3D_Addr,$21,%0010_0000 ,8,8)
  'ctrl_reg2 :2g, update continuously, big endian, no boot, 12bit
  i2c.writeLocation(ACC3D_Addr,$22,%0000_0000 ,8,8)
  'ctrl_reg3 :no filters
  repeat
    a:=i2c.readLocation(ACC3D_Addr,$28 ,8,8)<<8  'outxh
    a+=i2c.readLocation(ACC3D_Addr,$29 ,8,8)      'outxl
    if a>32000
      accel:=800*(a-65536)
    else
      accel:=800*a
' Excerpt from i2cobject.spin Copyright (c) 2006 James Burrows
```

The code object that we will use to measure tilt for our robot is called tilt.spin. It measures the gyroscope and accelerometer sensors, and fuses the result with a Kalman filter, which is implemented as fixed-point math in Spin code. See Fig. 6-8 for a schematic on how to connect the accelerometer and gyroscope to the Propeller.

FUZZY LOGIC CONTROL–THE BRAINS

I've spent a lot of time optimizing the DanceBot's control logic. At the highest level, the problem is quite simple: We steer and the robot balances. Our high-level interface

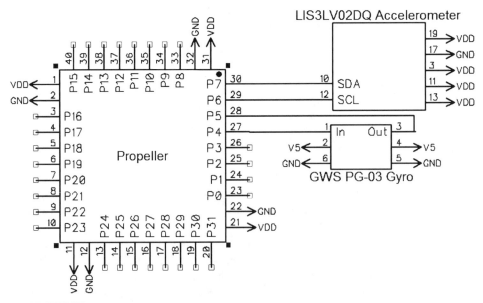

Figure 6-8   Schematic of tilt circuit.

should be as simple as those of a car's remote control, requiring us to provide just the speed and steering inputs. The robot should do all the hard work of analyzing inputs and adjusting the speeds of the two motors. However, a balancing bot drives differently from a car; steering is easier, while managing speed is much more difficult.

Let's start by seeing what we need to do to steer the robot. Since the robot has two wheels, we can drive the wheels at different speeds to steer it. This is similar to how tanks are steered and controlled. If we speed up the left wheel while slowing the right wheel, our robot will turn to the right without affecting the bot's balance. This differential steering can be carried so far that we're driving one wheel forward and the other backwards, resulting in the bot spinning about its own axis.

Accelerating a balanced robot to a new speed is more complex than it first appears. Let's start with a simple experiment. Balance on your toes and try to run. Most people will slowly lean forward and only start running when their body is tilted to a certain angle. If we started running right away, we would fall backwards because our feet would be too far in front of our center of gravity. Decelerating to a stop requires you to run a bit faster until you're tilted backwards, at which point you can gradually slow down. If your legs were stopped suddenly because they were "tripped," you would fall. For a balancing robot, things get even more complicated because the robot has no heel with which to start a lean and can't move its upper body to change its center of gravity. The only thing the bot can do is drive its wheels. So to lean forward in order to move forwards, the robot first has to drive backwards! Similarly, when the robot wants to slow down, it first has to speed up to lean itself backwards. Controlling these complicated motions requires a sophisticated control algorithm.

Fuzzy logic has received somewhat of a bad rap because it was once hyped as the silver bullet to all problems. It's not a silver bullet, but it makes some things easier, especially configuring a complex control function with human intuition. We'll use fuzzy logic to create smooth, continuous mapping functions defined by just two setpoints that relate the variable being mapped to human concepts. Figure 6-9 shows a simple fuzzy mapping function that is used by DanceBot to monitor its speed. The input to the function is a real number—in this case, the bot's measured speed. The output is a set of five numbers, indicating membership in each of five *speed classes:* fast-backwards, slow-backwards, stopped, slow-forwards, and fast-forwards. While different people might have different definitions for the "fuzzy speed classes" mentioned, we all have a good idea what they mean. It's also quite easy to talk logically about fuzzy classes—for example, "If you're falling forward fast, you need to drive forward fast." Fuzzy logic allows us to tune the bot with just two numbers per transfer function: the setpoints for "slow" and "fast" in the previous example. The resulting function will be continuous, without sudden jumps, and will gracefully top out at the maximum values, all without

Figure 6-9   Fuzzy mapping function to translate speed to "fuzzy speed classes."

cluttering our code with a slew of `if` statements. ViewPort comes with a fuzzy logic object called fuzzy.spin, which implements a complete fuzzy logic engine, and a control panel, which graphically displays the transfer functions and rule maps.

Now that we've understood enough about fuzzy logic for our application, we need to understand one more concept to build our balancing bot controller: the *cascading PID loop*. Basic PID (*proportional–integral–derivative*) controllers attempt to correct the error between a measured variable and a desired setpoint by calculating and then outputting a corrective action that adjusts the process to minimize the error. The PID controller has three parts that are summed to yield the output. The *proportional* part makes changes proportional to the error. The *integral* part eliminates the residual steady-state error in the system, but might lead to overshoots. And finally, the *derivative* part reduces the magnitude of overshoot and improves stability, but is sensitive to noise in the input signal.

In a cascading PID control, two PIDs are arranged with one PID controlling the set point of another. In our case, the outer loop controller will control the speed of the bot, while the inner loop controller, which reads the output of the outer loop controller as a setpoint, controls the speed of the motors driving the wheels. This type of control logic will let our bot achieve two goals: driving to a specified location and staying upright while only using one output, and driving the wheels a certain speed. See Fig. 6-10 for a diagram of the control logic.

The following Spin code implements the *hybrid fuzzy logic cascading PID controller* that keeps the bot balanced and allows us to steer and drive it like a car. It starts out by calculating the position error (`posE`) by subtracting the target position from the measured position. By minimizing this error with the top-level PID controller, we'll keep the bot on the desired position. The `posE` is then fuzzified into five values stored in the **A** register of the fuzzy logic engine using the position error map. By defuzzifying this register with the velocity map, we get a target velocity (`velT`), which is proportional to

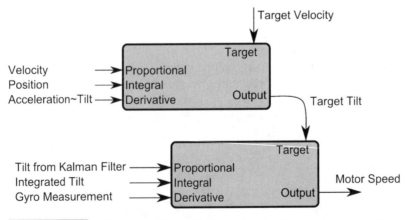

Figure 6-10   Diagram of DanceBot's control logic showing cascading PID controller.

the position error. We also defuzzify the A register using the tilt map. This yields an angle that's proportional to position error, which we'll use later as the derivative signal for the inner PID control. Calculating the velocity error (velE) from the measured velocity and target velocity (velT) is a straightforward difference. We fuzzify that result with the velocity map and defuzzify the result with the tilt map to get t2. Now we come to the top PID controller, where we use t1, t2, and tilt to calculate a target tilt for the inner PID controller. For the inner PID controller, we can calculate the proportional tilt error (tiltE), the integral tilt error (tiltI), and use the gyroscope (turn) signal as the derivative tilt error. Finally, we can calculate the output to drive the motor by summing the PID components of the inner controller.

```
posE:=-pos+posT               'posE=positive if on right of target
    f.fuzzifyA2(posE,FposE)    'A=fuzzy posE
    velT:=-f.defuzzify2(FvelT) 'velT=positive if need to move right
    t1:=f.defuzzify2(FtiltT)
    velE:=vel-velT                'velE=positive if going slower than target
    f.fuzzifyA2(velE,FvelE)    'A=fuzzy velE
    t2:=f.defuzzify2(FtiltT)
    '-t1-t2 is a pi controller, need d which is acceleration, use tilt
    tiltT:=-t1-t2-tilt/6
    f.share(@fuzz)
    tiltE:=tilt-tiltT          'tiltE=positive if tilted right of target
    tiltI+=tiltE/200           'tiltI=positive if tilted right for too long
    tiltI:= (tiltI<# maxTI) #> -maxTI
    f.fuzzifyA2(tiltE,FtiltE)  'A=fuzzy tiltE
    t1:=f.defuzzify2(Fmotor)
    f.fuzzifyA2(turn,Fturn)     'A=fuzzy turn=positive if turning right
    t2:=f.defuzzify2(Fmotor)
    motor:=-t1-t2/2'+tiltI/1500
```

ACHIEVING BALANCE WITH VIEWPORT

Parallax built a great tool to load programs written in assembly or high-level Spin language to the Propeller. This is fine for getting started writing simple programs, but I quickly determined that I needed a more powerful debugging tool to develop and configure my robot. This led me to develop an application I now call ViewPort. I started by dedicating one of the eight cogs to continuously share data stored in the Propeller's memory with a PC application. This let me monitor and change variables, all while the other seven cogs were running at full speed. When I added a module that sampled all 32 I/O pins at 80 MHz, other hobbyists became interested in using my application to debug their integration code, and ViewPort was born. Since then I've added other capabilities to the ViewPort application to turn it into a complete debugging package.

In other parts of the book, you'll see how you can use ViewPort to debug Spin programs with its visual debugger, how to stream video to debug computer vision problems, and how to use the virtual instruments to measure signals. In this section, we'll use

ViewPort Debugger

Monitor/Change Variables
Analyze Data
OpenCV Vision Engine
PhysX World Simulation
Integrated Debugger

Propeller with 8 Cogs

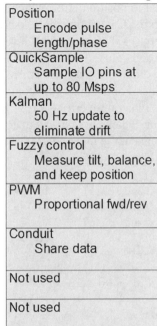

Position
Encode pulse length/phase
QuickSample
Sample IO pins at up to 80 Msps
Kalman
50 Hz update to eliminate drift
Fuzzy control
Measure tilt, balance, and keep position
PWM
Proportional fwd/rev
Conduit
Share data
Not used
Not used

DanceBot

Gyro
Accelerometer
Encoder
Camera

Motors

Figure 6-11   **System diagram with ViewPort, DanceBot, and six active cogs.**

ViewPort to make sure all parts of our robot are working correctly before we put it all together to start balancing. First, refer to Fig. 6-11 to see how all the pieces fit together. The diagram shows ViewPort graphically showing the state of the system on a PC. It's connected to the Propeller with a serial connection, where data can be sent to or from the bot. The Propeller is running code that we've discussed in this chapter on six of the eight available cogs. The "Motor" cog continually outputs direction and pulse width modulation data to the H-bridges to drive the motor, while the "Position" cog continually runs code from the motor.spin object to track where the robot is by monitoring the quadrature encoder. The "Kalman" cog continually reads the accelerometer and fuses this measurement with the gyroscope reading with a Kalman filter to produce a clean and steady tilt measurement. Finally, the "Conduit" cog sends data to and from ViewPort, while the "Spin" cog executes the main fuzzy logic code to keep the robot balanced. Now that we understand the system, let's take some measurements.

Let's start by looking at the states of the Propeller's I/O pins using ViewPort's LSA view. Since the Propeller's I/O pins are connected to sensors and actuators, looking at their state will tell us the raw signals that are sent to and from the Propeller. Refer to Fig. 6-12 to see 10 traces, which represent the communication with the accelerometer, gyroscope, motors, and encoders. The accelerometer clock (acCL) trace shows the pulses used to clock the data coming and going to the accelerometer over the accelerometer data (acDT) line using the I²C protocol. Notice that multiple bytes are transmitted

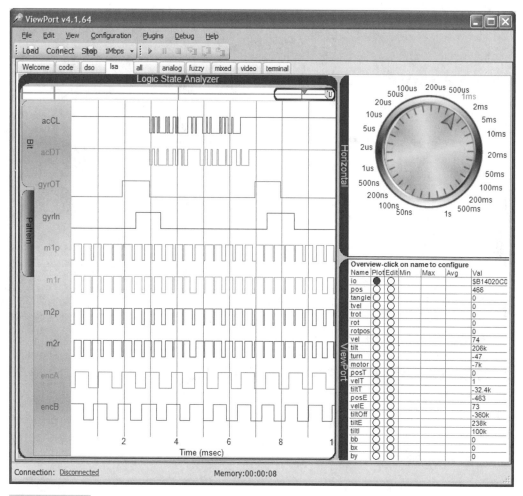

Figure 6-12   ViewPort measurement of input/output signals.

to/from the accelerometer to first query and then read in the two bytes of the sensor. The gyro out (gyrOT) trace shows the pulse output to the gyroscope, which is returned slightly later in the gyro in (gyroIn) trace. As the bot is rotated, this signal gets both longer and shorter. The next four traces (m1p, m1r, m2p, m2r) are the direction and pulse width modulation lines used to control the two motors. And finally, the encoders (encA, encB) show the pulses received from the quadrature encoder; remember their frequency and phase tells us how quickly and in which direction the wheel is turning.

Refer to the Fig. 6-13 to see the sensors and the Kalman filter in action. The screenshot of ViewPort's oscilloscope shows the values of three variables while I tilt the DanceBot back and forth. The blue trace shows the raw signal from the gyroscope—it's a clean signal that indicates how quickly I'm rotating the bot—and the signal is highest when the bot is being turned. The red trace shows what happens when I integrate the signal

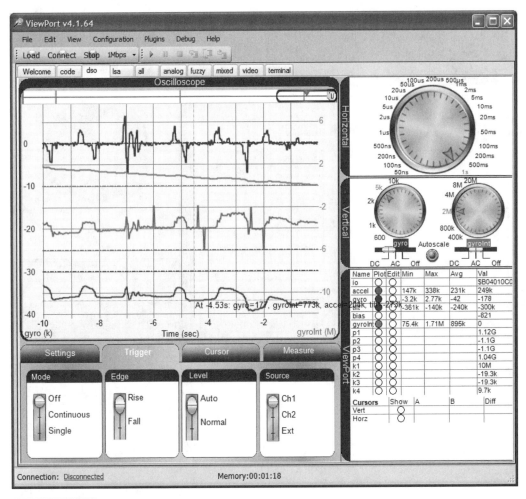

Figure 6-13   ViewPort measurement of the tilt system.

from the gyroscope; here, the signal is highest when the bot is turned all the way to one extreme. At first, the red trace accurately reflects the bot's tilt. However, over time, the integration errors add up and the signal has drifted so far as to be useless. The green trace shows the raw signal from the accelerometer—notice how noisy it is because it's continually registering small accelerations. However, averaged over time, it accurately indicates which way the bot is tilted. The purple trace is the output of the Kalman fused signal. It's clean and doesn't drift over time, perfect for controlling our robot.

Ensuring that both of these figures are correct helps us confirm that the robot is working correctly. With something this complex, it's almost impossible to track down the source of a problem to a loose wire if we don't have a tool that shows us what's going on! Now that our robot is wired correctly and we've confirmed that our sensors and calculations are correct, let's see the robot balance.

Controlling the DanceBot

The DanceBot is controlled like a car: It requires two channels of information. Channel 1, speed, controls how fast the bot should travel. Channel 2, turn rate, controls how quickly the bot should turn about its own axis. DanceBot manages its two motors to stay balanced and to achieve the desired input. Unlike a car, the bot is capable of turning in place. At first, I controlled my robot with remote control to get the hang of it. It took me much longer to get it to drive programmatically to where I wanted. Here's some code that continually drives the robot in a figure 8 by driving at a set speed while turning in one direction and then the other.

```
repeat
  repeat 200
  bal.do(10,10,0)
  repeat 200
  bal.do(10,-10,0)
```

I quickly realized that while programming the DanceBot to move by itself was fun, it would be much more fun if it could interact with others as well just by watching what they were doing. Read the next chapter to find out how I added computer vision to my DanceBot.

Summary

Although the DanceBot is quite complex, the Propeller's multicog architecture makes it easy to understand, build, and debug each of the bot's subsystems one at a time. In this chapter, we've integrated powerful sensors to the Propeller to measure the robot's position and orientation in the world. Coupled with our control logic, we were able to build a robot that balanced itself while we steered it like a normal RC car. In the next chapter, we'll add computer vision capabilities to the robot to make it interactive.

Exercises

In this section, we are going to explore a number of exercises with the Propeller-powered balancing robot.

1 Try changing the center of gravity of the robot. What happens when the center of gravity is made higher, for example, by attaching a heavy weight to the top of the bot? Does this make balancing easier or harder?

2 Try using different wheels on different surfaces. What happens when the robot's contact area with the ground gets smaller or larger?

3 Try running the robot up and down inclined surfaces. Does the robot stay balanced? Why?

7

CONTROLLING A ROBOT WITH COMPUTER VISION

Hanno Sander

Introduction

The Propeller is powerful enough to capture and analyze video. In this chapter, I'll look at two vision technologies that can control a robot with computer vision. Here's what we are going to cover in this chapter:

- Computer vision: Seeing and understanding the world one pixel at a time
- Using computer vision in our robot to interact with people
- Developing a vision engine that runs on the Propeller: PropCV
- Integrating with OpenCV for state-of-the-art computer vision

Resources: Demo code and other resources for this chapter are available for free download from ftp.propeller-chip.com/PCMProp/Chapter_07.

VISION-GUIDED ROBOTS

It's time to build more sophisticated robots. With today's technology, our robots should no longer require us to program their every move or steer them via remote control. Instead of stumbling through the world with touch or proximity sensors, our robots should stand up and watch where they're going. The future of vision-guided robots is bright. Let's see how we'll get there.

The Propeller provides a great entry into vision-guided robots. It's powerful enough to process video and has enough memory to perform simple vision algorithms. But most importantly, it's easy to integrate with all sorts of robotic sensors and actuators. The Parallax community has written Propeller objects for almost anything you would want to interface with. Integration is typically as simple as downloading an object and running the code in one of the Propeller's eight cogs. Once you've configured your robot, adding vision capabilities is a matter of choosing how much sophistication you require. ViewPort's PropCV vision engine provides simple vision capabilities and runs on as little as one of the Propeller's cogs. Yet, the OpenCV vision engine integrated into ViewPort gives you access to state-of-the-art computer vision—recently used to win the Defense Advanced Research Projects Agency (DARPA) competition by guiding a car autonomously through an urban setting.

Understanding Computer Vision

A touch sensor is easy to understand. When something touches the sensor, the switch closes and the robot can react to that event. A vision sensor is a bit more complex. First, the robot's camera, or set of cameras, needs to image the environment by focusing and registering the incoming light and turning it into a video signal. The video signal is digitized and stored in the processor's memory. Vision filters are applied in a vision chain to analyze the scene. At the end of the vision chain, the large amount of incoming video data has been processed into the nugget that tells the robot what it needs to know. See Fig. 7-1 for a diagram. For example, the robot can sense the location of a guide pattern, the presence of an obstacle, or even the location of a beer bottle. Once the location or presence of the desired object has been calculated, the robot can react appropriately—for example, drive to the beer bottle and pick it up with its gripper. Central to this whole process is the design of the vision filters.

Users familiar with photo-editing software like Adobe Photoshop have plenty of experience with filters. In Photoshop, I can apply the "brighten" filter to make my image brighter. So how does the program accomplish this behind the scenes? The picture is

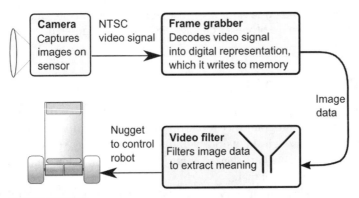

Figure 7-1   In a vision sensor, video is filtered to yield a nugget, which controls the robot.

made up of many individual pixels, thousands in each row with thousands of rows. Each pixel has a color that's specified by some number of bytes. For example, a 24-bit color image might use one byte to specify how red the pixel is, one byte to specify how blue, and one byte to specify how green for a total of three bytes. When the "brighten" filter is applied, the individual bits in each byte of every pixel are manipulated to make the image brighter—for example, by adding a value to each byte.

In computer vision, filters have to be applied continually, ideally as fast as new images come from the robot's camera. This can require significant computer processing power and memory to manipulate all the bytes that make up each image. Also, while it's easy to code a filter that will brighten an image, designing a filter that will find a bottle of beer in a cluttered room is much more difficult.

Decades of research have been spent trying to understand how the human brain processes vision. We've learned that the brain's vision system has many parts, each of which is specialized for some task. For example, some cells in our eyes are active only when we're looking at lines at a certain angle, and further along the chain, neurons are arranged to decode three-dimensional information from our two eyes. Some researchers have even found a neuron that recognizes just the subject's grandma! Almost half of the human brain is dedicated to processing vision. It takes that much processing power to do what we take for granted. Luckily, modern desktop computers and even the Propeller are getting to the point that we can apply them to computer vision.

PropCV: A Computer Vision System for the Propeller

In the next sections we'll build a complete computer vision system that will run entirely inside the Propeller. I originally built PropCV so that my kids could dance with my DanceBot, but it's now a part of ViewPort, so people can use it for all sorts of projects. One of my customers is even working on a project to make industrial robots weld more accurately. So, what's involved? We'll start by building a frame grabber to digitize video from a camera for storage in the Propeller's main memory. Then, we will continue by building different filters to process the video and recognize different objects.

FRAME GRABBER

I designed PropCV to work for any video source that can output a National Television System(s) Committee (NTSC) composite signal, which makes the solution highly flexible. I used a $20 grayscale camera that fits on my fingertip, while others use professional-quality camcorders with autofocus and a servo-driven zoom lens. I have chosen the C-Cam-2A camera for the DanceBot. It measures just 16 × 16 × 16 mm, uses less than 100 mW at 5 V, and only has one option: the gamma setting. (See Fig. 7-2 for the complete vision hardware for my robot.) The output signal consists of a 1 V peak-to-peak composite video signal when terminated into 75 ohms. Like other NTSC composite

Figure 7-2   C-Cam 2A grayscale camera and ADC.

sources, you can watch the camera's output on your TV by simply connecting it to the composite. Teaching the robot to understand what the camera sees is a bit harder, so we'll take it one step at a time.

To start, we have to digitize the analog signal. To sample slower waveforms with the Propeller, you would use delta-sigma modulation with a capacitor and resistor, but since we need to resolve the individual pixels in a frame, we need a faster solution. The ADC08100 is a 20–100 Msps, eight-bit analog-to-digital converter. Given the correct clock signal, it will output the digital equivalent of the input analog voltage on its eight digital data outputs. We'll use one of the Propeller's 16 hardware counters to clock the ADC at 10 MHz and read the result on pins 0..7 (see Fig. 7-3 for the schematic).

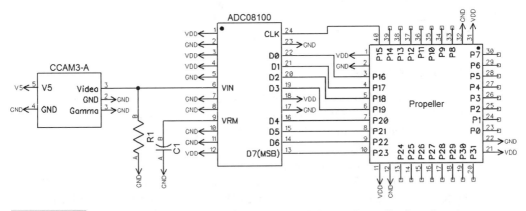

Figure 7-3   **PropCV schematic showing camera, ADC, and Propeller interface.**

Now that we've finished our vision hardware, we can use ViewPort to verify that the camera and ADC are working. We'll use a short Spin program to generate a 10 MHz clock signal for the ADC and use the QuickSample object to quickly sample the INA register. Our ViewPort configuration tells ViewPort to decode the ADC's digital data stream and show an analog presentation within ViewPort on a 50 μs timescale.

```
CON
 _clkmode        = xtal1 + pll16x
 _xinfreq        = 5_000_000
OBJ
 vp  :           "Conduit"        'Transfers data to/from PC
 qs  :           "QuickSample"    'Samples INA up to 80MHz
 Freq :          "Synth"
pub demoADC|a,frame[1600] 'Frame stores 1600 samples+configuration
 vp.register(qs.sampleINA(@frame,1 ))
 vp.config(string("var:io,adc(decode=io[0..7])"))
 vp.config(string("dso:view=adc,trigger=adc<15,timescale=50us"))
 vp.share(0,0)
 Freq.Synth("A",8, 10_000_000) "Drive ADC with 10MHz clock
 repeat
```

Refer to Fig. 7-4 for the waveform we've captured from the camera. The waveform represents one of the 425 scan lines. The signal starts with a horizontal sync pulse followed by the NTSC color burst and then 50 μs of data. In an NTSC signal, the beginning of a frame is indicated by a vertical sync, where the signal is at its minimum for a specified time. Each scan line is marked by a horizontal sync, where the signal is also at its minimum, but for a shorter time than the vertical sync. We won't worry about the color burst signal, as PropCV will only decode the grayscale information, not the color. The 50 μs of data represent the pixel's brightness by the reconstructed analog signal value. The peak in the analog signal indicates a bright object in the middle of the camera's view.

The following assembly code implements a simple, low-resolution frame grabber. It detects the horizontal and vertical sync marks and then packs the pixel data into memory. It captures 100 lines of video, each with 120 pixels with 16 gray levels. This consumes quite a bit of the Propeller's 32 KB of RAM:

100 lines · 120 pixels/line · 4 bit/pixel · 1 byte/8 bit = 6000 bytes or 1500 longs

We can distinguish between vertical and horizontal sync marks by the different amounts of time that the signal stays at its minimum. After detecting a vertical sync mark, the code initializes a new frame and processes every other video line. It detects the horizontal synch, skips past the color burst, and then samples the ADC's value every eight instructions. To store the pixels efficiently into memory, we pack eight 4-bit pixel values in a 32-bit long into the Propeller's global memory, which is accessible by all other cogs.

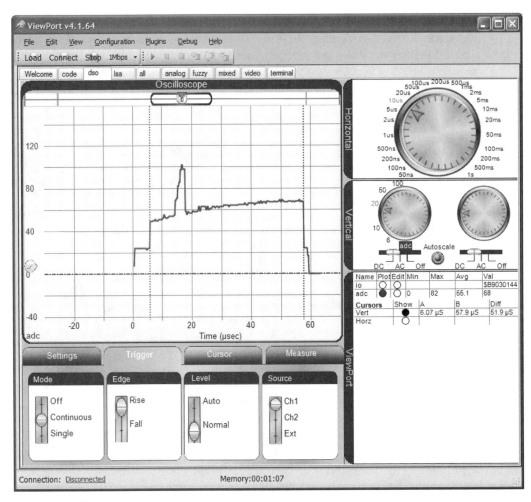

Figure 7-4 NTSC waveform captured by the Propeller.

```
doLoFrame       mov     pixPtr,par      'Start new frame
                call    #findVSync      'Find vertical sync
                mov     fieldCtr,#100   '15longs/line*100lines=1500 longs
:sampleField    mov     lineCtr,#15     '15 longs*8=120 samples
                call    #findHSync      'Find horizontal sync twice to skip a
                                         line
                waitcnt sB,#0
                call    #findHSync
:sampleLine     mov     sb,#7           'Pack 8 pixels/long
                mov     sL,#0
```

```
:sample8Pix
            mov        sA,ina           '8 instr/pixel
            and        sA,video
            add        sL,sA
            ror        sL,#4
            nop
            nop
            nop
            djnz       sb,#:sample8Pix
            mov        sA,ina
            and        sA,video
            add        sL,sA
            ror        sL,#4+VIDEOSTART
            wrlong     sL,pixptr
            add        pixptr,#4
            djnz       lineCtr,#:sampleLine
            djnz       fieldCtr,#:sampleField
doLoFrame_ret ret
findVSync
            waitpeq    syncval,sync 'Wait for sync
            mov        sA,cnt
            waitpne    syncval,sync 'Wait for exit
            mov        sB,cnt
            sub        sB,sA
            cmp        sB,vSync wc 'c set if sEnd-sStart<vsync
     if_c   jmp        #findVSync
            mov        sB,cnt
            add        sB,vblank
            waitcnt    sB,cycPline
findVSync_ret ret
   findHSync
            waitpeq    syncval,sync 'Wait for sync
            mov        sA,cnt
            waitpne    syncval,sync 'Wait for exit
            mov        sB,cnt
            sub        sB,sA
            cmp        sB,hSync wc 'c set if sEnd-sStart<hsync
     if_c   jmp        #findHSync
            mov        sB,cnt
            add        sB,burst
            waitcnt    sB,cycPline
findHSync_ret ret
```

The following Spin code uses the PropCVCapture object in HIVIDEO mode to continuously sample video from the camera at 30 fps with a resolution of 240 pixels · 200 lines · 4 bit/pixel. This uses even more memory, but provides a sharper image:

200 lines · 240 pixels/line · 4 bit/pixel · 1 byte/8 bit = 24,000 bytes or 6000 longs

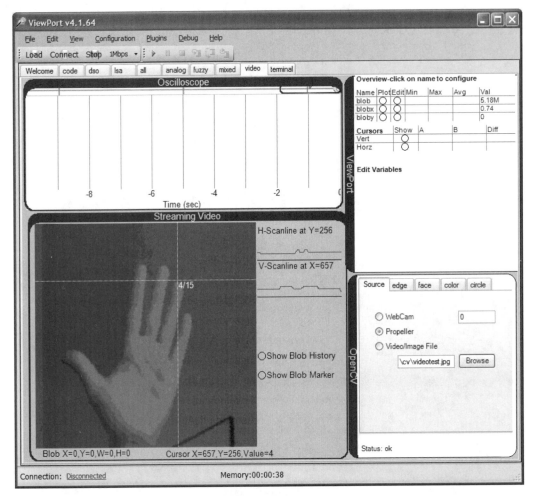

Figure 7-5 Picture imaged by the PropCVCapture object.

The program uses the Conduit object to stream the video to ViewPort, where I can watch the camera's output, look at horizontal and vertical scan lines, and measure the value of individual pixels. See Fig. 7-5 for our first picture of the world!

```
CON
  _clkmode      = xtal1 + pll16x
  _xinfreq      = 5_000_000
OBJ
  vp   : "Conduit"            'Transfers data to/from PC
  video: "PropCVCapture"      'Capture video signal
pub watchTV|videoFrame[6000],a,blob
  vp.register(video.start(@videoFrame,video#HIVIDEO))
  vp.config(string("start:video"))
  vp.share(@blob,@blob)
  repeat
```

Apply Filters and Track a Bright Spot in Real Time

Now that we can digitize video from a camera into the Propeller's memory, we need to do something useful with it. We have quite a bit of video data updated at 30 times a second. We need to filter this information to provide us with just two variables to control our robot.

We'll start by implementing a filter that identifies the location of the brightest spot in each frame of video. The following assembly code is passed some configuration about the location and size of the video to filter and then searches for the maximum value in our array of pixel-brightness values. Since our frame grabber has packed 8 four-bit pixels into each 32-bit long, we need to make sure to look at each individual pixel by first decoding it. This filter processes one pixel every five instructions. The filter processes only data, not sync marks or color bursts, so let's calculate if it's fast enough to process the incoming data.

$$1 \text{ pixel/5 instructions} \cdot \text{line/120 pixels} \cdot \text{frame/100 lines} \cdot 20 \text{ M instructions/sec}$$
$$= 333 \text{ frames/sec}$$

New frames are only coming in at 30 fps, so the Propeller is fast enough to apply multiple filters to incoming video in real time.

```
doMax
'setup ptrs to positions
                mov     dnp,src          '   src points to first pixel
                add     dnp,bytesNline 'dnp is one row below src
                sub     n,#15            'Number of pixels to inspect
                mov     dest,#0  'Location of pixel
                mov     sum,#0           'Brightest value so far

:loop           rdlong  old,src
                add     src,#4
                rdlong  dn,dnp
                add     dnp,#4

                mov     m,# 8
                mov     new,#0
:dodiffb        mov     tmp,old
                and     tmp,#15 'Decompress 4 bits out of a long
                mov     t1,dn
                and     t1,#15 't1=pixel down
                add     tmp,t1 'tmp=pixel value of two vertical pixels
                cmp     tmp,sum wc 'c if tmp<sum
```

```
if_c         jmp      #:notMax  'Found a new maximum
             mov      dest,sum   'So save the position and value
             shl      dest,#16
             add      dest,n
             mov      sum,tmp 'Reset max
:notMax      ror      dn,#4
             ror      old,#4
             djnz     m,#:dodiffb
             djnz     n,#:loop
             wrlong   dest,val
             jmp      #cmdLoop
```

We can build additional filters to invert an image, look for edges, and apply a threshold. Unlike Photoshop, we want to apply our filter continually to new frames coming from our frame grabber. We also want to chain multiple filters together, linking the output of one filter to the input of another. To do this, we'll implement the concept of a video buffer and a simple scripting language (see Fig. 7-6 for a diagram).

In the last section, we allocated one contiguous array in memory and used the PropCVCapture object to write new frames of video into that array at high resolution. From now on, we'll continue allocating one array in memory, but treat it as four separate video buffers. We'll name the first 1500 longs of the buffer "Quadrant 0" and display this in the upper-left inside area of ViewPort. Our frame grabber will fill this buffer with raw video from the camera if we use the VIDEO4 option. We'll name the remaining 4500 longs Quadrants 1, 2 and 3 and display them in the upper-right, bottom-left, and bottom-right areas, respectively. Using our scripting language, we'll filter frames to and from these video buffers. You can think of each of these buffers as a temporary video "variable" that we use for our video calculations.

We'll keep our scripting language simple. For each script action, we'll store one long that indicates which filter to apply. We'll use two additional longs to point to the first pixel in the source and destination video buffers. To make the system flexible, we'll allow filters to access any number of additional parameters to configure their behavior. The following assembly code implements the scripting engine, which continually applies our vision filters to the vision buffers.

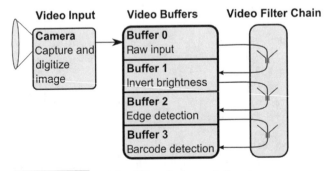

Figure 7-6   Video filter chain scripting language and video buffers.

```
                org 0
doEdit          mov         cmdPtr,par
cmdLoop         rdlong      action,cmdPtr
                add         cmdPtr,#4
                rdlong      dest,cmdPtr
                add         cmdPtr,#4
                rdlong      src,cmdPtr
                add         cmdPtr,#4
                mov         n,len
                add         action,#:jmpTable
                jmp         action
:jmpTable       jmp         #doEdit
                jmp         #dolimit
                jmp         #doinvert
                jmp         #dodiff
                jmp         #dochaos
                jmp         #domax
                jmp         #dopattern
                jmp         #docopy
```

Last, we'll need some methods to let other programs specify which filters to run in their engine. The following Spin code lets users programmatically add two types of filters: those taking a value argument and those that don't. The start method is used to start a cog, which will then continually cycle through the filters.

```
con
#0,None,Threshold,Invert,Difference,Chaos,Max2,Pattern,Copy 'Fuzzy maps
var
  long cmd[20]
pub start(tVptr,tmode)
{{start a cog to continuously apply vision filters, tvptr points to the video
buffer,
  tmode set to 0 for a 6000kb buffer, >=1 for a 1500kb buffer}}
  if tmode>1
    len:=1500
    len4:=6000
    llen:=60
  vptr:=tVptr
  cognew(@doEdit,@cmd)
pub filter(tDest,tSrc,tFilter)
{{add a filter to the filter chain, tDest=destination frame, tSrc=source frame,
tFilter=filter number}}
  cmd[cmdNum]:=tFilter
  cmd[cmdNum+1]:=vptr+tDest*len4
  cmd[cmdNum+2]:=vptr+tSrc*len4
  cmdNum+=3
```

```
pub filterValue(tDest,tSrc,tFilter,tVal)
{{add a value filter to the filter chain, tDest=destination frame, tSrc=source
frame, tFilter=filter number, tVal=parameter value}}
  cmd[cmdNum]:=tFilter
  cmd[cmdNum+1]:=vptr+tDest*len4
  cmd[cmdNum+2]:=vptr+tSrc*len4
  cmd[cmdNum+3]:=tval
  cmdNum+=4
```

The following Spin code integrates the vision filter chain with the frame grabber we've developed previously. This time, we'll allocate 6000 longs for our vision data and select the VIDEO4 mode from the PropCVCapture object to capture lower-resolution video into the upper-left quadrant. This allows us to use the other quadrants to view filtered versions of our data. To filter the video, we'll make several calls to "ve.filter" to configure and set up the filter chain. We'll use the invert, threshold, difference, and max filters that come with PropCVFilter. Once set up, the filters will run in their own cog, periodically updating the video data and the variable indicating where the brightest spot is.

```
CON
  _clkmode       = xtal1 + pll16x
  _xinfreq       = 5_000_000
OBJ
  vp   : "Conduit"              'Transfers data to/from PC
  video: "PropCVCapture"        'Capture video signal and view in Viewport
  ve:    "PropCVFilter"         'Applies vision algorithms to video
pub vision|videoFrame[6000],blob,b1,x,y,w,h
  vp.register(video.start(@videoFrame,video#VIDEO4))  '1 video buffer
  vp.config(string("var:blob,b1,x,y,w,h"))
  vp.config(string("dso:view=blob"))
  vp.config(string("video:view=blob"))
  vp.config(string("start:video"))
  vp.share(@blob,@b1)
  ve.start(@videoFrame,video#VIDEO4) 'Apply vision
  ve.filter(1,0,ve#invert)           'Invert pixels from 0 to 1
  ve.filter(2,0,ve#difference)       'Calculate difference from 0 to 2
  ve.filtervalue(3,0,ve#threshold,10)'Only show brightest spots
  ve.filtervalue(0,0,ve#max,@b1)     'Find maximum'
```

Figure 7-7 shows a screenshot of the video streamed from the Propeller. Notice the original image in the upper-left quadrant, the negative of the image in the upper-right quadrant, then the horizontal edge filtered image in the bottom-left quadrant, and the thresholded image in the bottom-right quadrant. Also, the marker represents the location of the brightest spot as found by the "findMax" filter.

We now have two channels of information updated at 30 times a second with which we can drive the two control channels of a robot: speed and direction. We use the x position of the spot to control the turning rate of the robot. If the spot is in the middle, we don't need to do anything. However, if it's on the left side of the image, the algorithm

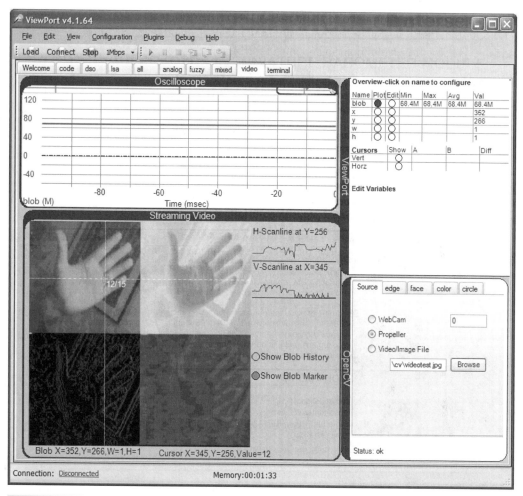

Figure 7-7   Multiple video buffers showing the effects of negative, edge, and threshold filters.

tells the robot to turn left until the spot is in the center and the robot is facing the source of the spot. The robot's speed is controlled using a similar technique: using the vertical position of the spot. The algorithm's goal is to keep the spot's position centered in the image. So when the spot is too low, the robot is instructed to move forward, which brings it closer to the spot's source, and because our camera is looking up at the spot, will raise the spot in the image. Conversely, if the spot is too high, our robot is too close, so it's commanded to drive backwards. Translating this algorithm into code is simple: We just scale and offset the x,y location of the spot to control the robot.

To illustrate the high-speed tracking ability of this filter, I can use ViewPort to display the streamed video with a superimposed trail showing the position returned from the filter over the last minute. Figure 7-8 shows a screenshot with the grayscale image as seen by the camera and a trail showing the path the bright source took.

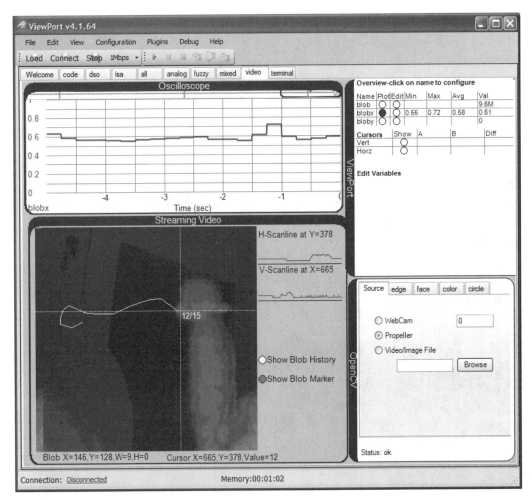

Figure 7-8   **Tracking an object at high speed.**

Following a Line with a Camera

In the previous section, we controlled the behavior of the robot by shining a bright light at the camera. This works in some environments where we can control the lighting and ensure that no other objects reflect or create light to the camera that's brighter than our flashlight. It's also an active method, where we have to power the flashlight. We'll now look at a filter that is less restrictive and uses a passive method to steer a robot.

Traditional line-following robots use two phototransistors to stay on a line. One possible strategy works by ensuring that one detector is on the dark line while the other is on the lighter background. More sophisticated robots use additional detectors to determine the robot's exact position on the line, to look ahead, or even to recognize junctions.

Figure 7-9   DanceBot with camera configured to follow lines.

In this section, we'll build a filter that uses our existing frame grabber to perform following a line with a camera.

Since our frame grabber gives us too much information, we need to design a filter that will steer a robot in the middle of a line. We start by mounting the robot's camera so that its field of view is from below the horizon to just in front of the robot (see Fig. 7-9). Now, we can stream the video to ViewPort and analyze what the video of a properly programmed robot would do. It becomes apparent that a good strategy is to average the location of the darkest pixel in each line. This is quite robust, easy to program, and gives a good control signal to the robot. The following assembly code filter is used for this task. Integrating this filter with the rest of the DanceBot is straightforward. We just use the average position of the line to control the direction of the robot while it's moving along at a set speed.

The following assembly code filter averages the location of the darkest pixel in each line. Watch a video of my DanceBot using this filter to follow a black line here: http://mydancebot.com/viewport/videos.

```
doLowest
               rdlong    t1,cmdPtr    't1 points to seexy will be written with
                                       sum of pos of lowest item
               add       cmdPtr,#4
               mov       t2,#0        't2 is sum of pos


               mov       n,linesNpanel 'Loop over panel
:loopLines     mov       x,longsNline  'Loop over line
               sub       x,#2
               mov       val,#15       'Reset val
:loop          rdlong    old,src       'Loop over longpixels
               add       src,#4
               mov       m,#8
               mov       new,#0

:limit         mov       tmp,old
               and       tmp,#15
               cmp       tmp,val wc   'c set if v1<v2
if_c           add       new,#15
if_c           sub       val,#1
if_c           mov       t3,x          't3 is pos of lowest item
               ror       new,#4
               ror       old,#4
               djnz      m,#:limit

               add       dest,#4
               djnz      x,#:loop 'Loop over longpixels
               add       t2,t3
               add       src,#8
               add       dest,#8
               djnz      n,#:loopLines 'Loop over lines
               wrlong    t2,t1
               jmp       #cmdLoop
```

Track a Pattern

We've gone from tracking an active, bright spot to following a passive line on an artificial background. Now it's time to track a passive pattern in the real world. Our goal for this section is to develop an algorithm and pattern that will steer the robot in practically any environment. As an experiment, take a look around you and imagine what type of pattern would stand out in our typical cluttered world—in computer vision terms, we're looking for an appropriate fiduciary pattern. In most places, a barcode-like pattern of repeated black and white lines should do well. This pattern is relatively rare in most settings, and we can use a chain of simple vision filters to locate it, even in visually cluttered scenes.

First, we need to build a filter that's sensitive to horizontal edges in our video. Looking at relative changes in value improves the robustness of our algorithm, especially when lighting conditions change. The horizontal Sobel filter is quick to implement and does a good job of detecting vertical edges. To compute it, just replace each pixel with the absolute value of the difference of its horizontal neighbors. When we run the horizontal Sobel filter on our pattern, we notice that area corresponding to the pattern turns white; the filter finds many edges close together in that area.

Next, we need to design a filter that highlights these areas where multiple edges occur close together. The algorithm I chose keeps a running total of the last eight pixels. When we chain this filter with the horizontal Sobel filter, it has the effect of finding large areas that contain strong horizontal edges caused by multiple black and white stripes.

For our last filter we'll reuse the "findmax" filter we designed two sections ago. By chaining these three filters together, we find the location of the maximum running total of vertical transitions—in other words, we find a black/white striped pattern. We can now use the location of the found object to steer our robot without any active input. When used with the DanceBot, we end up using all eight cogs (see Fig. 7-10 for a diagram of the complete system). Two cogs are used by PropCV to capture and process

ViewPort Debugger

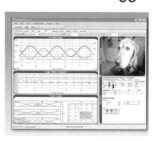

Monitor/Change Variables
Analyze Data
OpenCV Vision Engine
PhysX World Simulation
Integrated Debugger

Propeller with 8 Cogs

| Software Frame Grabber
NTSC->ADC->4 Pins
Memory |
| Image Processing
Frame->Variables |
| Position
Encode pulse
length/phase |
| QuickSample
Sample IO pins at
up to 80 Msps |
| Kalman
50 Hz update to
eliminate drift |
| Fuzzy Control
Measure tilt, balance,
and keep position |
| PWM
Proportional fwd/rev |
| Conduit
Share data |

DanceBot

Gyro
Accelerometer
Encoder
Camera

Motors

Figure 7-10   Complete DanceBot system diagram.

video to steer the robot, while the remaining six cogs are used to balance the robot with its multiple sensors (see Chap. 6 for more details).

To make our pattern robust at different distances, we can repeat it at various scales. When we now place this pattern on our belt, we can start dancing with our DanceBot. Stepping closer to the robot makes the image of our pattern move up in the robot's field of vision—this causes the robot to move backwards and maintain a set distance from us. Stepping to one side of the robot causes the pattern to move horizontally, which commands the robot to turn and face us. While dancing with my robot, I discovered some interesting behavior that I hadn't planned on. When I jumped up or crouched down, the robot would change its set distance to me. When I turned around, thereby covering the pattern, the robot also turned around. It no longer detected the pattern and went into search mode, where it turned on its axis. This level of vision-guided interaction is still quite novel, but it's all possible with a simple camera, an ADC, and a couple of the Propeller's eight cogs.

State-of-the-Art Computer Vision with OpenCV

The PropCV objects give us a good foundation to perform computer vision with the Propeller. We've explored all the components that make up a vision sensor: the frame grabber, the chain of vision filters, and the multiple vision buffers to store our images. The Propeller is fast enough to do simple vision processing in a single cog with its own memory; however, we need more power to perform more sophisticated computer vision. So, let's use our knowledge to tackle the state-of-the-art vision system that's included in ViewPort: the OpenCV plug-in.

OpenCV is a library of computer vision functions that was originally developed by Intel in 1999. It's now available to everyone as an open-source project on sourceforge.net and has corporate support from Willows Garage. It's been used successfully in video surveillance systems, video games, and robotics. Stanley, an autonomous car created by Stanford University, used OpenCV to find its way through 132 mi of California desert to win the 2005 DARPA Grand Challenge.

The goal of the OpenCV project is to advance vision research by providing open and optimized code for vision infrastructure. Rather than reinventing the wheel, researchers can reuse and complement vision algorithms contained in OpenCV. Developers can add features to OpenCV and even develop commercial applications.

The library is written in the C language and includes more than 500 functions that cover the entire space of computer vision. Besides image manipulation functions, the library can be used to find objects like human faces, understand gestures, construct 3-D models from stereo cameras, track motion, and even do statistical-based machine learning. The functions are performance-optimized to take advantage of advanced multimedia instructions found in modern processors. In the following sections we'll explore how OpenCV implements several of these functions and then use them in our Propeller Spin program.

FINDING COLORED OBJECTS

Since OpenCV supports color video sources, one useful filter we can employ is a color blob filter. This filter finds the area containing the largest number of pixels that match our target color. To understand the color filter, however, we need to understand two different color spaces.

Earlier in this chapter we learned that pixels make up an image and that each pixel in a 24-bit color image is typically defined by three bytes that define the red, green, and blue levels of the pixel. This, the RGB color space, is the most common color space used by electronic systems like computers, cameras, and webcams. It's an additive color model where different proportions of the three primary colors are added together to reproduce the broad array of colors we see on our monitors. For example, to produce a yellow pixel, we would add equal amounts of red to equal amounts of green—using more of both would make the pixel brighter. Other color spaces exist and are appropriate for different uses. For example, the subtractive color space CMYK (cyan, magenta, yellow, and black) is used in the printing industry, and HSV (hue, saturation, and brightness) is used in televisions.

Video coming from a webcam is typically in the RGB format. Searching this stream of data for pixels that exactly match a chosen color is easy: We just have to match the three bytes of each incoming pixel to our target RGB values. However, this solution doesn't solve our problem. To accommodate changes in lighting, shadows, and color, our filter should match colors that we humans would judge as being similar to the target color. Doing this is much more efficient when we first convert each pixel's color from the RGB to the HSV color space. Fortunately, converting from RGB to HSV is straightforward and OpenCV provides us with a high-speed filter to do that.

Once our pixels are in the HSV color space, we can restrict our search to pixels whose HSV values lie within a specific range. For example, to search for bright, highly saturated blue pixels, we could look for pixels whose hue lies between 150 and 170 with saturation and brightness values higher than 200. We can generalize this to find all brightly colored pixels—for example, toys—by looking for pixels with saturation and brightness values higher than 200, ignoring the hue altogether.

OpenCV allows us to chain filters together just like our PropCV system. The first filter results in a frame where only the pixels that match our criteria are set to white. Then, we pass that image to a blob-finding filter, which finds the location and size of the best blob that fits our pixels. The result is a filter that's highly robust for finding colored objects—something that's useful as long as we have a color camera and are looking for colored objects.

FINDING SPECIFIC OBJECTS

Let's look at another interesting OpenCV filter: the Viola Jones object detector. This filter is used to accurately find objects that statistically resemble objects that the detector has been trained for. As configured in ViewPort, it's quite good at finding all types of human faces: with or without glasses, different hairstyles, and from different ethnic groups.

The Viola Jones detector looks for features in the frame. Specifically, the detector evaluates different parts and scales of the image to match their features with those identified in the training set. These features are rather simple when we look at them: They consist of a small number of rectangular areas where pixels are darker or brighter than the surrounding area. Taken by itself, a single feature isn't enough to detect an object, but by combining multiple features that match in the right location and at the right scale, the algorithm efficiently finds objects.

The key to using the Viola Jones detector successfully is a good dataset of images from which the features are learned. In theory, the filter is able to detect a wide variety of objects when you take the time to build a clean learning set. So far, it has been used mostly to find human faces.

Unlike the color blob finder, this filter only looks at the grayscale component of the video. Color frames sourced from a color webcam are first processed by a grayscale filter before the object detector is run. Since only grayscale frames are needed, you can use this filter on video sourced from the PropCV frame grabber.

FINDING SHAPES

Faces are quite complex objects. Let's look at a filter that finds a simple shape: a circle. A circle filter is quite useful because spheres viewed from any direction have the shape of a circle. Also, balls are quite common in many of the sports we play. A technique to accurately track where they are in a robotic system can be useful!

It's easy to confirm the location and radius of a circle once we've found it. All we have to do is look at the pixels that are one radius away from the circle's center and make sure they mark the edge of the object we're seeking. This filter typically preprocesses the frame with an edge detector filter, so the circle's edge will ideally be completely white. However, finding the location and radius is inefficient in our traditional Cartesian system, where pixel values are arranged geometrically by their x and y locations. There are so many possibilities! Luckily, we can transform our frame data from the Cartesian system into a three-dimensional space using the Hough transform, where this process becomes much easier. This mathematical technique is a linear transformation that maps edge points to a specific point in the three-dimensional location/radius space. Once the transform has been applied, finding the circle is a quick matter of finding the maximum in the location/radius space.

Enough math! Now let's get back to the Propeller and find some objects with OpenCV!

OpenCV and Propeller Integration

I designed ViewPort's architecture to allow myself and others to add functionality to the program through the use of plug-ins. I developed the OpenCV plug-in to make it easy for Propeller users to add state-of-the-art computer vision to their robotic creations

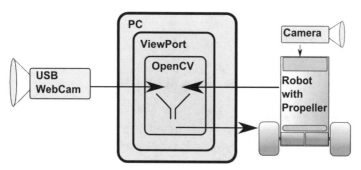

Figure 7-11   Propeller integrated with ViewPort running OpenCV filters.

without having to learn C or set up a development environment for OpenCV. People have been doing computer vision research with OpenCV for quite some time, but integrating it with real-world devices like sensors and actuators wasn't easy. With the plug-in and the Propeller, people have the best of all worlds: easy integration with all sorts of real-world sensors and actuators with the Propeller and state-of-the-art vision algorithms from OpenCV, all presented with a simple interface inside of ViewPort.

Figure 7-11 shows a diagram of an OpenCV-enabled Propeller application. To perform sophisticated computer vision, the OpenCV code is run on the host PC to take advantage of the faster processor and larger amount of memory available there. There's a big difference between the 32 KB of RAM available on the Propeller and the multiple gigabytes available on most PCs! Running on the host PC also allows us to tap OpenCV's various input options to process video from a variety of sources. We can still process the grayscale video captured by the Propeller-based PropCV frame grabber, but we can also process video from USB webcams, Ethernet cameras, and video files. Configuring OpenCV's input source and filter options can be done through ViewPort's graphical interface or programmatically in Spin code. The results of OpenCV-filtered video is shown inside of ViewPort, and the location of found objects is continually sent to the Propeller, where it can be used to steer robots.

Let's take a look at a Propeller program that incorporates OpenCV. The following Spin code allocates four variables: pos, x, y, and sum. It shares these variables with the PC running ViewPort using the vp.share method from the "Conduit" object. One line of code configures the video widget to take "webcam0" as a source, use "face" as the mode, and pass the result to the variable called "pos." This line is powerful! It's telling ViewPort to look for a human face with its webcam. And, most importantly, it's configuring ViewPort to continually update Spin variable "pos" with the location of that human face. We can use the location of the face to control our robot, make our system interactive, or, like the Spin code shows, calculate using the x and y coordinates.

```
CON
 _clkmode       = xtal1 + pll16x
 _xinfreq       = 5_000_000
OBJ
  vp   : "Conduit"              'transfers data to/from PC
pub findblob|pos,x,y,sum
  vp.config(string("var:pos,x,y,sum"))
  vp.config(string("dso:view=[x,y],timescale=1s"))
  vp.config(string("video:source=webcam0,mode=face,result=pos"))
  "source should be webcam0..1000,prop,or a file like: \cv\videotest.jpg
  "mode should be face,edge,color,circle
  vp.config(string("start:video"))
  vp.share(@pos,@sum)
  repeat
    x:=pos & $3ff
    y:=pos >>10 & $3ff
    sum:=x+y
```

Okay—time to go into more detail. First, the result returned from the OpenCV plug-in contains the location and size of the desired object, all packed into a single long. The x and y coordinates are stored in the lower 20 bits and range from 0,0 (lower left) to 1023,1023 (upper right), no matter what camera you use. The height and width are stored in the upper 12 bits and range from 0,0 (small) to 63,63 (the whole camera's width and the whole camera's height). This makes it easy to write one program that will work for people, regardless of what hardware they use. Most people will use OpenCV with their computer's built-in webcam or an inexpensive USB webcam. However, you can use the "source" option to set OpenCV's input to the PropCV frame grabber, to an AVI file on your PC, or to other video devices. OpenCV filters will run at the resolution and color mode of the source input. Processing a 24-bit megapixel video stream on the PC is no big deal. Let's look at the memory consumption for a frame of that:

$$1 \text{ Mpixel} \cdot 24 \text{ bit/pixel} \cdot 3 \text{ byte/24bit} = 3 \text{ MByte: a lot for the Propeller,}$$
but little of a PCs gigabytes of RAM!

Want to see how well OpenCV finds your face? All you'll need is a webcam and Propeller connected to your PC running ViewPort. When you load the previous code (Tutorial #15 in ViewPort) to the Propeller, you'll see a color video from the camera in ViewPort's "video" view. Your head should be marked with a red frame, and the x and y variables on the Propeller should indicate the scaled position (see Fig. 7-12 for a screenshot). It should be straightforward to drive a hobby servo with the x signal so that the servo tracks your face. To find different objects, or to use a different video source, choose the appropriate controls in the "openCV" panel.

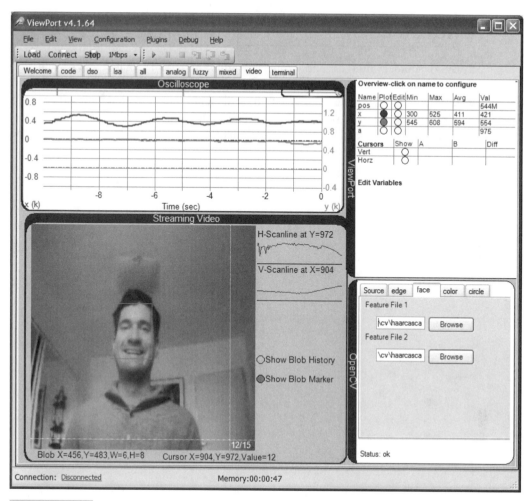

Figure 7-12   Author's head detected by OpenCV's face detector.

Summary

I've had a lot of fun building the vision-guided DanceBot with the Parallax Propeller and ViewPort. The Propeller's unique architecture of eight identical cogs made it easy to split the problem of guiding a balancing robot with vision into manageable pieces. Depending on the performance required, I could either write the code in the high-level, object-based Spin language or dive down to assembly to write completely deterministic code. The PropCV objects introduced in this chapter are a great start for adding simple computer vision capabilities to any Propeller robot. The OpenCV plug-in for ViewPort allows state-of-the-art computer vision applications to be developed with the Propeller.

Exercises

1 Build another filter for the PropCV engine, and verify that it does what you intended.

2 Build a robot that others can program by sketching the intended path with a laser. While it's being "programmed," your robot should track the location of the brightest spot. Your robot should then drive a path that resembles the one taken by the laser light.

3 Experiment with the OpenCV filters. Mount the webcam on a servo controlled by the Propeller. Have the camera follow your face, a red ball, or even a circle.

8

USING MULTICORE FOR NETWORKING APPLICATIONS

Shane Avery

Introduction

In this chapter we will show how to make use of the multiple cores in a Propeller to create networked applications. We will explore the reasons why having multiple cores provides the programmer with a new paradigm for creating applications. The platform for this will be the HYDRA hobby video game development system with the EtherX Ethernet add-in card running a simple game called "Button Masher." Since the focus of this chapter is networking, before we get to the "Button Masher" example, we will take a look at Ethernet and Internet protocols such as TCP/IP and UDP/IP. Here's what we will cover in this chapter:

- Ethernet and Internet protocols
- EtherX add-in card for the Propeller-powered HYDRA
- Creating a simple networked game

Resources: Demo code and other resources for this chapter are available for free download from ftp.propeller-chip.com/PCMProp/Chapter_08.

Ethernet and Internet Protocols

Before we begin talking about the EtherX card and how to use it, we need a basic understanding of computer networks. Computer networking is a complicated field of study in computer science which encompasses different protocols and topologies that take years of studying to grasp. However, we will narrow our focus to the Ethernet link layer and the common Internet protocols such as TCP/IP.

ETHERNET

Ethernet was developed by Xerox PARC in the 1970s as an implementation for a *local area network* (LAN) and was standardized by the *Institute of Electrical and Electronics Engineers* (IEEE). In theory, Ethernet can pass any kind of data the user would want. The Ethernet standard not only defines what needs to be passed from device to device to be valid data, but also defines the wiring and physical standards. It has become the de facto standard for local area networks, replacing older standards like Token Ring and *fiber distributed data interface* (FDDI).

A valid Ethernet frame contains between 64 and 1518 bytes. The first 14 bytes of the frame comprise the header, which contains a destination MAC, source MAC, and Ethertype. Then there is the payload data followed by a 4-byte checksum. A *MAC address* is a 6-byte unique address for every Ethernet device made. You may also hear this referred to as the *physical address*. The Ethertype indicates to the Ethernet device a little about what kind of data the payload is, because in theory, the payload could be anything. Internet protocols such as TCP/IP will have an Ethertype of 0×800. The last four bytes are used to compute the CRC32 checksum. This is just fancy talk for an algorithm that verifies that the data that was sent was sent correctly. Figure 8-1 shows an Ethernet frame.

INTERNET PROTOCOLS

For Internet protocols, the first piece of data in the payload is the Internet Protocol (IP) header. This IP header contains a lot of information about this particular packet of information, much of which we don't need to know for our Propeller programming. As it pertains to the EtherX card, the two things we care about most are the protocol fields and the IP address fields. The protocol fields indicate the type of data that appear after this IP header. In this chapter, the two types of protocols we will focus on are Transmission Control Protocol (TCP) and User Datagram Protocol (UDP). The other key data in the IP header that we care about are the IP addresses. Every computer

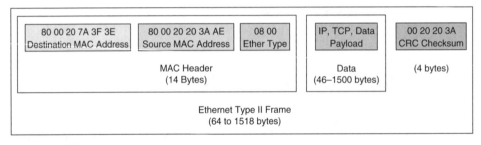

Figure 8-1   Ethernet frame.

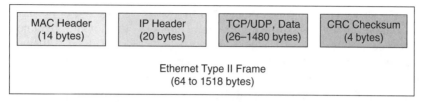

Figure 8-2   Ethernet frame with IP, TCP, or UDP header and data.

connected to the Internet has a unique IP address (this is not strictly true, like with a MAC address, but we will make this statement for now). The IP address is sort of like a phone number. Computers called *routers* have an idea (or know exactly) where an IP address is located on the Internet. In this way, these packets of data will hop along from router to router until they reach their destination address. So now we have an Ethernet MAC header followed by an IP header that looks like Fig. 8-2.

The IP header in all its glory is shown in Fig. 8-3.

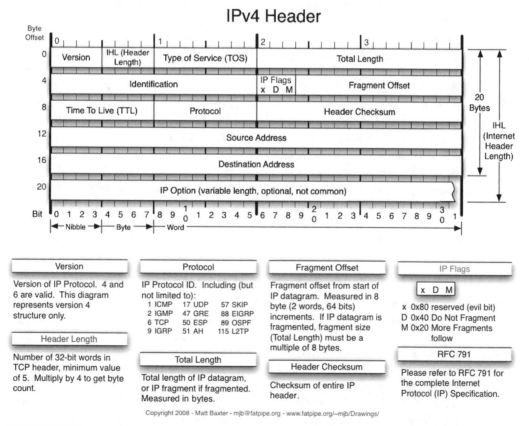

Figure 8-3   IP header field.

Now that we have IP addresses, we know where to go, but we need some more information about what to do with the data. Think of your computer right now. Lots of Internet data comes and goes. There's your web browser, mail client, instant messaging, BitTorrent, and so on. How does your computer know that a piece of data it received from the Internet came for your web browser and not your e-mail? The answer is a number called a port number. Your web browser works on a specific port, your e-mail on a different port, and so on. Another header is added to help tell Internet devices about the port number, and this layer is a TCP or UDP header. Once again, these headers contain much more data than just the port numbers, but for the purposes of working with the EtherX card, this is what we will focus on.

> **NOTE:** As a side note, the layer following the IP header doesn't need to be TCP or UDP. It could be something else, such as *Address Resolution Protocol* (ARP), *Domain Name System* (DNS), or PING.

What's the difference between TCP and UDP? Both have port information, but TCP guarantees that the data will get there, whereas UDP does not. That means if a packet of data is lost, TCP will continue to resend until the packet reaches its destination. Also, TCP provides flow control, meaning it will try to find the optimum speed to send the data (as fast as possible without losing too much data). UDP simply sends the data and hopes for the best.

Which should you choose? Typically, UDP is chosen when speed matters. It takes more time to verify that packets have arrived and to resend when needed, as is the case with TCP. Typically, in UDP, data is constantly sent, so if you lose a data packet, it's not that big a deal because another packet will come along soon. A game like Microsoft's *Halo* would be an example of a UDP game, as the game needs to constantly update player position, orientation, and action. If UDP packets begin to get lost, there is a momentary hiccup, but the game will happily continue on once it gets the next packet. TCP-based games are used when data loss matters more than time. A networked chess game is a good TCP example because you would only send the data when a player makes a move, and that data must arrive or else the opponent would never know it's his or her turn. A good rule of thumb for games is that real-time games use UDP and turn-based games use TCP.

UDP AND TCP: WHICH DOES THE INTERNET CHOOSE?

Any real-time streaming data will be UDP, and this includes things like streaming video and audio, which is why you notice hiccups sometimes when streaming data on the Internet. Those hiccups are the periodic packets being lost between the streaming server and your computer. The time it would take to straighten out what packets were lost would take too long and would leave you way behind in the stream. It's best to just ignore the hiccup and move on. File transfers, on the other hand, need to be perfect and thus use TCP, which is why when you look at a web page, download an MP3, or chat with friends, the data is always correct.

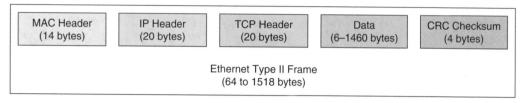

Figure 8-4 shows the Ethernet frame now with the TCP header and the data.
Figure 8-5 shows the TCP header fields.

Figure 8-6 shows the Ethernet frame now with the UDP header and the data.
Figure 8-7 shows the UDP header fields.

The best way to learn about the networking fields is to actually see them. WireShark
is a free program that not only acts as a packet sniffer that grabs every bit of Internet
traffic coming and going on your computer, but also breaks down each header for

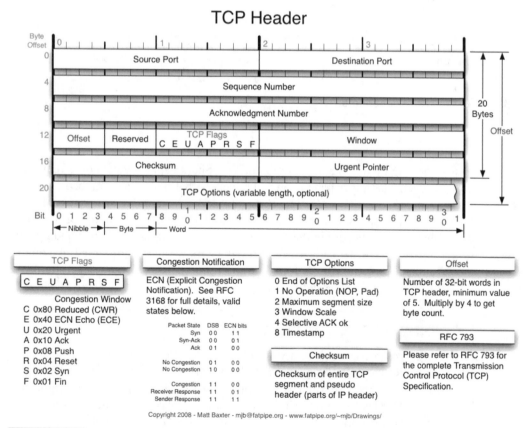

Figure 8-5   TCP header field.

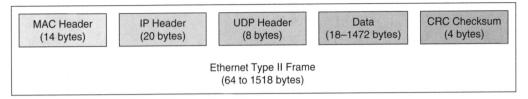

Figure 8-6 Ethernet frame with IP header, UDP header, and data.

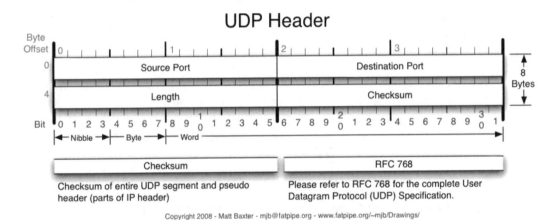

Figure 8-7 UDP header field.

inspection. Figure 8-8 is a screen grab of WireShark capturing data on a request to a web server. The upper pane shows the packets that have been grabbed (it shows these in real time as they come into the computer). The middle pane shows a header breakdown of the packet. The lower pane shows a complete hex dump of the entire data. Clicking on a header in the middle pane will highlight the corresponding data in the hex dump. In the screen capture, the TCP header is highlighted. A copy of the WireShark program can be found on the ftp.propeller-chip.com FTP site in the PCMProp/Chapter_08/Tools/ directory.

ORIGINS OF THE INTERNET

The origins of Internet are military in nature. The first Internet was developed by the Advanced Research Projects Agency (ARPA, which now is called DARPA—they added the word "defense") and was a simple packet-switching network that initially only linked up UCLA, Stanford, UCSB, and the University of Utah. It's hard to imagine that at one time the entire Internet consisted of four computers.

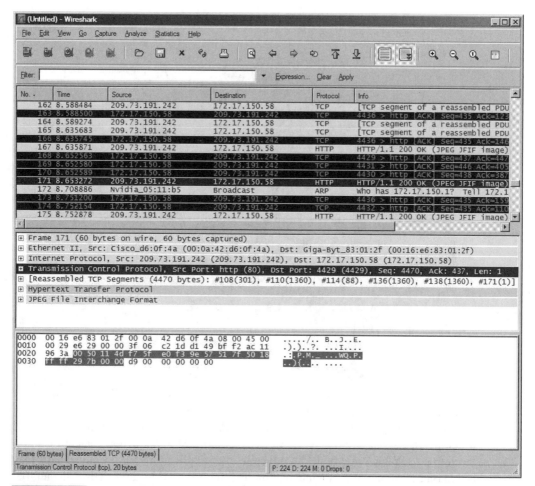

Figure 8-8   Screen capture of WireShark.

EtherX Add-in Card for the Propeller-Powered HYDRA

The platform that we will use to demonstrate the networking application for the Propeller is the HYDRA Game Development Kit with the EtherX card. The HYDRA Game Development Kit was built around the Propeller and was designed by André LaMothe at Nurve Networks, LLC. The HYDRA has many of the same features as the Propeller Demo Board, including PS/2 mouse, PS/2 keyboard, VGA, and NTSC/PAL video, just to name a few. Unique to the HYDRA are the NES connectors (for the gamepad), the HYDRA-Net (for networking two HYDRAs directly), and an expansion connector. If you'd like to learn more about the HYDRA, a fantastic book about it was written by

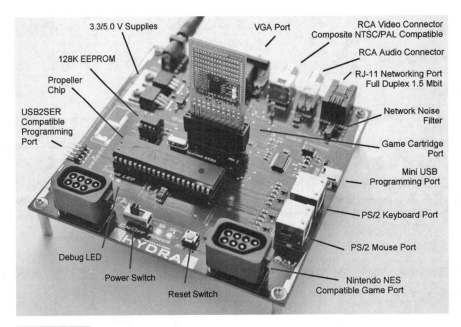

Figure 8-9   **HYDRA.** (*Courtesy of André LaMothe.*)

André LaMothe and is included with every HYDRA sold. Figure 8-9 shows a picture of the HYDRA Game Development Kit.

Since this chapter deals mainly with networking, we will focus on the EtherX card developed by Avery Digital. While we will try our best to show how the card is interfaced to the HYDRA and how to write networked applications for it, some details may be left out. The complete user manual for the EtherX card is located in the Chapter_08/ Docs/Documentation/ directory.

The EtherX card connects to the HYDRA via the expansion slot. The guts of the EtherX card is the W5100 Ethernet chip from Wiznet. This chip contains a hardwired TCP/IP stack, which offloads a lot of the Internet protocol from the HYDRA. For example, if the HYDRA wants to send a simple TCP/IP packet (assuming the connection between server and client has already been established), the HYDRA would need to send the packet and wait for an acknowledgement (ACK) from the receiver. If the ACK doesn't come in a timely manner, the HYDRA would need to resend. Furthermore, if the HYDRA were receiving multiple data packets, the data could come in out of order, which the HYDRA would need to stitch back together. There is also the implementation of the sliding window protocol for transmitting and retransmitting data to deal with. You get the picture. The complications just go on and on. The W5100 Ethernet chip handles all this for us. We simply initialize the device and then send or receive data. Figure 8-10 shows the EtherX card, and Fig. 8-11 shows the EtherX card inserted into a HYDRA.

Figure 8-10   EtherX.

Figure 8-11   EtherX inserted into the HYDRA.

ETHERX CARD ELECTRICAL INTERFACE

Before we dive head-first into the electrical interface of the EtherX card, we'll talk about the physical interface from the EtherX to the HYDRA. As previously stated, the EtherX card connects to the HYDRA via the expansion header, which not only provides Propeller I/O and power, but also USB signals, HYDRA-net signals, and EEPROM signals (so that an expansion card may be used for Propeller configuration). In our case, we only use the I/O and power. A table showing the EtherX interface to the HYDRA expansion slot is shown in Table 8-1.

The interface from the HYDRA to the W5100 is SPI, which stands for *Serial Peripheral Interface* (which was originally developed by Motorola). It's one of two popular modern serial standards, including I²C (which stands for *Inter-Integrated Circuit*) by Philips. SPI is a clocked synchronous serial protocol that supports full duplex communication. SPI needs three wires (data in, data out, clock), ground, and potentially chip select lines to enable the slave devices. The downside to SPI is that every slave device connected to the SPI bus needs its own chip select (also called slave select). Figure 8-12 shows a simple diagram between a master (left) and a slave (right) SPI device and the signals between them, which are:

- SCLK—Serial Clock (output from master)
- MOSI/SIMO—Master Output, Slave Input (output from master)
- MISO/SOMI—Master Input, Slave Output (output from slave)
- SS—Slave Select (active low; output from master)

SPI is a full duplex protocol, which means that as you clock data out of the master into the slave, data is clocked from the slave into the master. This is facilitated by a transmit-and-receive bit buffer that constantly recirculates, as shown in Fig. 8-13.

TABLE 8-1 ETHERX INTERFACE TO HYDRA EXPANSION SLOT

ETHERX CARD PIN	HYDRA PIN	FUNCTION
I/O_0 Pin 1	P16/Pin #21	OPMODE 2
I/O_1 Pin 2	P17/Pin #22	OPMODE 1
I/O_2 Pin 3	P18/Pin #23	OPMODE 0
I/O_3 Pin 4	P19/Pin #24	SCLK—SPI Clock
I/O_4 Pin 5	P20/Pin #25	SCS—SPI Chip Select
I/O_5 Pin 6	P21/Pin #26	MOSI—Master out/Slave in
I/O_6 Pin 7	P22/Pin #27	MISO—Master in/Slave out
I/O_7 Pin 8	P23/Pin #28	RST—W5100 Reset
3.3V Pin 14	POWER	3.3 V—Power
GND Pin 20	POWER	Ground

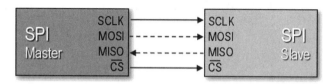

Figure 8-12   SPI electrical interface. (*Courtesy of André LaMothe.*)

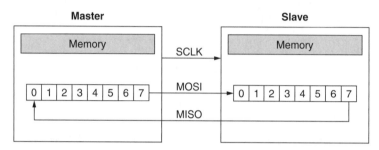

Figure 8-13   Circular SPI buffer. (*Courtesy of André LaMothe.*)

The use of the circular buffers means that you can send and receive a byte in only eight clocks rather than clocking out eight bits to send and then clocking in eight bits to receive. Of course, in some cases, the data clocked out or in is "dummy" data, meaning when you write data and are not expecting a result: The data you clock in is garbage, and you can throw it away. Likewise, when you do an SPI read, typically, you would put a $00 or $FF in the transmit buffer as dummy data, since something has to be sent, so it might as well be predictable.

Sending bytes with SPI is similar to the serial RS-232 protocol: You place a bit of information on the transmit line and then strobe the clock line (of course, RS-232 has no clock). As you do this, you also need to read the receive line since data is being transmitted in both directions. This is simple enough, but the SPI protocol has some specific details attached to it regarding when signals should be read and written—that is, on the rising or falling edge of the clock, as well as the polarity of the clock signal. This way, there is no confusion about edge, level, or phase of the signals. These various modes of operation are:

Mode 0—The clock is active when HIGH. Data is read on the rising edge of the clock and written on the falling edge of the clock (default mode for most SPI applications). The clock phase polarity here is zero and Fig. 8-14 shows a timing diagram for this.

Mode 1—The clock is active when HIGH. Data is read on the falling edge of the clock and written on the rising edge of the clock. The clock phase polarity here is one and Fig. 8-15 shows a timing diagram for this.

Mode 2—The clock is active when LOW. Data is read on the rising edge of the clock and written on the falling edge of the clock. The clock phase polarity here is zero and Fig. 8-14 shows a timing diagram for this.

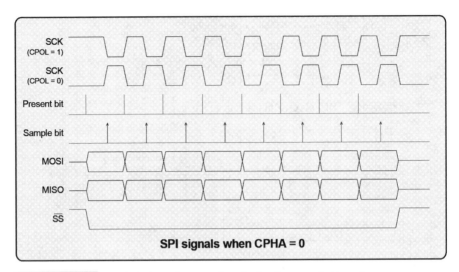

Figure 8-14   SPI timing diagram for clock phase polarity = 0. *(Courtesy of André LaMothe.)*

Mode 3—The clock is active when LOW. Data is read on the falling edge of the clock and written on the rising edge of the clock. The clock phase polarity here is one and Fig. 8-15 shows a timing diagram for this.

Notice that there is no timing-related logic in the method. Since the SPI protocol is totally synchronous, the master in charge of the clock can run it as fast (to maximum

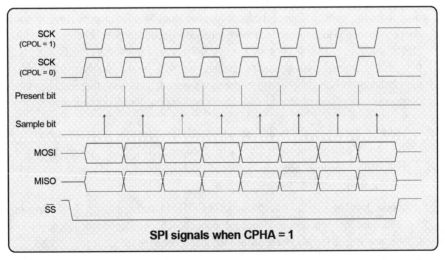

Figure 8-15   SPI timing diagram for clock phase polarity = 1. *(Courtesy of André LaMothe.)*

speed) or as slow (static, if desired) as he wants. Also, notice the data is sent out most significant bit (msb) to least significant bit (lsb).

For the EtherX card, we will operate in SPI Mode 0. This is the only mode that the W5100 operates in. The HYDRA has no built-in support for SPI, so we must bit-bang the protocol ourselves. The bit-banging of the SPI protocol lives as assembly code that fits completely in a cog.

Communication with the W5100 is exclusively on a register addressed basis. This means that any read or write to the W5100 will be to a register that has an address. All registers will have a 16-bit address and 8-bit data. Every read or write to the W5100 will be four bytes long. The first byte is the command (read or write), the next two bytes are the address, and the last byte is the data. Writing a byte from the HYDRA to the W5100 on the MOSI line will only be valid for a write. During a read, the final byte that you need to read will be on the MISO line. The SPI format is shown in Fig. 8-16.

W5100: BUS INTERFACE?

The W5100 actually has two interfaces: an SPI and a bus interface. The bus interface is actually far more efficient in transferring data to or from the W5100. This bus interface requires 15 address pins, 8 data pins, and 4 control pins. The HYDRA was unable to accommodate this because there weren't enough I/O pins mapped to the expansion connector. Keep this in mind if you are designing the W5100 into a new project. Also, the next-generation chip called the W5300, which promises data rates of more than 70 Mbps, contains only a bus interface (no SPI).

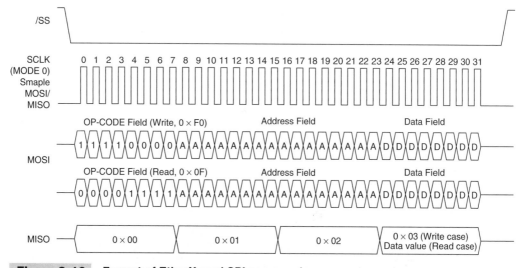

Figure 8-16   Format of EtherX card SPI commands. (*Courtesy of WIZnet Co., Ltd.*)

ETHERX CARD APPLICATION PROGRAMMING INTERFACE

Now that we understand electrically how the EtherX communicates with the HYDRA, we should mention the driver that provides a clean interface for the code written on the Propeller. The driver can be found in every example in the Chapter_08/Source/ directory. It is also located on the Avery Digital website at www.averydigital.com/products.html. Figure 8-17 shows the overall driver architecture and its relationship to the application and hardware.

As you can see from the figure, between the application and the EtherX card there are three levels. The bottommost layer is the hardware interface layer, which is the physical interface that has been discussed already in the electrical interface section. The next layer is the SPI assembly code layer, which resides in a separate cog within the Propeller. This layer handles the bit-banging of the SPI protocol, as shown in Fig. 8-16.

The next layer (read and write) provides another layer of abstraction for the driver by providing two Spin methods that will read and write data given a register address. The application layer has direct access to these two methods, which would allow the user application complete access to all registers within the W5100.

The final layer of abstraction contains Spin methods that perform the most common operations used in applications. These methods add another level of abstraction to open sockets, read and write data, and initialize the W5100. In truth, they are unnecessary because all they do is call the read method and write method with appropriate address and data which the application could do for itself since it has access to the read method and write method. They exist to make application code easier to read and write.

The user application needs to call a method to initialize the W5100 by providing a MAC, IP, subnet, and gateway address. After that, the W5100 programming model

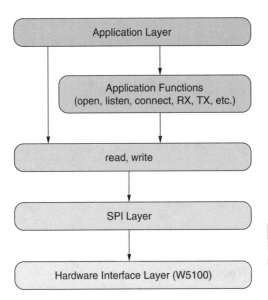

Figure 8-17 Simplified illustration of the EtherX software model.

TABLE 8-2 API LISTING
START AND STOP FUNCTIONS
start—Starts the SPI layer assembly code in a new cog.
stop—Stops and frees the cog containing the SPI layer assembly code.
READ AND WRITE FUNCTIONS
read (a0, a1)—Reads a byte from address a0, a1 from the W5100 via SPI.
write (a0, a1, dout)—Writes byte dout at address a0, a1 to the W5100 via SPI.
APPLICATION FUNCTIONS
init (gway, subnet, ip, mac)—Initializes the W5100's gateway, subnet, IP, and MAC.
open (type, source_port, dest_port, dest_ip)—Opens the socket.
close—Closes the socket.
listen—Used when TCP server should listen for connection from client.
connect—Used when TCP client should establish a connection with server.
con_est—Used to determine if a TCP connection has been established.
TX (dataptr, size)—Transmits size bytes from main memory pointed to by dataptr.
RX (dataptr, size, block)—Receives size bytes to main memory pointed to by dataptr.
read_rsr—Returns the number of bytes the W5100 has stored in its internal receive buffer.
MODE FUNCTION
mode(opmode)—Changes the operating mode of the W5100.

closely resembles socket programming. A socket is opened on a particular port. When opening the socket, the user determines if it should be UDP or TCP. If TCP, there is a listen call (if we intend to be a server) and a connect call (if we intend to be a client). UDP is connectionless, so this is not needed. Data is then transmitted or received by TX or RX method calls. Table 8-2 shows a listing of the API methods.

SPI Assembly Layer   The lowest level of the driver is the SPI layer. We'll discuss in detail here how it works, but all details aside, all it really does is send and receive bytes using SPI. The SPI layer is written completely in assembly code and is designed to live wholly in a cog with its 396 bytes.

The first thing that the SPI layer does is set the correct pins to input or output as needed to enable communication between the HYDRA and the W5100 via SPI. Once that is done, we take the W5100 out of reset. Since there is no power-up reset on the W5100, the HYDRA needs to reset the W5100 sometime after power-up. We take care of this by pulling down the reset pin of the W5100 on the EtherX card, which will hold the W5100 in reset mode when the device is turned on. It is up to the HYDRA to take the W5100 out of reset by setting the reset line to the W5100 to logic high. Thus, the W5100 is held in reset until the SPI assembly code portion of the driver is started in a new cog. After that, the SPI layer looks at global variables to determine what to do next.

Five global variables are used to interface the Spin methods of the driver to the SPI layer. Since the variables are global, they reside in main memory and are accessible by all cogs, including the one where our SPI layer is located. The first global variable is the SPIRW variable. The main loop of the SPI layer will constantly look at this variable to determine if it should take any action. As long as the variable is equal to zero, it will do nothing. When the variable is equal to one, the SPI layer will read a byte from the W5100 on the EtherX card. This will cause the SPI layer to send 0x0F as a first byte to the W5100 to indicate a read operation (see Fig. 8-16). When the variable is equal to two, the SPI layer will write a byte to the W5100 on the EtherX card. This will cause the SPI layer to send a 0xF0 as a first byte to the W5100 to indicate a write operation (see Fig. 8-16). After reading or writing a byte, the SPI layer will clear this variable back to zero.

If you look at the assembly code for the SPI layer, the first part simply loops, waiting for the SPIRW variable to not be zero. Once the variable is not zero, the SPI layer determines, based on the SPIRW variable, if the first nibble should be 0x0 (for a read) or 0xF (for a write). If the value should be 0x0, it sets the MOSI line logic low and pulses the SCLK line four times. If the value should be 0xF, it sets the MOSI line logic high and pulses the SCLK line four times. The assembly code then does the same thing for the second nibble (which should be 0xF for a read and 0x0 for a write). This would complete the first byte sent to the W5100 to indicate whether we will read or write a byte from the W5100. Note that most SPI interfaces allow for full duplex (read and write at the same time), but the W5100 command structure only allows one read or one write per SPI transaction, thus making it half-duplex.

The next two variables, called add0 and add1, are the address that the SPI layer will read from or write to. The interface of the W5100 is an addressable register interface. This means that every read or write will be to a register in the W5100 and every register has an address. As you can see in Fig. 8-16, two bytes are needed to address the register you want to read or write. Whether the operation is a read or write, the address portion of the transaction is exactly the same. Thus, after sending the first byte to indicate whether the operation is a read or a write, the SPI layer will read the two address variables and send them to the W5100, regardless of whether we are reading or writing.

The assembly code to accomplish sending the address to the W5100 acts as a shift register. The assembly looks at the most significant bit of the address variable and sets the MOSI line logic high if the most significant bit of the address variable is high. Similarly, if the most significant bit of address variable is low, the MOSI is cleared to low. We then pulse the SCLK line to clock in that bit. The SPI layer will shift the address variable left one bit and repeat the process until the entire address variable is clocked into the W5100. The assembly code will do this twice, as there are two address variables.

Note: Using the term "most significant bit" here is a bit of a misnomer. Technically, the global variables are 32 bits. This means that the most significant bit is the 32nd bit of the variable. When we say "most significant bit" here, we mean the most significant bit that we care about, which is the eighth bit because we send data to the W5100 one byte at a time.

After the address is sent to the W5100, the SPI layer will shift out (just like the address variables) the value in the next global variable, called dataout, at the same time it shifts in the data from the MISO line to the global variable called datain. It does this whether the operation is a read or a write. In the case of a write, datain will contain 0x03 as per the W5100 documentation. In the case of a read, the MOSI line will output whatever data is located in the dataout, which is fine since the W5100 will ignore whatever this value is during a read operation. In other words, even when we want to read data from the W5100, we will output a byte on the MOSI line as though we are writing. The W5100 knows that we are performing a read operation, so it will ignore it. Keep in mind that whenever you perform a write operation to the W5100, the previous value of the global variable datain will be wiped out and now contains 0x03.

The last thing the SPI layer will do is set the SPIRW variable back to zero. The SPI layer was never intended to be directly accessible to the application layer. To change this, you just need to make the global variables (SPIRW, add0, add1, datain, and dataout) visible to both the application and the SPI layer. However, in the current version of the driver, two small wrapper Spin methods called read and write are used to interface to the SPI layer; we will discuss these next.

Information: The assembly source code is omitted here in the text since assembly is a little obfuscated by its very nature. Nonetheless, if you truly want to understand how the driver implements SPI specifically for the W5100, check out the source code which is included every example in the Chapter_08/Source/ directory. Use the force, read the source!

Read and Write Layers   The two main methods for interfacing to the SPI layer are the read and write methods. These are two simple wrapper Spin methods that are easy to use and understand. The read method in its entirety is shown here:

```
PUB read(a0, a1) : ret_val

" Call this method to read a byte from the W5100 via SPI.
" The arguments are two bytes that contains address byte 0
" and address byte 1. See W5100 data sheet for what registers
" these actually address. The method will return the byte
" that came via SPI from the W5100.

'Read data from the address specified in the arguments
add0 := a0                      'Set the arguments to the global
add1 := a1                      'variables values
SPIRW := SPI_RD                 'Set SPIRW to the read value
```

```
'Wait until the driver clears SPIRW. This indicates that it is done.
repeat until SPIRW == SPI_DONE

'The driver wrote the result to datain. Thus, that is the value
'that we will return.
ret_val := datain
```

You can see that the `read` method has two input parameters: a0 and a1. These represent the address you want to read from. The `read` method will then set these address parameters to the global variables add0 and add1. It will then set the SPIRW global variable to 1, which is equal to the constant SPI_RD (set in the CON section of the driver).

The `read` method then waits until the SPIRW global variable is set to 0, which is equal to the constant variable SPI_DONE (set in the CON section of the driver). Remember that the SPI layer will clear the SPIRW global variable to 0 when it is done with the SPI transaction. Thus, this method will block (which means it will wait, doing nothing) until the SPI layer completes the transaction. It then returns the value that is stored in the datain global variable, which is where the SPI layer writes the result of the transaction.

The `write` method is just as simple as the read method, as shown here:

```
PUB write(a0, a1, dout)

  " Call this method to write a byte from the W5100 via SPI.
  " The arguments are three bytes that contains address byte 0,
  " address byte 1 and the data we wish to write. See W5100
  " data sheet for what registers these actually address.
  " Note that the datain global variable will be overwritten
  " during this process.

  'Write data to the address specified in the arguments
  add0 := a0                    'Set the arguments to the global
  add1 := a1                    'variables values
  dataout := dout
  SPIRW := SPI_WR               'Set SPIRW to the write value

  'Wait until the driver clears SPIRW. This indicates that it is done.
  repeat until SPIRW == SPI_DONE
```

You can see that the `write` method has three input parameters: a0, a1, and dout. These represent the address you want to write to and the data to be written. The `write` method will set the address parameters to the global variables add0 and add1. It will then set the dout parameter to the dataout global variable and will set the SPIRW global variable to 2, which is equal to the constant variable SPI_WR (set in the CON section of the driver). The `write` method then waits until the SPIRW global variable is set to 0, which is equal to the constant variable SPI_DONE (set in the CON section of the driver). Remember that the SPI layer will clear the SPIRW global variable to 0 when it is done with the SPI transaction. Thus, this method will block (which means it will wait, doing nothing) until the SPI layer completes the transaction.

That's it! These methods are meant to be used by the application layer to read and write bytes to and from the W5100. If you wanted to, you could interface to the HYDRA EtherX card entirely just using these two methods. In fact, if you wanted to configure or use the W5100 beyond what the HYDRA EtherX driver provides in the application methods, you would need to use the read and write methods. In actuality, all the other methods that are provided for the application layer in the EtherX driver use these two methods. We will now explore these other application methods.

Application Methods Layer   While you could simply use read and write to use the EtherX, there are still a lot of sticky details about what data you need to send and to what addresses to actually make the EtherX card useful. Application methods exist to provide a means for you to use the EtherX card without actually needing to know the gritty details of how the W5100 works. This is great for getting an application running and learning about Ethernet and in the Internet in general, but I encourage you to look at the methods and the W5100 documentation itself to understand what is going on.

Mode Method   One of the only application methods that does not use the read and write methods is the mode method. This allows you to change the mode of the W5100 to change the speed and half/full duplex of the W5100. By default, the Ethernet interface of the EtherX operates at 10 Mbps and at full duplex. Pull-up and pull-down resistors on the EtherX card set the mode by default, so if you never call the mode method, the card will power up and come out of reset as 10 Mbps, full duplex. However, you can change this if you are so willing. Be careful, though, as the EtherX card can become unstable at 100 Mbps. It is no big deal that the default of the EtherX operates at 10 Mbps, as the SPI interface to the W5100 is so much slower than 10 Mbps that you would never notice the difference—in addition, all Ethernet devices are backward-compatible to 10 Mbps. But if you wanted to operate the device at 100 Mbps, this is the method you would call to do so.

If you choose to call the mode method, it must be called before you start the EtherX driver. This is because one of the first things the EtherX driver does is take the W5100 out of reset; once the W5100 is out of reset, you shouldn't change the mode of the device. Thus, the mode method first checks to see if the EtherX driver has already been started and if so, it will do nothing. The method is shown here:

```
PUB mode(opmode) | temp

  " Call this method with the value wanted for opmode before
  " the start method is called. If the start method has
  " already been called then this method will do nothing.

  if (cogon == 0)              'Be sure driver not already running
    DIRA |= $70000             'Set pins 16-18 to output
    temp := opmode << 16       'Shift our arg 16 bits left
    OUTA &= $FFF8FFFF          'Clear all outputs on pins 16-18
    OUTA |= opmode             'Set only required pins on pins 16-18
```

Functionally, all the mode method does is set the pins that map to the W5100 mode pins to output and set them according to the opmode input parameter. To make the W5100 100 Mbps and full duplex, call the mode method with opmode equal to 0.

Start and Stop The only other two methods that don't use the read and write methods are the start and stop methods. You should recognize these, as they are used in practically all drivers for the HYDRA.

We'll discuss the stop method first, as it is the most simple. It simply stops the SPI layer and frees the cog it runs on. There, done!

The start method starts the SPI layer (written in assembly), which handles the actual bit-banging of the SPI data to the W5100 in a new cog and returns the new cog number that the SPI layer lives in. If you choose to call the mode method, it should be called before the start method. All other Spin method calls (including read and write) should be called after the start method has been called.

The code for start and stop is shown here:

```
PUB start : okay

  " This is the public start method. It starts
  " a new cog at the assembly entry point

  'Start the SPI code in a new COG
  stop
  okay := cogon := (cog := cognew(@entry, @SPIRW)) > 0

PUB stop

" Stops driver—frees a cog

  if cogon~
      cogstop(cog)
```

Init The init method should be called right after the start method. It initializes the W5100 by telling it what our gateway, subnet, IP address, and MAC address are. The init method is shown here:

```
PUB init (gway, subnet, ip, mac)

  'Init the registers
  'Gateway
  write($00,$01,byte[gway])
  write($00,$02,byte[gway+1])
  write($00,$03,byte[gway+2])
  write($00,$04,byte[gway+3])
```

```
'Subnet
write($00,$05,byte[subnet])
write($00,$06,byte[subnet+1])
write($00,$07,byte[subnet+2])
write($00,$08,byte[subnet+3])

'MAC
write($00,$09,byte[mac])
write($00,$0a,byte[mac+1])
write($00,$0b,byte[mac+2])
write($00,$0c,byte[mac+3])
write($00,$0d,byte[mac+4])
write($00,$0e,byte[mac+5])

'IP
write($00,$0f,byte[ip])
write($00,$10,byte[ip+1])
write($00,$11,byte[ip+2])
write($00,$12,byte[ip+3])
```

You can see that the method takes the four arguments to the `init` method, which are pointers to arrays, and writes their values to the corresponding register addresses of the W5100. Keep in mind that the arguments are array pointers and you would fill in the values on the arrays in your application before calling this method. Also, you need to make sure that the arrays are `byte` arrays, not the default 32-bit arrays. An example of calling this method from your application might look like the following from the SimpleTCPClient program that is included in the Chapter_08/Source/SimpleTCPClient/Spin/ directory.

```
VAR
        byte   dip[4]
        byte   subnet[4]
        byte   ip[4]
        byte   gateway[4]
        byte   mac[8]
        byte   buffer[256]

OBJ

        term   : "tv_terminal_010.spin"    ' Instantiate the terminal object
        eth    : "W5100_drv_011.spin"      ' Instantiate the W5100 driver

PUB begin | size

  'Start the tv terminal.
  term.start
```

```
'Start the W5100 driver.
eth.start

'Display a title string and indicate we are
'waiting for the ethernet to come up.
term.out($02)
term.pstring(string("HydraEtherX TCP Client Test",13,13))
term.out($01)
term.pstring(string("Waiting for Eth",13))

'Wait for a while for the ethernet to come up.
waitcnt(cnt + $17d78400)

'Indicate that we are done waiting.
term.pstring(string("Done waiting",13,13))

'Fill out the arrays needed for initialization.

'Destination IP is IP we are sending to
dip[0] := 192
dip[1] := 168
dip[2] := 1
dip[3] := 2

'Fill out the rest of the arrays needed for initialization.
'W5100 IP address.
ip[0] := 192
ip[1] := 168
ip[2] := 1
ip[3] := 100

'Gateway.
gateway[0] := 192
gateway[1] := 168
gateway[2] := 1
gateway[3] := 1

'Subnet.
subnet[0] := 255
subnet[1] := 255
subnet[2] := 255
subnet[3] := 0

'W5100 MAC address.
mac[0] := $00
mac[1] := $70
mac[2] := $6e
mac[3] := $69
```

```
mac[4] := $73
mac[5] := $0a

'Initialize the W5100.
eth.init(@gateway, @subnet, @ip, @mac)
```

Notice that the `gateway`, `subnet`, `ip`, and `mac` arrays are byte arrays; to pass the address of the pointer to the `init` method, you use the @ symbol. This example would set your IP to address 192.168.1.100, your gateway to 192.168.1.1, and your subnet to 255.255.255.0, and your MAC address would be 00.70.6e.69.73.0a. Technically, you should apply to IEEE for your own MAC address, especially if you intend to produce a commercial product based on the W5100. But in reality, for educational purposes, all you need to do is be sure that the MAC address isn't the same as any other Ethernet device in your local network.

Again, as you see here, you should call the `init` method after the `start` method and before you call the `open` method, which we will discuss next.

Open and Close   Socket programming has been around for a long time as an attempt to standardize network interfaces. The W5100 loosely models the socket programming scheme. It supports up to four independent sockets at a time. The EtherX driver only supports one socket at a time, but there is nothing stopping you from modifying the driver to support all four sockets. Sockets need to be "opened" in the W5100, and that is why we have the `open` method call. To open a socket on the W5100, we need to tell it whether the socket is a UDP socket or a TCP socket, what the source and destination ports are, and the destination IP address. Actually, if you set up the W5100 to be a TCP server, the destination IP address is not known at the time when you open the socket—that value can be anything, as it will be ignored. The `open` method is shown here:

```
PUB open (type, source_port, dest_port, dest_ip) | temp, temp2

  " This method will open either a TCP or UDP socket based on
  " the type argument. The source_port and dest_port are the
  " source and destination port numbers. The last argument
  " is meant to be a four byte array that stores the destination
  " IP. We could have just made this a long arg but this
  " format is more human friendly.

  'Set the socket type so that other method know if we are
  'TCP or UDP
  socket_type := type

  'Configure socket for UDP for TCP (TCP is the default)
  if(socket_type == UDP)
    write($04,$00,2)
  else
    write($04,$00,1)
```

```
'Configure source port
temp := (source_port >> 8) & $FF
temp2 := source_port & $FF
write($04,$04,temp)
write($04,$05,temp2)

'Configure dest port
temp := (dest_port >> 8) & $FF
temp2 := dest_port & $FF
write($04,$10,temp)
write($04,$11,temp2)

'Dest IP
write($04,$0c,byte[dest_ip])
write($04,$0d,byte[dest_ip+1])
write($04,$0e,byte[dest_ip+2])
write($04,$0f,byte[dest_ip+3])

'Open socket
write($04,$01,1)
```

You can see that the method has four parameters passed into it. The first is the type, which is either 0 or 1. A 0 means UDP and a 1 means TCP. The CON section defines UDP to be 0 and TCP to be 1 for easier readability. The next two arguments are the source and destination ports. The final argument is a pointer to a byte array of the destination IP. This value doesn't make sense if we intend to be a TCP server, as we cannot possibly know the IP address of the client that may attempt to connect to us. Thus, if we intend to be a TCP server, this value can be anything because it will be ignored. Notice that the open method will set the socket type, source and destination ports, and then the destination IP address before actually opening the socket. If you choose to not use the open method and instead open a port yourself using the write method, be sure not to open the socket until these are set. Do not attempt to change the socket type, ports, or destination IP after the socket has been opened. If you need to change one of these parameters, "close" the port first.

The close method will close a socket. After the close method has been called, the socket will no longer send and receive data on that socket. Once closed, however, you can change the socket type, source and destination ports, and destination IP address. Then if you wanted, you could reopen the socket and those new parameters will take effect. The following is the code for the close method:

```
PUB close

  " Close the socket.

  write($04,$01,$10)
```

Listen, Connect, and Connection Established   If you choose open the socket as a TCP socket, you need to determine if you will be a TCP server or a TCP client. TCP will establish a connection before data is exchanged. Two methods will work to establish a connection—which one you use will be based on whether you want the W5100 to be a server or a client. If you choose to implement your socket as a UDP socket, there is no need to use the listen, connect, or con_est method calls because UDP doesn't establish a connection before sending and/or receiving data. Following is the code for the listen, connect, and con_est methods:

```
PUB listen

  '' This method will initiate a TCP listen. Call this method
  '' for a TCP server to listen for a connection.

  if(socket_type == TCP)
    write($04,$01,2)

PUB connect

  '' When TCP this method will establish a connection with a server.

  if(socket_type == TCP)
    write($04,$01,4)

PUB con_est : ret_val

  '' Call this method to determine if a TCP connection has been
  '' established. The method will return true if a connection
  '' has been established.

  if(read($04,$03) == $17)
    ret_val := true
  else
    ret_val := false
```

If the EtherX card will be a TCP server, it will call the listen method after it calls the open method. This method will tell the W5100 that it will be a server and that it should wait for a client to try and establish a connection with it. After you call the listen method, you will need to call the con_est method, which returns true if a connection has been established with a client and false if a connection has not been established. Following is a snippet of code from the SimpleTCPServer code that shows how to use listen and con_est. It can be found in the Chapter_08/Source/SimpleTCPServer/Spin/ directory.

```
'Initialize the W5100.
eth.init(@gateway, @subnet, @ip, @mac)
```

```
'Open the socket for TCP communication.
eth.open(TCP, PORT, PORT, @dip)

'Listen for a connection.
eth.listen

'Display message.
term.pstring(string("Listening...",13))

'Wait for connection to be established.
repeat while eth.con_est == false

'Display connected message.
term.pstring(string("Connected to PC",13,13))
```

You can see the order of the methods that need to be called to set up the W5100 for data exchange as a TCP server in the previous snippet. Call the init, open, and listen methods in that order and then call con_est as/when needed to determine when and if a connection has been made. If the con_est method returns true, a connection has been established with a client and you can then call the RX and TX methods to exchange data with the client. Do not attempt to receive or transmit data before a connection has been established.

If the EtherX card will be a TCP client, it will call the connect method after it calls the open method. This method will tell the W5100 that it will be a client and that it should attempt to establish a TCP connection with a TCP server location at the IP address specified in the destination IP address field that we specified in the open method call. Once the connect method has been called, you need to call the con_est method to determine if a connection has been made. Do not attempt to receive or transmit data before a connection has been established. The con_est method will return true when the connection has been established and will return false if a connection has not been established. Following is a snippet of code from SimpleTCPClient that shows how to use connect and con_est. It can be found in the Chapter_08/Source/SimpleTCPClient/Spin/ directory.

```
'Initialize the W5100.
eth.init(@gateway, @subnet, @ip, @mac)

'Open the socket for TCP communication.
eth.open(TCP, PORT, PORT, @dip)

'Display message.
term.pstring(string("Connecting...",13))

'Connect to the PC.
eth.connect
```

```
'Wait for connection to be established.
repeat while eth.con_est == false

'Display connected message.
term.pstring(string("Connected to PC",13,13))
```

You can see that the processes for setting up the W5100 as a TCP server and as a TCP client differ by only one line. You use listen for a TCP server and connect for a TCP client. Other than that, the process is the same. Whether you choose to implement your socket as a UDP, TCP client, or TCP server, the next thing you need to do is send and receive data. This is accomplished using the RX and TX methods, which we will discuss next.

Transmit   Once the socket has been opened (and a connection established if you are a TCP socket type), you can send and receive data. The TX method is called to transmit data; its code is shown here:

```
PUB TX (dataptr, size) | tptr, offset, startadd, a0, a1, counter, temp

  '' Call this method to send data via the W5100.

  'Read the offset so we know what the starting address is
  'of the TX buffer.
  tptr := read($04,$24)          'Read the transmit pointer
  tptr := tptr << 8
  tptr += read($04,$25)
  offset := tptr & $7FF          '2K of buffer determines mask of $7FF
  startadd := $4000 + offset     'Add offset to the starting address

  'Write the data to the W5100 internal memory buffer
  repeat counter from 0 to size-1.
    a1 := startadd & $FF
    a0 := (startadd & $FF00) >> 8
    write(a0, a1, byte[dataptr])
    offset++
    offset := offset & $7FF      '2K of buffer determines mask of $7FF
    startadd := $4000 + offset   'Add the offset to the starting address
    dataptr++

  'Update the offset counter and write it back to the W5100.
  tptr += counter
  temp := (tptr & $FF00) >> 8
  write($04,$24,temp)
  temp := tptr & $FF
  write($04,$25,temp)

  'Tell the W5100 to write the data.
  write($04,$01,$20)
```

You can see that the method takes two arguments. The first is a pointer to a byte array that contains the data you want to send. The second argument specifies how many bytes you actually want to send. The driver writes the number of bytes you specify in the size argument from the address in main memory, starting at the address pointed to by the dataptr argument to the internal buffer of the W5100. Once the bytes have been written to the internal buffer of the W5100, the EtherX driver instructs the W5100 to transmit that data. Keep in mind that the default size for the internal buffer for socket 0 (the socket that the EtherX driver operates on) is 2048 bytes. This means that you can't write more than 2048 bytes' worth of data at one time using the TX method.

GOT A PROBLEM WITH BEING LIMITED TO 2048 BYTES?

No problem. Check out the W5100 data sheet (it's included in the Chapter_08/Docs/DataSheets/ directory) and read about how to configure the W5100. Specifically, you want to modify the size of the socket buffer. Learn what register's address you need to write to and what data you need to write to get the buffer size you want; then write that value to the address you need using the write method. That's what it's there for!

Another "gotcha"—besides the fact that, by default, you can't transmit more than 2048 bytes at a time—is that you need to be sure you don't specify the size argument to be any larger than the buffer that is pointed to by the dataptr argument. There is no automatic checking for an array that is out of bounds.

Receive The TX method is nice in that it operates the same no matter what your socket type is—TCP or UDP. It is used the same way no matter what. However, the RX method is called to receive data from the W5100, and it behaves differently, depending on your socket type (UDP or TCP). The RX method is shown here:

Note: Lines marked with the "Ð" symbol should appear on the same line.

```
PUB RX (dataptr, size, block) : ret_size | offset, startadd, a0,Ð
a1, counter, rdptr, temp, tempsize Ð

  " Call this method to receive data via the W5100.
  " The method will return the number of bytes read.
  " Set block == true if you want the method to block waiting for data.
  ..
  " This method will return the actual number of bytes read.

  'Block waiting for the W5100 to tell us that there is data.
  'Technically we are reading the number of bytes that have
  'been received.
```

```
if(block == true)
  repeat
    temp := read($04,$26)
    temp := temp << 8
    temp += read($04,$27)
  while temp == 0            'Wait as long as we have received zero bytes

'If the user wants a non-block RX call then we will return immediately
'with a size of zero if there is no data to receive.
else
  temp := read($04,$26)
  temp := temp << 8
  temp += read($04,$27)
  if(temp == 0)
    return 0


'Compute the starting address.
rdptr := read($04,$28)          'Read the receive pointer
rdptr := rdptr << 8
rdptr += read($04,$29)
offset := rdptr & $7FF          'Socket 0 has 2K of buffer and that
                                ' determines mask here of $7FF
startadd := $6000 + offset      'Add the offset to the starting address
                                ' of socket 0


'Determine how many bytes we need to read.
tempsize := read_rsr
if(tempsize > size)
  ret_size := size
else
  ret_size := tempsize


'Now we read the data from the W5100 and write it to the array
'pointed to by dataptr.
repeat counter from 0 to ret_size-1
  a1 := startadd & $FF
  a0 := (startadd & $FF00) >> 8
  byte[dataptr] := read(a0,a1)
  dataptr++
  offset++
  offset := offset & $7FF       'Socket 0 has 2K of buffer and that
                                ' determines mask here of $7FF
  startadd := $6000 + offset    'Add the offset to the starting address
                                ' of socket 0


'Need to increment the rdptr by the number of bytes actually read.
rdptr += ret_size
```

```
'Then write the value of the new pointer back.
temp := (rdptr & $FF00) >> 8
write($04,$28,temp)
temp := rdptr & $FF
write($04,$29,temp)

'Tell the W5100 that we have read the data.
write($04,$01,$40)

'Issue the read command to the W5100 which just updates registers.
'We will block waiting for the W5100 to finish.
'This takes very little time but we will check it just to be sure.
repeat
  temp := read($04,$01)
while temp <> $00
```

As you can see, this is the largest and most complex method in the EtherX driver. You can see why we saved it for last. Still, when you boil it down, all it really does is read some data from an internal buffer in the W5100 and write that data to main memory in the HYDRA. This method takes in three arguments. The first, dataptr, is the memory address of the byte array that the EtherX driver will write the data to that it reads from the W5100.

The next argument, size, stipulates the maximum number of bytes that the RX method should write to the memory pointed to by dataptr. For example, the receive buffer that you have allocated in the HYDRA may only be 32 bytes in size and the W5100 may have received 40 bytes from the Internet. In this case, you want to be sure that the EtherX driver doesn't write 40 bytes' worth of data into a buffer that you have only allocated 32 bytes for. The next argument, block, we will skip for a second. The method will return the actual number of bytes written to the buffer pointed to by the dataptr argument. As an example, let's again say you had a 32-byte receive buffer set up in the HYDRA that was pointed to by the dataptr argument and the W5100 had only 16 bytes' worth of data that it had received. The EtherX driver will write 16 bytes to the buffer pointed to by the dataptr argument and return a value of 16. In our first example, where the W5100 contained 40 bytes but only wrote 32 bytes to the buffer, the return value would be 32.

The third argument in the method, block, is a Boolean variable that will instruct the RX method what to do if there is currently no data waiting for us in the W5100. If there is no data, we can do one of two things. The first is to wait until data is available—this is called *blocking*. The program will "block," which means it sits there and does nothing, waiting for data to arrive. Set the block argument to true if you want the RX method to "block" and wait for data to arrive. The second thing you could do if there is no data waiting for us in the W5100 when called is to return immediately and somehow indicate to the application that there was no data available at the time. In our case, we would return immediately and set the return value to 0 (which, remember, indicates how many bytes have been read from the W5100 into the HYDRA main memory pointed to by the dataptr argument). To set the RX method to "nonblocking," set the block argument to false.

BLOCKING OR NONBLOCKING?

Which should you use? Remember that a blocking call will wait until data is available, which, in theory, could be a long time. Simply ask yourself what would happen to your application if it paused and waited around for too long. Would you miss some user input (i.e., Gamepad sample)? Would you be unable to update the video or the music? If a method pausing for too long would cause a problem in your application, use nonblocking. If your application will never do anything other than look for data from the W5100, it's better to use blocking. The concepts of blocking and nonblocking are important when writing code for things like device drivers and operating systems.

Sometimes it's helpful to know before we call the RX method if data is available. The EtherX driver has a Spin method to read the receive size register called read_rsr. If this method returns a non-zero number, there is data in the W5100 for us to read.

The other thing we need to address here is the difference between a call to RX when we have a socket type of TCP versus a socket type of UDP. Assuming there is already data in the W5100 internal buffer waiting for us to read, when our socket type is TCP and we read data from it, we will read only the data that was received by the W5100. When we have a socket type of UDP, every received packet of data will have an eight-byte header appended to it by the W5100. This header consists of the IP address of the received packet (four bytes), the source port (two bytes), and the data size (two bytes). So if you specified the socket type as UDP and were expecting your PC to send you 32 bytes, you would, in fact, read 40 bytes from the W5100 when you read the data from its internal buffer.

Just as with the transmit buffer of the W5100, the receive buffer of the W5100 can only store 2048 bytes by default. Don't let the W5100 get too much data into it before you read, or your data will be corrupt. Just as with the transmit buffer, you can configure the W5100 to have a larger receive buffer using the write method provided by the EtherX driver.

Receive Size Register   The last application method is the read_rsr method, which will read the receive size register. It will return the number of bytes that the W5100 has stored in its receive buffer for you to read.

If this method returns a non-zero number, there is data in the W5100 for us to read. This read_rsr method is also handy in determining if we have read all the data out of the W5100. For example, if we make a call to the RX method and specify that we should read and store 32 bytes from the W5100 to the main memory of the HYDRA pointed to by the dataptr argument and the return value is 32, we have a problem. We don't know if there are more bytes in the W5100 for us to read. Remember, we specify a maximum amount of bytes to be written to the main memory in the HYDRA. If we store that maximum amount of data, there could still be more in the W5100 waiting for us to read it. But how could we know? The answer is to call the `read_rsr` method. If, after we call RX with a size argument of 32 and we get a return value of 32 and we call the `read_rsr` and it returns 0, then we know we got all the data. If it returns a non-zero value, we know there is still data left over in the internal W5100 buffer for us to read out. The code for this method is shown here:

```
PUB read_rsr : ret_val

  '' Read the receive size register.
  '' This will return the value in the receive size register
  '' in the W5100.
  '' A non-zero value indicates that there is data to be read.

  ret_val := read($04,$26)
  ret_val := ret_val << 8
  ret_val += read($04,$27)
```

Creating a Simple Networked Game

To show how to use the Propeller to implement networked applications in a multicore environment, we will create a simple networked game, since the HYDRA is a game development platform, after all. Our new game will be called "Button Masher." The game is on a split screen, where the player's score appears on the left and the opponent's score appears on the right. Whenever the player hits any button on the gamepad, his score is increased. Conversely, whenever the opponent hits a button, his score is increased. The first person to hit a button five times is the winner. The entire source code for "Button Masher" is found in the Chapter_08/Source/ButtonMasher/Spin/ directory. Also, most readers will not purchase two HYDRAs to network together (although it is a blasty blast), so a C program was created that will send an updated score to the HYDRA whenever you press ENTER. That way, you can simulate another HYDRA using your PC. You can find this program executable and source in the Chapter_08/Source/ButtonMasher/C/ directory.

A key thing here is that when the player presses a button, he increments his score and sends his new score to the opponent in an Ethernet packet (using the UDP protocol). Of course, when the opponent presses a button, his score is incremented and he sends an Ethernet packet to the player.

So, what will we need to implement this game? Of course we need a means to display the graphics of the game on the screen. We will also need to monitor the gamepad and, when a button is pressed, increment our score and send the new score to the opponent. Last, we need to monitor the EtherX card to see if the opponent has sent us his updated score.

TRADITIONAL MICROCONTROLLER APPROACH

For some perspective, let's look at how your traditional low-power/cheap micro-controller would implement this funtastic game. The majority of the time is spent drawing the image on the TV screen. The vertical sync is typical when anything other than video in a game (sampling the gamepad, sound, game logic, network-ing, etc.) is dealt with. If you are unfamiliar with game programming, the *vertical sync* is the time on the TV when the electron gun swings from the bottom of the screen to the top.

There is some time during the horizontal sync as well (although small) if the pro-grammer is clever. Again, if you are unfamiliar with game programming, *horizontal syncs* are when the TV's electron gun swings from the far right of the tube to the far left of the tube to draw the next line. The horizontal sync time is often spent determin-ing what data to write during the next line, so it is usually not practical to do any game logic, sound, gamepad sampling, etc. during this time. It is important to note that the timing for writing an image to a TV is precise. Even one clock-cycle difference can result in distortion of the image or no image at all. Thus, *interrupts cannot be used while drawing an image because it will affect the video timing.*

While there is nothing wrong with these implementations, the game programmer is limited in that the only time he or she has to work with is during the vertical sync. Things also start to get complicated when we consider conditions like what happens when both player and opponent have a score of four and push the button within the same video frame. Remember that we won't be able to check gamepad and EtherX until the vertical sync. So if both happened within the same frame, we would have no way of knowing who actually won. It would appear as though they both happened at the same time, and the game logic would be forced to declare a tie.

MULTICORE APPROACH USING THE PROPELLER

The Propeller, by contrast, would not need to wait until the vertical sync to sample the gamepad, check the EtherX card, play a new sound, etc. All of these things happen in a different cog in parallel with one another and completely independent of each other. Let's see the "Button Masher" code and discuss some of the important points of the implementation.

Real quick, let's break down what cogs will handle what.

Cog 0—Boots into the Spin interpreter by default and will start the drivers for our I/O peripherals and handle our game logic

Cog 1—Gamepad (NES controller) driver

Cog 2—EtherX SPI driver

Cog 3—TV driver (NTSC)

Cog 4—Graphics driver; provides methods like drawing shapes and displaying text on the TV

Cog 5—Monitors the gamepad buttons and transmits the new score via EtherX card to the opponent when a button is pressed

Cog 6—Constantly polls the EtherX card to determine if there is a score update from the opponent

Here's the Cog 0 game logic code. This is the code after we have started all of the drivers, initialized the video, initialized the EtherX card, etc.

```
'Initialize game variables
winloss := 0
player_score := 0
opp_score := 0
score_change := 0

'Create a new lock to manage the EtherX card between cogs
eth_lock := locknew

'Start the gamedpad monitor in a new COG
cognew(MonitorButtons, @cog_stack[0])

'Start the ethernet monitor in a new COG
cognew(MonitorEthernet, @cog_stack[128])

'//////////////////////////////////////////////////////////
'Begin the game loop
'//////////////////////////////////////////////////////////
'Draw the playing field.
gr.clear

gr.colorwidth(3,3)
gr.textmode(6,6,6,0)
gr.text(50,60,@pscore)

gr.colorwidth(3,3)
gr.textmode(6,6,6,0)
gr.text(170,60,@oscore)

repeat
  if(score_change == 1)

    'Clear the score_change variable
    score_change := 0

    'Display the new scores
    gr.clear

    gr.colorwidth(3,3)
    gr.textmode(6,6,6,0)
```

```
    pscore[0] := $30 + player_score
    gr.text(50,60,@pscore)

    gr.colorwidth(3,3)
    gr.textmode(6,6,6,0)
    oscore[0] := $30 + opp_score
    gr.text(170,60,@oscore)

    if(winloss == 1)
       gr.colorwidth(1,2)
       gr.textmode(4,4,6,0)
       gr.text(25,140,@vic)
    elseif(winloss == 2)
       gr.colorwidth(2,2)
       gr.textmode(4,4,6,0)
       gr.text(40,140,@def)
```

The `repeat` keyword indicates the main loop of the game. The `score_change` variable is a global variable that will be changed by Cog 5 when the player hits a button or by Cog 6 when the opponent sends us a score update because he has pressed the button. All we do is clear the screen and draw the new scores on the screen. We then check another global variable called `win_loss`, which is initialized to 0 and set to 1 by Cog 5, if we won, or set to 2 by Cog 6, if we lost. If we won, in addition to displaying the score, we display "Victory;" if we lost, we display "Defeat."

You can see in the code how we start executing methods in a new cog using the `cognew` keyword. Both the `MonitorButtons` method and the `MonitorEthernet` method will begin executing in new cogs. Actually, the Spin interpreter will be loaded into two new cogs, which will start executing Spin instructions in locations in main memory. Here are the code listings for these two methods:

```
PUB MonitorButtons

  'This routine will run in another cog and monitor the gamepad.
  'It runs as a continuous loop waiting for player to push any button.

  repeat

    'Wait for the player to stop pushing on the pad
    repeat until (pad.read & $ff == 0)

    'Debounce button
    waitcnt(cnt + $10000)

    'Wait for the player to push something on the pad
    repeat until (pad.read & $ff <> 0)

    'Increment player score
    player_score += 1
```

```
'Determine if we have won
if (winloss == 0 and player_score == 5)
  winloss := 1

'Grab the ethernet lock and send new score to opponent
repeat while (lockset (eth_lock) == true)
eth.TX (@player_score, 4)
lockclr (eth_lock)

'Set the score_change variable
score_change := 1

'Debounce button
waitcnt (cnt + $10000)

PUB MonitorEthernet | UDP_Header0, UDP_Header1, data

'This routine will run in another cog and monitor the network.
'It runs as a continuous loop waiting for data from the opponent
'over the EtherX card.

repeat

  'Grab the ethernet lock
  repeat while (lockset (eth_lock) == true)

  'If read_rsr > 0 then there is data in the EtherX card for us.
  if (eth.read_rsr > 0)

    'We are making a dangerous assumption here that the
    'only data to have arrived is the data from the other
    'HYDRA and we KNOW that it will send 1 words of data.
    'Note here however, that we read 3 words. This is because
    'when the W5100 reads UDP data the UDP header is included
    'in the data. The UDP header is 8 bytes (two longs) so we
    'read two longs of data which we junk. Then we read the data
    'we care about.
    eth.RX (@UDP_Header0, 4, false)
    eth.RX (@UDP_Header1, 4, false)
    eth.RX (@data, 4, false)

    lockclr (eth_lock)

    'Set the opponents score to the data received
    opp_score := data
```

```
'Determine if we lost
if(winloss == 0 and opp_score == 5)
  winloss := 2

'Set the score_change variable
score_change := 1

else
  lockclr(eth_lock)
```

There's one last topic to cover concerning multicore networking. You'll notice that we've created a `lock` in Cog 0 using the `locknew` keyword. Resource management is a huge concern in multicore processing. As an example, let's assume that two cores want to write to a serial port (or SPI in our case). First, the two cores cannot write at the same time. This is especially true in an architecture like the Propeller (with no dedicated serial logic such as a UART), where the cores need to bit-bang the I/O. What is needed is a mechanism for one core to know if a particular resource is being used by someone else. One way to solve this problem is with something called a semaphore.

Using our serial port example, if a core wanted to send data out to a serial port, it would first look to see if the semaphore corresponding to the serial port were available. If so, the core would "grab" it. Once the core has grabbed the semaphore, the core keeps it until giving it up, and then the core can send data out the serial port, knowing that another core will not stomp on the output. After the data has been output on the serial port, the core needs to "release" the semaphore so that another core can use the serial port if it needs to.

IF A CORE GRABS A SEMAPHORE, IT MUST RELEASE IT AT SOME POINT

This is important, because if another core is blocking (not running anything else) while waiting for that semaphore to be available and that semaphore never becomes available because the previous owner never released it, you have a problem.

The Propeller calls these semaphores "locks." Although they have a different name, the concept is the same. Notice in the code that when a button is pressed, we need to send the updated score to our opponent via the EtherX card. But we need to make sure the `MonitorEthernet` method is not talking to the EtherX first. So we wait to grab the `eth_lock`, which is a lock that we created specifically for managing communication with the EtherX card. We know we have grabbed the lock when `lockset` returns true. Also notice that when we are done sending the updated score to the opponent, we immediately release the lock with `lockclr`. Similarly, in the `MonitorEthernet` method, we grab the lock before we check to see if there is any data from the opponent and release it as soon as possible to give the `MonitorButtons` method an opportunity to access the EtherX if needed.

Summary

So now we can see how to effectively use the multiple cores of the Propeller to implement a network application. In this chapter we have shown how to take an off-the-shelf Ethernet chip, interface that chip physically and electronically to the HYDRA game development platform, and write an API in both assembly and Spin to communicate with it. We showed how to write code for a simple network application while making use of the multiple cores available to us and managing resources using semaphores. We even explored a little how this multicore solution is preferable to low-power/cheap microcontrollers. Hope you enjoyed the chapter!

Exercises

1 Modify the "Button Masher" game to stream UDP packets at a constant rate, as opposed to only when the player or opponent pushes a button. This way, even if a packet is lost, it won't matter too much because another packet will be along shortly. This will turn "Button Masher" into an actual real-time game.

2 Modify the "Button Masher" game to communicate via the HYDRA-Net port instead of Ethernet.

3 Use the last cog to add sound to "Button Masher." This can be as simple as a sound that is played when the game ends.

4 Make "Button Masher" more time efficient. Currently, the `MonitorButtons` and `MonitorEthernet` methods are Spin methods that are interpreted. This means that they execute one Spin instruction every time the Hub comes around to that cog which is slow. Rewrite `MonitorButtons` and `MonitorEthernet` to be assembly instructions that fit within one cog (they must be fewer than 512 bytes).

9

PORTABLE MULTIVARIABLE GPS
TRACKING AND DATA LOGGER

Joshua Hintze

Introduction

In this chapter we will use the Propeller to sample and log data from a GPS receiver module and a miniature barometric pressure sensor. The received data will be logged to a Secure Digital (SD) card that is also attached to the Propeller. A dataset will be collected from the sensors and post-processed using free online software tools and also a spreadsheet program like Microsoft Excel, Google Docs, or OpenOffice.org. We will cover the following topics:

- Learn what multicore means and why it's important
- Overview of the hardware connections between the Propeller and the sensors
- How to interface to a GPS receiver and barometric pressure sensor
- How to log sensor data to an SD card
- Conversion equations between pressure and altitude
- How to plot waypoint tracks using online tools and Google Earth

Let's begin by discussing the advantages of using a multicore processor like the Propeller to capture sensor data versus a sequentially operating single-core microcontroller.

Resources: Demo code and other resources for this chapter are available for free download from ftp.propeller-chip.com/PCMProp/Chapter_09.

SINGLE CPU VERSUS MULTICORE SENSOR READING

A vast number of methods are available for interfacing a microcontroller to a sensor. This can range from simple asynchronous serial communication (see Fig. 9-1) using three wires—TX, RX, and ground—to other forms of clocked serial communications, such as *Serial Peripheral Interface* (SPI) and *Inter-Integrated Chip* (I^2C). In contrast to serial communications, some sensors and peripherals may be interfaced using parallel data and control lines. For example, the Wiznet Ethernet Chip (discussed in Chap. 8) supports sending and receiving Ethernet data through a 16-bit/8-bit parallel address/data bus. Finally, there are more exotic methods conceived by manufacturers where careful reading of the datasheet or hardware reference manual is required for interfacing.

Once a device is physically connected to a microcontroller, software must be written to send and receive data and commands. This task is generally trivial when interfacing one sensor to a microcontroller, but it can grow in orders of magnitude in complexity when talking with multiple sensors. The reason why is because different sensors and peripherals operate at diverse frequencies. While one sensor outputs its custom data once every second (1 Hz), another sensor could be transmitting its data 10 times a

Full Duplex UART

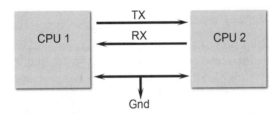

Serial Peripheral Interface (SPI)

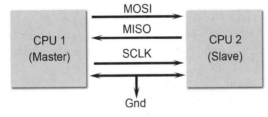

Inter-Integrated Circuit (I^2C)

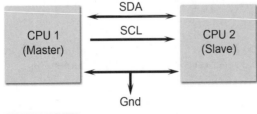

Figure 9-1   Interface technologies.

second (10 Hz). What is worse is when both sensors send their payload at exactly the same time, requiring the microcontroller to capture and act upon the data arriving at its input pins simultaneously.

Since microprocessors and microcontrollers have historically been *single-core* (meaning a single CPU contained inside that acts upon a sequential set of instructions), this problem has been solved by adding peripheral hardware for many of the standard interfaces. These hardware peripherals capture and buffer data into separate memory buffers without CPU intervention. For example, a typical *Universal Asynchronous Receiver/Transmitter* (UART) that is capable of receiving serial data will contain a shift register that clocks in each bit of data into an 8- or 16-bit register. Once enough data bits have been clocked in, the serial data is then copied to another register and a flag is set in the CPU to indicate that serial data has been received. The CPU must react to this flag by interrupting its current flow of program execution and then copy the received serial data from the saved buffer into main memory. If it does not act in a timely fashion, the received serial data could be lost. See Fig. 9-2 for an overview of this process.

Now the previous example is grossly simplified; hardware peripherals may contain multiple levels of *First-In-First-Out* (FIFO) buffers so that the CPU does not get interrupted as often. The peripheral may also contain *Direct Memory Access* (DMA) hardware capability so that it may directly transfer received data into the larger main memory without any CPU intervention. Regardless of the process, a single CPU microcontroller almost always becomes interrupted when new sensor data is required to be

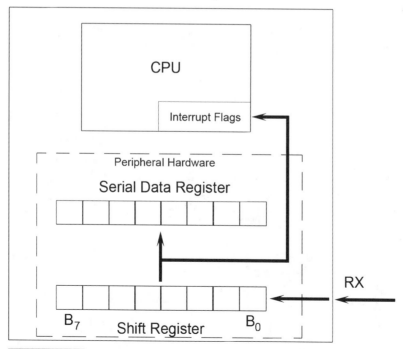

Figure 9-2   Serial UART hardware peripheral.

read and saved. This can become a time-consuming operation, as any CPU registers that are currently being operated on, including any status registers, must be stored; this is called a *context switch*. If a processor is continuously being interrupted to service sensor data requests, a large amount of time is spent switching between processes; in the extreme condition where a CPU may spend all of its time jumping back and forth between interrupts, that it can become deadlocked and no real work is performed.

Putting aside CPU time wasted from context switching, debugging software programs that have multiple processes and threads of execution to capture data using a single CPU resource can become a logistical nightmare. Trying to trace execution paths with debugging equipment while the CPU is being interrupted can become problematic. However, difficult as these challenges sound, this has been the method of operation for decades until multicore processors appeared.

When using a multicore processor such as the Parallax Propeller, we can dedicate one core, or cog in Parallax terms, to each individual sensor. This alleviates the burden of a single processor from being interrupted to handle different data rates. It also makes it easier to communicate to sensors that use nonstandard communication protocols because the entire cog can dedicate 100% of its time to reading in the data. This is especially beneficial for slow devices such as PS2 protocol keyboards and mice. Finally, debugging the communication is easier since the program operation will not randomly jump to an interrupt routine and instead will execute sequentially within each cog.

Overview of the Sensors

As discussed in the introduction, we will be connecting a GPS receiver module and a barometric pressure sensor (which also contains a built-in temperature sensor) to the Propeller. Data will be read in and displayed on a TV screen, and it will also be stored to a SD card for post-processing. Figure 9-3 shows the overall hardware connections, and Table 9-1 shows the descriptive pin connections.

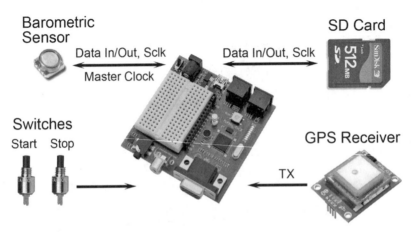

Figure 9-3   Hardware connections overview.

TABLE 9-1   PROPELLER PIN CONNECTIONS

PROPELLER PIN	PERIPHERAL	SIGNAL NAME DESCRIPTION
P1	GPS TX	Receive pin from the GPS transmit
P2	SD nCS	SD card chip select (low asserted)
P3	SD DI	SD card data in
P4	SD SCK	SD card clock in
P5	SD DO	SD card data out
P6	Barometer SCK	Barometer serial clock in
P7	Barometer DO	Barometer data out
P8	Barometer DI	Barometer data in
P9	Barometer Master Clock	Barometer master clock input (32.768 kHz)
P12	NTSC Out 0	NTSC Output 0
P13	NTSC Out 1	NTSC Output 1
P14	NTSC Out 2	NTSC Output 2
P15	NTSC Out 3	NTSC Output 3
P16	Start Switch	Start push button switch input
P17	Stop Switch	Stop push button switch input
P18	LED Mounted	LED indicator for SD card mounted.
P19	LED GPS Sats	LED indicator enabled when satellite lock is 3 or greater
P20	LED Heartbeat	LED indicator toggles when recording a data segment
P21	LED Pressure Good	LED indicator for barometer temperature is in valid range
P22	LED Recording	LED indicator enabled when logging is on

To make this experiment easier to produce, we will use the Parallax Propeller Demo Board since it already has a number of hardware connections for TV video out, prototyping areas, and light-emitting diodes (LEDs) on board. Plus, it already has all the necessary components for operation, such as power supplies, EEPROM for program storage, clock inputs, and hardware-interfacing mechanical connections. For a GPS receiver, we will be using a Parallax GPS receiver module (Item Code: 28146). There is also a barometric pressure sensor module MS5540C by Intersema that we will cover in depth later on. Finally, a standard Secure Digital card is attached for data logging purposes.

As you can see from Table 9-1, we have only one data connection to the GPS. This is because on startup, the GPS is wired to output raw data at a constant rate that the Propeller will read in serially.

Next the Propeller is connected to an SD card using an SPI interface, meaning there is a Serial Clock, Data Out, and Data In connection from the Propeller to the SD card. The barometric pressure sensor uses a slightly modified SPI interface, but it also contains all of the same communication pins as does the SD card.

One additional input into the barometric pressure sensor is the Master Clock pin input. Since the barometric pressure sensor contains its own built-in microcontroller for sampling the pressure sensor and communicating with outside devices (i.e., the Propeller), it requires an input clock for the CPU operation. The required clock rate is 32.768 kHz, and we will synthesize that frequency using the Propeller's counter registers and output it from a pin on the Propeller into the input pin on the pressure sensor. This saves us from having to purchase an oscillator chip.

WHY DO WE USE 32.768 kHZ?

The 32.768 kHz signal is a common clock frequency that you may see often, especially in designs that call for a *real-time clock* (RTC). The reason why is that with a real-time clock, we are interested in keeping track of the time of day, and one of the easiest methods is to have an input crystal that oscillates at 32,768 cycles per second. This is not a random number, but rather it is 2^{15}. If we were to take this clock rate and input it into a 16-bit counter with outputs labeled Out[15:0], bit 15 would transition from a 0 to a 1 exactly one second after being reset. The 15th bit, or the *most significant bit* (MSB), could then be fed back into the reset of the counter so that the counter is self-resetting. Also the 15th bit could be fed into a much larger counter that keeps track of the number of seconds that have transpired since startup or last reset. Even though the pressure sensor does not have any RTC circuitry, at least none that we can read, the developers of the sensor probably choose this as its operating speed since many circuit boards do have this clock source available, and possibly to make calculations easier.

The Propeller is connected through a binary weighted resistor network to produce NTSC television output for data monitoring and status information. However, when road testing the sensors, a television is not immediately available so we will make use of an array of LEDs that can signal any potential problems and toggle an LED on and off for each sample recorded to the SD card. Take a look at the completed circuitry in Fig. 9-4.

GPS RECEIVER

GPS stands for *Global Positioning System*. It was created as part of a United States Department of Defense (DOD) grant and is currently being supervised by the United States Air Force in Colorado. Initially developed in the early 1960s for the military, it became available for civilians in 1995.

A GPS receiver operates by receiving signals from a constellation of satellites that are orbiting the Earth (see Fig. 9-5). Thirty-two satellites orbit the Earth, allowing for

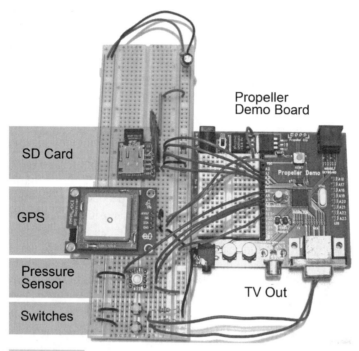

SD Card

GPS

Pressure
Sensor

Switches

Propeller
Demo Board

TV Out

Figure 9-4   Propeller Demo Board connected to experiment sensors.

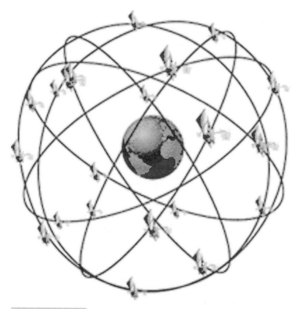

Figure 9-5   GPS satellite constellation.

complete coverage with a little bit of overlap. Each satellite continuously transmits three messages at 50 bps for 30 s, yielding a message length of 1500 bits. Encoded in the first message is the time of week, the current week of the year, and the satellite health information. In the second message, the satellite broadcasts its current position, also known as its *ephemeris*. Finally, in the third message is a table of all the other satellites and the course positional information. The table of satellites is known as an *almanac*.

Each data transmission is modulated into a higher-frequency signal before being sent to earth. The two most common signals, L1 and L2, operate at 1575.42 MHz and 1227.60 MHz, respectively. The L1 signal carries the navigation message, a course-acquisition (C/A) code, and an encrypted precision (P) code. L2 contains another P code transmission. Civilian GPS receivers can only decode the course-acquisition code, and without further augmentation can only achieve a positional accuracy of around 3 m under low-atmospheric error conditions. The P code can achieve ten times better accuracy, or 30 cm, without additional help. Since the military controls the GPS satellites, they could potentially disable the C/A code in times of war or other extenuating circumstances, leaving only P code receivers that contain the correct decryption keys capable of operating.

A civilian GPS receiver geolocates itself by receiving the L1 signal through its antenna and then it demodulates and decodes the data stream. Upon receiving a message, it records the current time and compares it to the original time that the signal was sent encoded in the message. If $T(s)$ is the time the message was sent and $T(r)$ is the time the message was received, we can use the following to calculate the distance the GPS receiver is from the satellite:

$$\text{Distance} = [T(r) - T(s)] \cdot c, \text{ where } c = \text{speed of light}$$

We can think of the position of the GPS receiver as a point on a sphere, with the center of the sphere being the position of the satellite and the radius of the sphere equaling the distance calculated from the previous equation. If the GPS receiver is currently receiving transmission from three satellites, it may then use geometric trilateration to determine its current position as the intersection of the three spheres (see Fig. 9-6 for an illustration). A large problem arises if there are even slight differences between the receiver's and the satellite's clocks. For example, if there was a mismatch of 100 ns and we use the speed of light as the propagation speed of the radio signal through the ionosphere, this would lead to an error of

$$100 \text{ ns} \cdot c = 100 \cdot 10 - 9 \cdot 299{,}792{,}458 \text{ m/s} = 299.79 \text{ m of error!}$$

Thus, the GPS receiver either needs to have a really expensive and accurate clock onboard or it can use positional information from more than three satellites to correct its clock and accurately produce a position. In practice, four satellites are required for a 3-D position fix unless other information, like current elevation, is known a priori.

GPS receivers are manufactured by many different vendors. This means that the information communicated by them can come in a form of a proprietary protocol. Fortunately

Figure 9-6   GPS trilateration.

for us, most GPS receivers support a dual-mode transmission that is capable of transmitting standard human-readable ASCII text. This is known as the *National Marine Electronics Association 0183* (or NMEA for short). The NMEA protocol describes a number of serial data "sentences" that describe the current status of the GPS. Take a look at an example set of NMEA sentences:

```
$GPGGA,123519,4807.038,N,01131.000,E,1,08,0.9,545.4,M,46.9,M,10,3*47<CR><LF>
$GPGSA,A,3,04,05,,09,12,,,24,,,,,2.5,1.3,2.1*39<CR><LF>
$GPRMC,123519,A,4807.038,N,01131.000,E,022.4,084.4,230394,003.1,W*6A<CR><LF>
```

The first thing to notice is that each sentence begins with a "$" and ends with a carriage return <CR> and a line feed <LF> ASCII character. Also, each field in the message is separated by a comma. Near the end of each NMEA sentence is an asterisk "*" that signals that the following two bytes after the asterisk is an error-checking code. To compute the error-checking code on your own, you would eXclusive-OR (XOR) all the bytes between the "$" and the "*" and then test it against the transmitted error-checking code before accepting the message.

Each sentence type can be identified by its first field: GPGGA, GPGSA, GPRMC, etc. There are many sentence types, but they all begin with GP. Once you have identified the sentence type, you can then parse out the individual elements that are separated by commas. For example, if we look at a GPGGA sentence, it could be separated as follows:

`$GPGGA,123519,4807.038,N,01131.000,E,1,08,0.9,545.4,M,46.9,M,10,3*47<CR><LF>`

GGA	GLOBAL POSITIONING SYSTEM FIX DATA
123519	Fix taken at 12:35:19 UTC
4807.038,N	Latitude 48 deg 07.038' N
01131.000,E	Longitude 11 deg 31.000' E
1	Fix quality: 0 = invalid
	1 = GPS fix (SPS)
	2 = DGPS fix
	3 = PPS fix
	4 = Real-time kinematic (RTK)
	5 = Float RTK
	6 = Estimated (dead reckoning) (2.3 feature)
	7 = Manual input mode
	8 = Simulation mode
08	Number of satellites being tracked
0.9	Horizontal dilution of position
545.4,M	Altitude, meters, above mean sea level
46.9,M	Height of geoid (mean sea level) above WGS84 ellipsoid
10	Time in seconds since last DGPS update
3	DGPS station ID number
*47	The checksum data; always begins with *

For additional information regarding GPS operation and the NMEA standard, see the following:

http://en.wikipedia.org/wiki/Global_Positioning_System

http://electronics.howstuffworks.com/gadgets/travel/gps.htm

http://aprs.gids.nl/nmea/

Reading the GPS Sensor In this experiment we will be using the Parallax GPS receiver module. The receiver module has four pins:

- GND: Ground reference supply
- VCC: 5 V input supply voltage

■ SIO: Serial data input/output (in our case, we will only use it as an output)
■ /RAW: Input pin that determines which communications mode to use

The /RAW input pin allows a user to select whether to use a binary command-based communications system when connected to VCC, or, when connected to GND, it continuously streams raw NMEA sentences. Since we are only interested in the NMEA sentences for this experiment, we will connect the /RAW pin to GND. This means we will receive serial GPS data at a baud rate of 4800 bps with 8-bit ASCII, no parity, and one stop bit.

To read in the serial NMEA sentences, we will make use of Parallax's Object Exchange (OBEX) Web site portal:

http://obex.parallax.com

The Object Exchange Web site contains hundreds (maybe even thousands by now) of user-submitted Spin, assembly, or C code objects for use by the general public under the MIT license. When searching the keyword "GPS," a number of NMEA parsers are already listed on the Web site. We elected to use a simple Spin language GPS parser that initially displays the GPS information on a VGA terminal. This file was originally created by Perry James Mole. The two files of interest to us for our experiment are:

■ GPS_IO_Mini.spin
■ FullDuplexSerial_Mini.spin

The FullDuplexSerial_Mini.spin is a slightly modified version of the Parallax full-duplex serial driver, with the only modifications being that some of the transmitter components are removed to save space in main memory.

The parsing of the NMEA strings takes place in the GPS_IO_Mini.spin object, specifically in the routine readNMEA listed here:

```
PUB readNEMA
  Null[0] := 0
  repeat
  longfill(gps_buff,20,0)
    repeat while Rx <>= "$"      ' Wait for the $ to ensure
                                 ' we are starting with
      Rx := uart.rx             ' a complete NMEA sentence
    cptr := 0

repeat while Rx <>= CR          '  Continue to collect data
                                '  until the end of the NMEA
                                '  sentence
      Rx := uart.rx             '  Get character from Rx Buffer
```

```
  if Rx == ","
    gps_buff[cptr++] := 0 '  If "," replace the
                          ' character with 0
  else
    gps_buff[cptr++] := Rx '  Else save the character

 if gps_buff[2] == "G"
   if gps_buff[3] == "G"
     if gps_buff[4] == "A"
        copy_buffer(@GPGGAb, @GPGGAa)

 if gps_buff[2] == "R"
   if gps_buff[3] == "M"
     if gps_buff[4] == "C"
        copy_buffer(@GPRMCb, @GPRMCa)

 if gps_buff[0] == "P"
  if gps_buff[1] == "G"
   if gps_buff[2] == "R"
    if gps_buff[3] == "M"
     if gps_buff[4] == "Z"
        copy_buffer(@PGRMZb, @PGRMZa)
```

'Excerpt from GPS_IO_MINI.spin Copyright 2007 Perry James Mole

The readNMEA method is called when a new cog is started as the main entry point. It begins by creating a null string initially and then enters into an infinite repeat loop. Inside the repeat loop it clears out the gps_buffer byte string with a call to longfill(...). Next it loops until we have received a "$" ASCII character from the serial port's RX pin using the following code:

```
repeat while Rx <>= "$" ' Wait for the $ to ensure
                        ' we are starting with
  Rx := uart.rx         ' a complete NMEA sentence
```

Once the $ is located, we know that we have the beginning of the NMEA sentence and that any character read in afterwards is part of this sentence up until the <CR><LF> characters. The next section of code reads in and stores the characters from the serial port into the gps_buff and replaces any occurrences of a comma with a 0 value. The reason why we are replacing commas with zeros will become apparent shortly. Once a <CR> has been encountered, the repeat loop ends, and the gps_buff should contain a complete NMEA sentence.

After the sentence has been read in, we can check which type of message it was by comparing the first couple of characters. This Spin code currently contains support for GPGGA, GPRMC, and PGRMZ sentences. The astute reader might have noticed that PGRMZ does not start with GP, even though we mentioned that all NMEA sentences begin with it. This is because GPS receiver manufacturers can add their own proprietary NMEA sentences, and in this case, the PGRMZ is a Garmin GPS receiver's proprietary altitude sentence.

Regardless of the message being received, the `copy_buffer(...)` method is called next, with arguments dependent on the sentence. The `copy_buffer(...)` method is listed here:

```
pub copy_buffer (buffer,args)
        ' Copy received data to buffer
        bytemove (buffer,@gps_buff,cptr)
        ptr := buffer
        arg := 0
        repeat j from 0 to 78          ' Build array of pointers
          if byte[ptr] == 0            ' to each record
            if byte[ptr+1] == 0        ' in the data buffer
                long[args][arg] := Null
            else
                long[args][arg] := ptr+1
            arg++
          ptr++
```

The first thing the `copy_buffer(...)` method does is perform a `bytemove(...)` copy of the data from the `gps_buff` to the first argument passed in. Next, it sets up a pointer to the beginning of the buffer and zeros out an argument counter. Finally, a repeat loop is entered that loops over the entire buffer, looking for those zeros that we replaced commas with. When a 0 is encountered, it saves a copy of the memory location to the `args` array that was passed in as the second argument to the `copy_buffer(...)` method. When execution returns from the method, we have a copy of the original received NMEA sentence, with commas replaced with zeros and an array of long pointers to each of the original comma-separated sections.

NULL-TERMINATE?

A question that you might be asking yourself is, "Why did we replace the commas with zeros when we could have just as easily searched for commas in the `copy_buffer(...)` method?" The reason why is because the standard convention for storing strings that have variable length is to NULL-terminate, or end them with a 0 value. Having strings NULL-terminated helps other methods determine when to stop reading characters from memory without needing to know the length of the string beforehand.

Now that we have pointers to the individual fields of the NMEA sentence string, we can return them to the parent Spin object that created the GPS_IO_Mini.spin object with methods like the following:

```
pub GPSaltitude
  return GPGGAa[8]
```

```
pub time
    return GPGGAa[0]

pub latitude
    return GPGGAa[1]

pub date
    return GPRMCa[8]
```

I have slightly modified the GPS_IO_Mini.spin object to add methods that return the degree and minute portions of the latitude and longitude separately so that it is easier to convert within a spreadsheet program. The modified source for the GPS_IO_Mini. spin object can be found here:

ftp.propeller-chip.com/PCMProp/Chapter_09/Source/GPS_IO_mini.spin

BAROMETRIC PRESSURE SENSOR

A barometric pressure sensor measures atmospheric pressure, which is the pressure exerted on a unit of area from a mass of air above the surface of Earth. When submerging into water, the deeper you travel beneath the surface, the greater the pressure will be built up from the water overhead. When rising from far down in the depths of water toward the surface, the pressure is alleviated. This is not at all different from the pressure felt from air when going from sea level to higher altitudes or vice versa. In fact, air does have mass—and a considerable amount of it. A cubic yard of air at sea level when the temperature is at 70°F weighs almost two pounds!

Since the air pressure diminishes as you travel into the upper atmosphere, we can use this knowledge to compute a reasonable altitude measurement. This is where a barometric pressure sensor is useful because it can measure the pressure felt upon it and produce a reading we can then use. There are many different applications for a pressure sensor, one of which is an altimeter for a large or small aircraft. Many people wonder, "Why not just use a GPS for calculating the altitude?" and the answer would be "You can." However, up until the mid 1990s, GPS receivers were not available for civilian aircraft, so the altimeter based upon atmospheric pressure was the method for calculating altitude. Also, under certain environmental conditions, the GPS receiver may not be capable of receiving the satellite radio waves—for example, in a deep canopy of trees, inside most buildings, or underground.

In our experiment, we will add a small *microelectromechanical systems* (MEMS) piezoresistive barometric pressure sensor. MEMS devices are a new, fascinating method for creating mechanical devices on silicon substrates that were originally meant for electronics. Through the fabrication of a silicon chip, a "micromachining" process can produce both mechanical and electrical components, making for orders-of-magnitude-smaller mechanical sensors and actuators.

The barometric pressure sensor we will be using is an Intersema MS5540C miniature barometer module, shown in Fig. 9-7. This unique device includes both a piezoresistive

Inter**s**ema

MS5540C (RoHS[*]) MINIATURE BAROMETER MODULE

- 10–1100 mbar absolute pressure range
- 6 coefficients for software compensation stored on-chip
- Piezoresistive silicon micromachined sensor
- Integrated miniature pressure sensor 6.2 x 6.4 mm
- 16 bit ADC
- 3-wire serial interface
- 1 system clock line (32.768 kHz)
- Low voltage and low power consumption
- RoHS-compatible & Pb-free*

DESCRIPTION

The MS5540C is a SMD-hybrid device including a precision piezoresistive pressure sensor and an ADC-Interface IC. It is a miniature version of the MS5534C barometer/altimeter module and provides a 16 bit data word from a pressure and temperature dependent voltage. MS5540C is a low power, low voltage device with automatic power down (ON/OFF) switching. A 3-wire interface is used for all communications with a micro-controller.

Compared to MS5534A the pressure range (measurement down to 10 mbar) has been improved. The MS5540C is fully software compatible to the MS5534C and previous versions of MS5540. In addition, the MS5540C is from its outer dimensions compatible to the MS54XX series of pressure sensors. Compared to the previous version the ESD sensitivity level has been improved to 4kV on all pins. The gel protection of the sensor provides a water protection sufficient for 100 m waterproof watches without any additional protection.

FEATURES

- Resolution 0.1 mbar
- Supply voltage 2.2 V to 3.6 V
- Low supply current < 5 µA
- Standby current < 0.1 µA
- -40°C to +85°C operation temperature
- No external components required

APPLICATIONS

- Mobile altimeter/barometer systems
- Weather control systems
- Adventure or multimode watches
- GPS receivers

BLOCK DIAGRAM

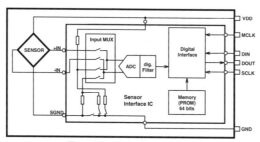

Fig. 1: Block diagram MS5540C.

---

[*] The European RoHS directive 2002/95/EC (Restriction of the use of certain Hazardous Substances in electrical and electronic equipment) bans the use of lead, mercury, cadmium, hexavalent chromium and polybrominated biphenyls (PBB) or polybrominated diphenyl ethers (PBDE).

DA5540C_003   June 16th, 2008
0005540C1193 ECN 1118

Figure 9-7   Intersema barometer module datasheet. (*Courtesy of Intersema; for the latest version consult the MEAS Web site at www.meas-spec.com.*)

barometer and an integrated chip (IC) that samples the analog output and converts it to digital value, called an analog-to-digital converter (ADC). It also provides a communications interface such that external devices can simply read the values preconverted without having to sample the analog voltage. In addition to the piezoresistive sensor, the MS5540C contains an internal temperature sensor that we will use in our altitude conversion equations, which will be discussed shortly. The first page of the Intersema MS5540C datasheet can be seen in Fig. 9-7.

Calculating Altitude from Pressure In the experiment code we will communicate to the barometer and access the compensated pressure and temperature values, and store those to the SD card, along with the GPS-received telemetry. As part of our post-analysis, we will look at the differences between the altitude given by the GPS and the barometer. For now, let's look at the math on how to compute your altitude given a pressure reading.

The following is the *barometric formula.* It is the formula used to calculate pressure at differing altitudes less than 86 kilometers (km) above sea level.

$$P = P_b \cdot \left[\frac{T_b}{T_b + L_b \cdot (h - h_b)} \right]^{\frac{g0 \cdot M}{R^* \cdot L_b}}$$

where P_b = Static pressure (Pa)
  T_b = Standard temperature (K)
  L_b = Standard temperature lapse rate (K/m)
  h = Height above sea level (m)
  h_b = Height at bottom of layer b (meters; e.g., h1 = 11,000 m)
  R^* = Universal gas constant for air: 8.31432 N · m/(mol · K)
  $g0$ = Gravitational acceleration (9.80665 m/s²)
  M = Molar mass of Earth's air (0.0289644 kg/mol)

The constants P_b, T_b, L_b, and h_b are different, depending on which region of the atmosphere you are in and are listed in Table 9-2.

In our experiment we will be gathering the data while driving a car around a residential neighborhood, so we can safely use the values from altitudes extending from 0 to 11,000 m above sea level.

P_b = 101325 Pa

Standard Temperature = 288.15 K

Temperature Lapse Rate = −0.0065 K/m

For additional information regarding the derivation of the barometric formula, see the following Web site; however, the formula is derived from the ideal gas law (pV = nRT).

http://en.wikipedia.org/wiki/Barometric_formula

TABLE 9-2 BAROMETRIC FORMULA CONSTANTS				
SUBSCRIPT b	HEIGHT ABOVE SEA LEVEL (m)	STATIC PRESSURE (Pa)	STANDARD TEMPERATURE (K)	TEMPERATURE LAPSE RATE (K/m)
0	0	101325	288.15	−0.0065
1	11,000	22632.1	216.65	0.0
2	20,000	5474.89	216.65	0.001
3	32,000	868.019	228.65	0.0028
4	47,000	110.906	270.65	0.0
5	51,000	66.9389	270.65	−0.0028
6	71,000	3.95642	214.65	−0.002

Using some math, we can solve the barometric formula for *(h)* and substitute in the known variables, arriving at the following *altimeter calibration formula*:

$$h = \frac{\left[1 - (P/P_{\text{ref}})^{0.19026}\right] \times 288.15}{0.00198122}$$

Thus, given a pressure, we can calculate our height above sea level. The equation requires the current sea-level–corrected pressure (P_{ref}) for the region in which the sensor is operating. An aircraft would normally get this pressure value from the radio tower, but we can find it by going to a weather Web site like www.weather.com. Note, however, that the typical units given are in inches of mercury when the equation calls for kilopascals. The conversion is:

1 in of mercury = 3.3860 kPa

Reading the Pressure Sensor   Unfortunately, an object wasn't already created for the MS5540C barometer module on the Parallax object exchange; therefore, we will make our own. Investigating the datasheet, we can see that the communications protocol is similar to SPI communications; however, there are subtle differences. Regarding the pin connections, we have a data input and data output and a clock that is controlled by an external "master" device. However, there is no chip select, which is common on SPI devices.

Another difference between this device's communication and SPI is that there are no established data lengths; instead, the barometer uses the notion of start flags (three bits high) and stop flags (three bits low) with commands embedded in between. This means that we cannot use an SPI object from the Object Exchange Web site either. Instead, we will need to control manually the signals to the chip; this is known as *bit-banging*. We will not cover any more details of the communication in this book since the datasheet dedicates a few pages to it; instead, we will discuss how to read and convert the barometer's registers into a temperature-compensated pressure. The MS5540C datasheet and Spin source code object to read the pressure and temperature values from the MS5540C can be found on the FTP site at this location:

Chapter_09/Docs/ms5540c-pressure_sensor.pdf

Chapter_09/Source/abs_pressure_01.spin

Referring to Fig. 9-8, we can see a flow chart of operations, provided from the manufacturer to execute in order, enabling us to read the pressure and temperature output. We will refer often to this figure while describing the steps to conversion shown in Fig. 9-8.

The first step is to read the factory 16-bit calibration words out of the barometer's memory and convert the words into calibration coefficients, which will be used later in conversion calculations. The coefficients are necessary for calculating the correct pressure and temperature because the output of the sensor is affected by changes in temperature. Also, the sensor is affected by light, especially direct sunlight, so it's best to create an enclosure around the sensor when using it during the daytime or under changing lighting conditions. The Spin code for reading the coefficients is show here:

```
cmd := %111_010101_000
calib1 := read_calib_word(cmd)
cmd := %111_010110_000
calib2 := read_calib_word(cmd)
cmd := %111_011001_000
calib3 := read_calib_word(cmd)
cmd := %111_011010_000
calib4 := read_calib_word(cmd)


' Retrieve the calibration bits from the words
c1 := calib1 >> 1
c2 := (calib4 & $3F) | ((calib3 & $3F) << 6)
c3 := calib4 >> 6
c4 := calib3 >> 6
c5 := (calib1 & %0001) << 15 | (calib2 >> 6)
c6 := calib2 & $3F
```

The read_calib_word(...) method is a private method in the abs_pressure_01.spin object, and it transmits the appropriate command and returns the calibration 16-bit word. After the calibration words are received, we use a combination of AND-bit masking, bitwise-ORing, and bit shifting to extract the calibration coefficients.

The next step is to read in the 16-bit data words that contain the current uncompensated pressure and temperature values. Again, we have created a private method that handles the nonstandard communication and have called it read_data_word(...). An example of its use is as follows:

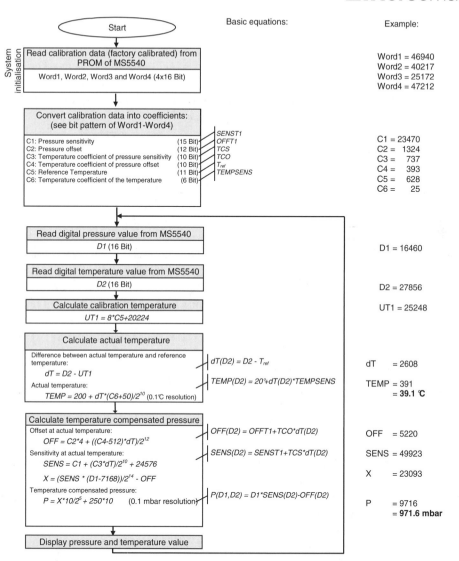

NOTES
1) Readings of D2 can be done less frequently, but the display will be less stable in this case.
2) For a stable display of 0.1 mbar resolution, it is recommended to display the average of 8 subsequent pressure values.

Figure 9-8   **Steps to reading the pressure sensor.** (*Courtesy of Intersema; for the latest version, consult the MEAS Web site at www.meas-spec.com.*)

```
' Read in the data words for temperature and pressure
d1 := read_data_word(%111_1010_000_00) 'Value for data 1
d2 := read_data_word(%111_1001_000_00) 'Value for data 2
```

After we have received the current data words for pressure and temperature, we need to calculate the temperature at which the sensor was factory-calibrated. One thing you may have noticed from Fig. 9-8 is that this calibration temperature value is constant and will not change, so to save calculation time, we have moved this outside the conversion loop. The relevant Spin code is shown here:

```
calib_temp := 8*c5 + 20224
```

Notice that the equation uses integer and not floating-point numbers even though the calibration temperature can be a fraction of a degree Celsius. At first, it may not be apparent, but the microcontroller multiplies any fractional number by 10. This means we can use integer math for all of our calculations, which is handy since the Propeller does not have dedicated hardware for floating-point processing.

Now it is time to calculate the actual temperature that the barometer is currently at while also determining the difference between the calibration temperature and current temperature and storing it as the variable dt. The code for this step is as follows; remember that the output variable temperature is 10 times degrees Celsius.

```
' Compute temperature difference from calibration
dt := d2 - calib_temp
'Actual temperature in .1 degrees C
temperature := 200 + (dt*(c6+50))>>10
```

Finally, we can calculate the temperature-compensated pressure by first calculating some intermediate variables: offset, sensitivity, and x. Then we use the intermediate values to compute the final temperature-compensated pressure in units of 10 times millibars. The code is shown here:

```
' Calculate temperature compensated pressure in .1 mbars
offset := (c2<<2) + (((c4-512)*dt)>>12)
sensitivity := c1 + ((c3*dt)>>10) + 24576
x := ((sensitivity * (d1-7168))>>14) - offset
pressure := ((x*10)>>5) + 2500
```

The abs_pressure_01.spin object requires one new cog for its operation and provides public methods for accessing the last read pressure and temperature that is read from the sensor four times a second (4 Hz). For our experiment, these variables will be logged to an SD card.

SECURE DIGITAL CARD

The Propeller has no built-in nonvolatile memory (memory that doesn't get erased when power is removed). The Propeller Demo Board does employ an external EEPROM used for programming the device, but we want to use something more portable, removable, and easily recognizable by a PC for analyzing the data log files. For this reason we have elected to use a *Secure Digital* (SD) card connected via SPI.

SD cards are essentially flash memory devices that are a collection of sectors. Each sector is usually 512 bytes, and there can be millions of sectors on very large SD cards. SD cards have no inherent file system. Thus, FAT16, FAT32, NTFS, Linux file systems, and so on have no idea how to read SD cards. Therefore, someone has to write a driver interface that translates, if you will, the SD card sectors into a file system. For example, the only operations you really need to implement on a FAT16/32 file system are the abilities to read and write a sector. The SD protocol and hardware supports this, so to support the FAT system, you have to write a driver that implements the FAT file system on the SD card and translates things like the directories, boot records, and so on. This is a complex business and outside the scope of this book. Fortunately for us, an object has been uploaded to the Parallax Object Exchange Web site that allows the Propeller to read and write files, among other things, to an SD card using the FAT16 file system. Before we discuss this SD card object; however, let's first talk about what would be required if you were to write your own SD card interface.

Let's start by looking at how SD commands are formatted. The first thing to remember is that everything is sent to and received from the SD card over the SPI interface one byte at a time; all commands, data, etc. are composed of single- or multiple-byte transactions. Thus, all the communications take place over the SPI write and read byte methods. The SD layer is on top of this.

The complete SD card specification is a monster of a document, not because of its length, but because of its lack of clarity. For those interested, you can read all about it in the SD_SDIO_specsv1.pdf located here:

Chapter_09/Docs/SD_SDIO_specsv1.pdf

Luckily, we aren't using the "SD protocol mode" but rather the "SPI protocol mode," which is much easier to deal with. Of course, it's not as fast as SD mode, nor does SPI mode have all the features of SD. But SPI mode allows us to read and write sectors of the SD card, and that's all that matters. Also, in SPI mode, both SD cards and MMC cards work in the same way, so the software, drivers, and our experiment discussed later would theoretically work with MMC cards as well.

In any event, all communication with the SD cards is done via the SPI interface, so when we talk about writing and reading, an SPI communications object interface is obviously used. This differs from "SD" mode communications, thus we want to place the SD card into "SPI" mode. This is actually the first step when initializing the SD card. Figure 9-9 shows how commands are formed for the SD card.

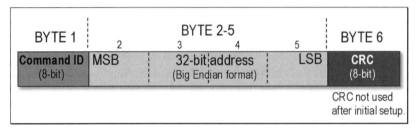

Figure 9-9 **Format of SD card commands in SPI mode.**

They are six bytes long and consist of the following parts:

Byte 1 : 8-bit command ID

Byte 2, 3, 4, 5 : 32-bit address

Byte 6 : 8-bit CRC checksum

The 32-bit address must be formatted in *big-endian* format—that is, high byte to low byte. This is the opposite of the Intel format, which is little-endian. Therefore, if you want to send $FF_AA_00_11 in big-endian, you would send $FF, followed by $AA, $00, and finally $11 to complete the address. Of course, the address word might not be important for the particular command; if not, always make it $00_00_00_00. Finally, the *CRC checksum byte* (CRC stands for cyclic redundancy check) is a byte that you, the sender, must compute based on bytes 1..5. The SD card, after receiving bytes 1..5, compares the CRC you send to it, and it computes its own; if they are not the same, the SD card will return an error. Also, when the SD card receives a command, it sends back a response along with a CRC. You are free to inspect the CRC if you wish. After you send the six-byte command to the SD card, it always responds with a one-byte response code. These response codes mean different things in different situations, but generally are used to catch errors. Figure 9-10 shows the bit encoding for the response codes.

CRC CALCULATION

When the SD is in SPI mode, it doesn't require CRC bytes; therefore, the CRC bytes can be anything. However, before the SD card is in SPI mode, it does require that the CRC be correct. If you're interested in how CRC bytes are computed, try this link:

http://en.wikipedia.org/wiki/Cyclic_redundancy_check

Note: SD and MMC cards use the CRC-7 method.

The next question, of course, is what commands does the SD card support? Table 9-3 lists the complete command set for SD SPI mode.

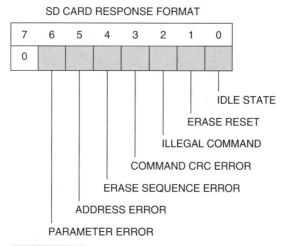

SD CARD RESPONSE FORMAT

IDLE STATE

ERASE RESET

ILLEGAL COMMAND

COMMAND CRC ERROR

ERASE SEQUENCE ERROR

ADDRESS ERROR

PARAMETER ERROR

Figure 9-10   **Response code binary bit encoding format.**

Notice CMD0. This command is very important and literally the first command that we need to issue to the SD card. This command tells the SD card to switch into SPI mode. Therefore, we are going to issue this command first, after which the SD card should respond with a response byte of $01 to indicate success. This brings us to Table 9-4, which is a short listing of SD SPI commands that must be supported and their respective response bytes.

Logging to the SD Card   Through the use of the SD SPI commands, a processor can read from and write to an SD card a sector at a time. The next step would be to implement software for reading a file system from those sectors. We will not cover file systems in this book; however, for further reading see the following:

Chapter_09/Docs/MS_fat_white_paper.doc

http://en.wikipedia.org/wiki/File_Allocation_Table

Instead of creating our own object for SD card support, we will be using the FSRW Spin object by Radical Eye Software. Again, this can be found on the Parallax Object Exchange Web site, or it can be found here:

Chapter_09/Source/CODE/fsrw.spin

Chapter_09/Source/CODE/sdspi.spin

The FSRW SD card object has a number of public methods for mounting, opening, reading, and writing to a file, which are documented in Table 9-5.

TABLE 9-3 COMMAND LIST FOR SD SPI MODE

CMD ID	ABBREVIATION	SDMEM MODE	SDIO MODE	COMMENTS
CMD0	GO_IDLE_STATE	Mandatory	Mandatory	Used to change from SD to SPI mode
CMD1	SEND_OP_COND	Mandatory		
CMD5	IO_SEND_OP_COND	Mandatory		
CMD9	SEND_CSD	Mandatory		CSD not supported by SDIO
CMD10	SEND_CID	Mandatory		CID not supported by SDIO
CMD12	STOP_TRANSMISSION	Mandatory		
CMD13	SEND_STATUS	Mandatory		Includes only SDMEM information
CMD16	SET_BLOCKLEN	Mandatory		
CMD17	READ_SINGLE_BLOCK	Mandatory		
CMD18	READ_MULTIPLE_BLOCK	Mandatory		
CMD24	WRITE_BLOCK	Mandatory		
CMD25	WRITE_MULTIPLE_BLOCK	Mandatory		
CMD27	PROGRAM_CSD	Mandatory		CSD not supported by SDIO
CMD28	SET_WRITE_PROT	Optional		
CMD29	CLR_WRITE_PROT	Optional		
CMD30	SEND_WRITE_PROT	Optional		
CMD32	ERASE_WR_BLK_START	Mandatory		
CMD33	ERASE_WR_BLK_END	Mandatory		
CMD38	ERASE	Mandatory		
CMD42	LOCK_UNLOCK	Optional		
CMD52	IO_RW_DIRECT	Mandatory		
CMD53	IO_RW_EXTENDED	Mandatory		Block mode is optional
CMD55	APP_CMD	Mandatory		
CMD56	GEN_CMD	Mandatory		
CMD58	READ_OCR	Mandatory		
CMD59	CRC_ON_OFF	Mandatory	Mandatory	
ACMD13	SD_STATUS	Mandatory		
ACMD22	SEND_NUM_WR_BLOCKS	Mandatory		
ACMD23	SET_WR_BLK_ERASE_COUNT	Mandatory		
ACMD41	SD_APP_OP_COND	Mandatory		
ACMD42	SET_CLR_CARD_DETECT	Mandatory		
ACMD51	SEND_SCR	Mandatory		SCR includes only SDMEM information

Note: The command code IDs start at $40 (64). For example, CMD0 = $40 + 0 = $40, CMD17 = $40 + 17 = $51.

TABLE 9-4   SUBSET OF SD SPI COMMANDS NEEDED TO IMPLEMENT SD CARD DRIVERS

COMMAND	MNEMONIC	ARGUMENT	RESPONSE (2)	DESCRIPTION
0 ($40)	GO_IDLE_STATE	None	$01	Resets SD card and places in SPI mode
1 ($41)	EXIT_IDLE_STATE	None	$01	Exits reset mode
17 ($51)	EAD_SINGLE_BLOCK(3)	Address	$00	Reads a block at byte address
24 ($58)	WRITE_BLOCK(4)	Address	$00	Writes a block at byte address
55 ($77)	APP_CMD(1)	None	$00	Prefix for application command
41 ($69)	SEND_APP_OP_COND(1)	None	$00	Application command

Note 1: The last two commands are not mandatory, but help differentiate if the card is SD or MMC. Only SD can reply to these commands.
Note 2: Refer to Fig. 9-10 for encoding of response byte.
Note 3: When reading a block, a data token of $FE will be received after the initial response code of $00. Then the next bytes will be the sector data.
Note 4: After the write command is sent, the SD card expects the data token $FE to follow, signifying the host is ready to send bytes.

TABLE 9-5   AVAILABLE METHODS FROM FSRW SD CARD OBJECT

PUBLIC METHOD NAME	DESCRIPTION
mount(basepin)	Resets the SD card connected to the base pin into SPI mode and attempts to read the FAT16 file system
popen(s, mode)	Opens a file named s for read, write, or append operation
pclose	Closes and flushes the header and write buffer to the previously opened file
pread(ubuf, count)	Reads from the previously opened file and stores into ubuf with count bytes
pwrite(ubuf, count)	Writes from ubuf into the previously opened file with count bytes
pputc(c)	Outputs a single byte, c, to the previously opened file
pgetc	Reads and returns a single byte from the previously opened file
pflush	Flushes any unwritten header metadata and write buffers to the previously opened file
opendir	Closes the previously opened file and prepares the read buffer for calls to nextfile
nextfile(fbuf)	Finds the next file in the root directory folder and copies its name in 8.3 string format into fbuf
SDStr(ptr)	Outputs the NULL-terminated string ptr to the previously opened file
SDdec(value)	Outputs the 32-bit long value to the previously opened file
SDhex(value,digits)	Outputs the 32-bit long value to the previously opened file with digits count of hexadecimal characters
SDbin(value,digits)	Outputs the 32-bit long value to the previously opened file with digits count of binary characters

Main Spin Object

The main Spin object is the code that ties all the sensor collection and logging together, and it runs out of Cog 0 on startup. The main Spin object is responsible for creating and initializing the GPS serial parser object, the barometer pressure sensor object, the TV terminal for display purposes, and the SD card file system object. On top of managing these additional objects, it is continuously reading two input pins that are connected to pushbutton switches that control starting and stopping the SD card file log. By using switches and allowing the user to start, stop, and restart, logging multiple data collections may be achieved on a single SD card. The source code to the main object may be found at the following location:

Chapter_09/Source/Main_01.spin

We will not be thoroughly explaining each portion of the source code in this chapter; however, it is thoroughly commented for easy understanding. Instead, we will talk about the operation as a whole. Take a look at the flow chart describing the inner workings of the main object in Fig. 9-11.

Referring to Fig. 9-11, the first thing the main Spin object does after startup is to initialize the GPS, barometer, and TV terminal objects. Each of these objects starts a separate cog for execution as they begin their data collecting and TV output operations. The next step is to test whether the Propeller can successfully communicate and read in the FAT16 file system of the attached SD card. This is a reasonable test because the SD card may not be inserted and the user wishes to only watch the TV terminal display.

Now if the SD card was not found, the TV terminal is updated with the message "Failed to mount SD." After a brief period—enough for users to see the TV terminal message—the screen is cleared and we enter into an infinite loop that reads the current states from the GPS and barometer objects and prints them to the TV terminal. If a user wishes to insert a card after startup, he would need to reset it for the SD mounting test. This would be a good starting place for improving the main object, but I leave that up to you.

If an SD card was found, the directory contents of the SD card in the root folder are read and displayed to the screen. Once again, the execution pauses so that users may read the directory listing, and then the main object enters into the infinite repeat loop. Likewise, when the SD card is not mounted, the GPS and barometer outputs are displayed to the TV terminal. Next the Propeller tests to see if we are currently recording. Remember the user must initiate the recording with a pushbutton switch. If the Propeller is recording, the SD card is written to using a combination of the FSRW methods SDStr(...), SDdec(...), and pflush().

It is important that we call pflush() after every write to the SD card in our application. This is for the sake of the user in case he or she forgets to stop the recording before ejecting or powering down the system. Without pflush() being called, the file header metadata would not be written to the SD card and a PC's operating system would not be able to recognize the file length and location without it.

Finally, regardless of whether the data is being recorded, we arrive at the input pushbutton switch test. If the SD card is currently being recorded, we test to see if the

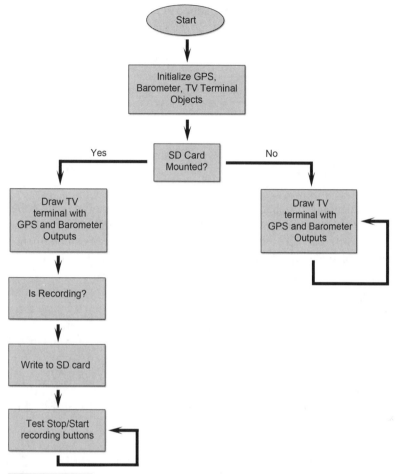

Figure 9-11   Flow chart of main Spin object.

STOP pushbutton switch is pressed. If so, we stop recording and close the file. Now if we were not recording to the SD card and the START pushbutton switch is pressed, we first look for a unique file name to open. To locate a unique file name, we search the root folder for a file named sensorXX.log, where XX is a number between 00 and 99. Once a suitable file name is found, we open it for writing.

Now a TV terminal is great for debugging purposes—for example, to see if the GPS receiver has three or more satellites or whether the SD card was mounted properly, and so on. However, for our experiment, we will be taking the hardware on the road for a test drive, and we will not be bringing along a TV. To aid us in debugging when a TV is not readily available, we will make use of a number of LEDs. The following lists the LED outputs we are using along with their description:

■ Mount SD: If the SD card was mounted properly, on startup, this LED will be lit.
■ GPS Satellites: If the GPS has three or more satellites, this LED will be lit.

- Heart Beat: Each time through the main object's infinite repeat loop, this LED will toggle. This is useful in helping the user see that all is working well.
- Pressure Good: This LED is lit when the temperature read from the barometer is within normal operating ranges, meaning we have established communication and are converting values properly.
- Recording: This LED is lit when the Propeller is in a state that is recording data to the SD card.

Experiment

In the experiment for this chapter, we will be taking the prototype data-collecting unit and traveling around a residential neighborhood. This neighborhood in particular consists of many different inclines with varying steepness. Since one of our main objectives is to compare and contrast the performance of the GPS altitude versus the barometric pressure sensor, having a nonflat terrain to travel over is ideal.

In addition to altitudes, we are interested in other measurements, such as the GPS reported speed. For our experiment we will be traveling between 0 and near 60 miles per hour. All collected data will be stored to the SD card and then postprocessed in a spreadsheet software program such as Microsoft Excel, Google Docs, or even OpenOffice.org. In order to facilitate easy importing into a spreadsheet program, we are going to output data captured in the main Spin object as comma-separated values. With the exception to the first line that serves as a header, each line of the ASCII text file log will contain a series of ASCII number strings separated by commas. An example of the text log file output is shown here:

```
30, 21.1403, N, 097, 54.2948, W, 00201.1, 000.0, 05, 040942, 210509, 072.3, 245, 9889
30, 21.1403, N, 097, 54.2948, W, 00201.1, 000.0, 05, 040943, 210509, 072.3, 245, 9889
30, 21.1403, N, 097, 54.2948, W, 00201.1, 000.0, 05, 040944, 210509, 072.3, 245, 9890
```

An explanation of each field, from left to right, can be found at the beginning of the log file, and for simplicity, is listed here:

- Latitude in degrees
- Latitude in minutes
- Northern or southern hemisphere
- Longitude in degrees
- Longitude in minutes
- Western or eastern hemisphere
- GPS altitude
- GPS speed
- Number of satellites locked on
- Time (Greenwich Mean Time)
- Date (DDMMYY format)

- GPS heading
- Barometer temperature (10 times Celsius format)
- Barometer pressure (10 times millibar format)

For this experiment, we have already gone ahead and collected a long dataset that can be found here:

Chapter_09/Source/Logs/SENSOR_5_20_2009.csv

Notice that we renamed the dataset so that it ends with a .csv file extension. This makes it easier for programs like Excel and Google Docs to import the data. When read into a spreadsheet program, all data fields will be separated into individual cells, where each row of the spreadsheet referenced a sampling of data in time.

DATA ANALYSIS

Once the data has been imported into a spreadsheet program, we can condense and convert some of the fields to prepare them for plotting or graphing. The first thing to convert is to change the latitude/longitude format from degrees and minutes to decimal degrees. For example:

70 degrees, 30.6 minutes = 70°, 30.6′ to 70.51 degrees = 70.51°

This conversion is as simple as taking the minute portion and dividing it by 60 and adding it to the degree portion. Also, we must take into account what hemisphere the data is collected in. If the data is collected in the southern hemisphere, we need to multiply the latitude decimal degrees by negative one. Likewise, if the GPS receiver was in the western hemisphere, we would need to multiply the longitude decimal degrees by negative one. We are not going to explain the details of entering formulas into a spreadsheet program here; instead, we have already imported and data-manipulated the CSV log file and stored it as a Microsoft Excel (XLS) file. This file is located here:

Chapter_09/Source/Logs/SENSOR_5_20_2009.xls

In addition to the latitude/longitude conversion, the Excel file converts the pressure from .1 millibars to kilopascals for use in the *altimeter calibration formula* we discussed previously. Now in order to convert the measured pressure into altitude above sea level, we first need to find out the local region's air pressure at *mean sea level* (MSL). This is known as the *altimeter setting*. Since this value is constantly changing depending on the weather, like a high- or low-pressure system moving in from a storm, we need to obtain this information beforehand. In fact, this is how aircraft operates because they will radio the tower before takeoff and adjust their altimeter accordingly. This information can be obtained from a local airport or more easily obtained from a weather Web site. For example, we recorded the local air pressure previous to the drive test at 32.02 in of mercury, or 101.65 kPa, from www.weather.com.

Figure 9-12   GPS altitude versus pressure altitude.

The Excel spreadsheet contains a column for the converted pressure sensor from kilopascals to altitude in meters above sea level. Now that we have this pressure converted, we can compare it to the GPS-reported altitude. Figure 9-12 shows an overlaid plot of the GPS altitude and the pressure sensor altitude.

Notice in Fig. 9-12 how the GPS-reported altitude and the pressure sensor's measured altitude are closely matched, especially near the end of the dataset. A possible explanation for the beginning of the dataset being mismatched is that when data collection had begun, the GPS was still acquiring and locking onto more satellites. As the receiver found these other satellites, its positional accuracy increased. One interesting experiment you might try is to adjust the spreadsheet cell containing the local region's air pressure. Notice how the overall altitude drops or rises when modifying this cell— this shows how important it is that airplanes receive the correct pressure information from the airport.

Plotting GPS Tracks   In addition to altitude, we have a number of other fields we can plot. We could plot the temperature reported from the pressure sensor over time or the number of satellites the GPS was locked onto. Probably the most interesting plot comes from plotting the GPS location tracks. A great Web site that assists with plotting GPS tracks is located at www.gpsvisualizer.com.

The GPS Visualizer Web site is a free do-it-yourself mapping Web site that allows users to upload their waypoint tracks and plot the data overlaid onto a number of different maps, including aerial photos, U.S. Geological Survey (USGS) topographical maps, county outlines, satellite photos, and much more. Data is uploaded as either plaintext (tab-delimited or comma-delimited) or Excel spreadsheet data, or you can even copy and paste into an edit box on the Web site.

We are now going to go through a simple step-by-step tutorial on how to format and upload GPS tracking data. For those who do not want to manually format their data or who want to skip to the plotting section, we have supplied preformatted comma-separated values (CSV) files located here:

Chapter_09/Source/Logs/GPSVisualizer_Lat_Long.csv

The GPS Visualizer Web site accepts many different data types, but we will choose to stick with a simple ASCII comma-separated file. In order for the GPS Visualizer Web site to correlate each column of data with its respective type, there needs to be a header row at the top. For example, to plot a simple GPS latitude/longitude track, we need the first row to look like the following:

```
Type, Latitude, Longitude
```

Capitalization and order do not matter as long as the data on subsequent lines follow the same order. The latitude and longitude fields need to be in *decimal degrees,* and the type field needs to contain the letter "T," which stands for "track."

✓ Organize your data in an Excel spreadsheet.

Next we need to copy the data from the Excel spreadsheet to the CSV file. When creating the CSV file, you will most likely start off with a separate spreadsheet, and when you are finished, save the output as a comma-separated values file. When you have finished, your file should look similar to the following snippet of data:

```
Type, Latitude, Longitude
T, 30.35233833, -97.90491333
T, 30.35233833, -97.90491333
T, 30.35233833, -97.90491333
T, 30.35233833, -97.90491333
```

✓ Save your Excel spreadsheet as a CSV file.

It is now time to load the data into the Web site for display. On the home page, there is an edit box for a file upload (see Fig. 9-13).

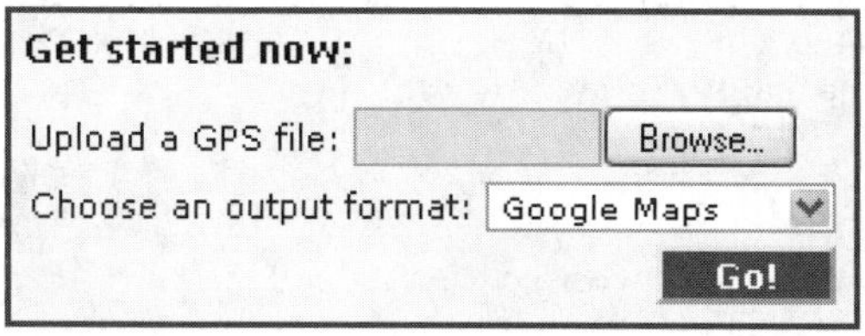

Figure 9-13   File upload box.

✓ Click the <u>Browse</u> button or click in the edit box, and a file browser window will open.
✓ Navigate and select your CSV file. In the Web site under <u>Choose an output format</u>:, select <u>Google Maps</u>.
✓ Now click the <u>Go</u> button, and after a few seconds the Web site will update with a GPS track overlaid onto an aerial photograph.

The aerial photograph is an interactive control that allows you to zoom in and out and pan around the GPS track. In the upper-right corner of the map you can select between different map types. You can even save the map and share it with others using another Web site called www.EveryTrail.com.

Now that you have the basics of plotting a GPS track, you can play around with some of the advanced options on the GPS Visualizer Web site. For example, we can add another field to the CSV file that contains the GPS speed. In this case, we would add another header called speed; then when we go to plot the data, the GPS track will be color-coded, as shown in Fig. 9-14. We have created a couple different CSV files with

Figure 9-14   **GPS Google Map track with speed.**

speed and altitude so that you can try out the advanced options on the GPS Visualizer Web site. The extra CSV files are located here:

Chapter_09/Source/Logs/GPSVisualizer_Lat_Long_Alt.csv

Chapter_09/Source/Logs/GPSVisualizer_Lat_Long_Speed.csv

One more feature of the GPS Visualizer Web site that I would like to point out is that they allow you to export your data as a *KMZ* or a *KML* file. KML stands for Keyhole-Markup Language, and it is essentially an XML file with tags recognized by their program. Keyhole Software created a mapping program called *Earth Viewer* that they sold mostly to the government until they were acquired by Google in 2004. After Google purchased the company, they renamed the software Google Earth and released a free version in 2005. The KML file allows users to input map markers, tracks, pictures, and even moving objects into Google Earth. A KMZ file is Zip-compressed container around the KML file. This means that we can upload our simple ASCII CSV files to the GSP Visualizer Web site and then download a KMZ file for viewing in Google Earth stand-alone software.

Summary

Throughout this chapter we have covered how to interface a Parallax Propeller to a GPS receiver and a barometric pressure sensor module. By using the Propeller's multicore CPU, we were able to isolate separate execution paths without the need for difficult threading and semaphore locking that you would find on a conventional CPU. This proved especially useful when we communicated with the barometric pressure sensor because of its nonstandard communications protocol. If instead we were to interface this sensor to a traditional single-CPU microcontroller, it would take away precious computation time from the main process in order to service the communications.

In addition to communicating with the sensors, we discussed how to store the log data to a Secure Digital (SD) card. An SD card is ideal for use with embedded processors because of its large capacity in a small footprint. Using Parallax's Object Exchange (OBEX) Web site, we were able to find suitable software that facilitated the complex SD card communication and file system handling.

After completing a test drive, the SD card was removed from the experimental hardware and inserted into a PC for post-analysis. We discussed equations to convert barometric pressure to altitudes and plotted the outputs in both a spreadsheet program and online mapping tools such as www.GPSVisualizer.com.

A number of improvements could be made to this experiment. First, we could always add sensors. We could add a barometric (absolute) pressure sensor, as well as a differential pressure sensor such as the Freescale Semiconductor MPX5050. By using a differential pressure sensor, we can connect a tube to one of the ports on the sensor and point it out toward the front of the car. The second port is a static reference.

By measuring the difference between the two pressures, we can compute the speed of the moving sensor. From here, we can add three (or two, depending if you have dual axis) rate gyros and a three-axis accelerometer (Hitachi H48C), and by the time we are done, we would have a full-blown autopilot system with the Propeller at the center.

Changes to the code could include outputting directly to a KML file so that a user could open up the SD card in a file explorer and directly import the track to Google Earth. Another option would be to skip an SD card altogether and connect the Propeller directly to a laptop computer while driving around and streaming the location-based data over a serial port. A program running on the laptop could then receive the data and save to a KML file while Google Earth is configured to automatically check for updates at a specified rate. By doing this, you have essentially created a moving map display!

Exercises

1 Modify the Main_01.spin so that the file extension on the output file is ".csv" instead of ".log."

2 Modify the Main_01.spin such that instead of outputting a comma-separated values file it outputs a KML file for use with Google Earth.

3 Add to the NTSC terminal display so that it shows the current logging status, such as "Logging: On/Off."

4 Add a digital compass—for example, the Hitachi HM55B compass module, available from Parallax, and record the logged heading. Then compare the compass's heading to the GPS's heading and explain any discrepancies.

5 Shine a flashlight onto the barometric pressure sensor and witness the change in pressure recorded. Try varying the intensity of the light by applying a piece of fabric over the flashlight before shining it onto the sensor.

10

USING THE PROPELLER AS A VIRTUAL PERIPHERAL FOR MEDIA APPLICATIONS

André LaMothe

Introduction

In this chapter, we are going to explore one of the most exciting uses of multicore processing; using the multicore processor as a *slave/server* to a host/client processor or computer. The multicore slave is issued commands via an *RPC* (remote procedure call) architecture, enabling the multicore processor (the Propeller chip, in our case) to execute commands on multiple processors with nearly zero load to the host/client issuing the commands. With this kind of architecture and a proper communications protocol(s), various virtual peripherals can be loaded into the multicore processor and executed on multiple cores, and the client/host can then communicate to these modules over a simple serial link and issue remote commands. Moreover, more advanced software could be developed that allows the host/client to request peripherals be loaded "on demand," creating a truly robust system.

Based on the model developed in this chapter, you will be able to not only control the Propeller chip over a simple RS-232 serial link, but also create a platform that can be used for *SPI* (serial peripheral interface) or other high-speed "chip-to-chip" low-level wired links with little modification. With that in mind, here's what's in store:

- Introduction, setup, and demo
- System architecture
- Remote procedure call primer
- Virtual peripheral driver overview
- Client/host console development

- Exploring the command library for the slave/server
- Enhancing and adding features to the system
- Exploring other communications protocols

Resources: Demo code and other resources for this chapter are available for free download from ftp.propeller-chip.com/PCMProp/Chapter_10.

Overview, Setup, and Demo

Before we start into the technical material of this chapter, let's first discuss the goal of the design, the hardware setup, the software, install anything needed, and, in general, get ready for the project. To enable as many people as possible to use this project "out of the box," it was decided to use a standard piece of hardware: the Propeller Demo Board Rev. C (Revisions A and B will work as well with slight changes to connections).

Referring to Fig. 10-1, the project requires an NTSC TV, VGA monitor, PS/2 keyboard, an audio amplifier (the NTSC TV's audio input will work fine), and the Parallax Demo Board Rev. C or better.

In addition, you are going to need a PC to run the serial communications software on. Technically, any PC that supports the FTDI VCP USB-to-serial drivers connection will work. The project simply needs to be compiled and downloaded to the Propeller

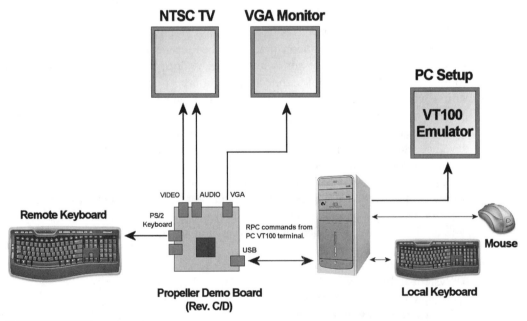

Figure 10-1 Overall system block diagram of what's needed for the setup.

Board; once this is accomplished, you don't need to use the Propeller Tool anymore (which is only officially targeted to PCs running Windows). Nonetheless, this project requires a standard RS-232 communications via a USB-to-serial connection. And since the USB chip on the Propeller Demo Board is an FTDI chip, your PC must support the driver. You can find out more and download the latest FTDI drivers from the Parallax and FTDI sites here:

www.parallax.com/usbdrivers

www.ftdichip.com/FTDrivers.htm

To compile and upload the code into the Propeller Demo Board, you must have a copy of the Propeller Tool (version 1.05.01 or better loaded on your system) and the PC running Windows connected to the Propeller Demo Board over a USB A to Mini B cable.

HARDWARE SETUP

The hardware setup for the project is rather straightforward. With the Propeller Demo Board in hand, make the following connections to the various media devices, as shown in Fig. 10-2:

1 Connect the Video output of the board to your NTSC TV's RCA video input.
2 Connect the Audio output of the board to your NTSC TV's RCA audio input or headphones. (Note: Older versions of the Propeller Demo Board have RCA jacks rather than headphone jacks, so you won't need a stereo-to-RCA converter cable with older versions.)
3 Connect the VGA out of the board to the VGA monitor.
4 Connect a PS/2 keyboard to the board.
5 Connect the Propeller Demo Board to the PC via the mini-USB cable.

Of course, you need to have the Propeller Demo Board 9 V wall adapter plugged in and powered on. Also, remember that step 4 is not only so the PC can ultimately communicate with the Propeller Demo Board via a serial terminal, but also so you can use the Propeller Tool to program the board itself; thus, the single USB serial connection is used for both programming and experimenting.

> **Tip:** Later, you might want to add another serial port to the Propeller Demo Board with an extra Parallax USB2SER adapter (Part #28024). That way, you don't have to play games with the single communications port for both programming and serial communications, and you can simply connect the extra serial port on another pair of free Propeller pins. The problem is that both the Propeller Tool and serial terminal programs like to "hold" the serial line. The Propeller Tool seems to release it after programming, but early versions didn't. However, most serial terminal programs must be closed before releasing the serial line, even though they are not connected.

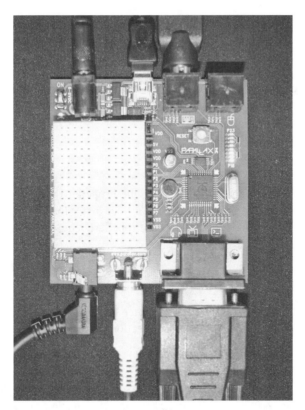

(a) Top view of board

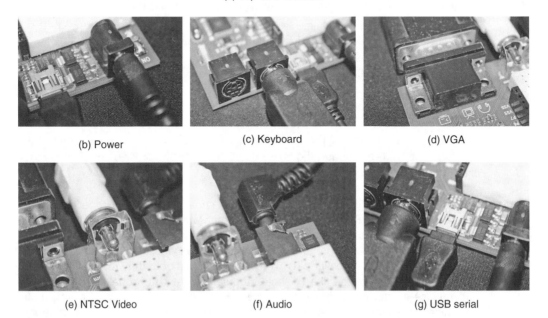

(b) Power

(c) Keyboard

(d) VGA

(e) NTSC Video

(f) Audio

(g) USB serial

Figure 10-2   **Making the hardware connections.**

SOFTWARE SETUP AND INSTALLATION

If you haven't installed the Propeller Tool, do that now. You can download the latest version of the Propeller Tool from the Downloads link at this Parallax site:

www.parallax.com/Propeller

Assuming you have the Propeller Tool installed and set up (refer to Chapter 2 for detailed instructions), confirm you have communications with the Propeller Demo Board with the Propeller Tool's Run → Identify Hardware command, and you can compile and download code to the board. Once you have that set up, the only remaining piece of the puzzle is to load a serial terminal program (the Parallax Serial Terminal doesn't support VT100 emulation, so we need something a little more complete), since this is what is used to communicate with the *CCP* (Console Command Program) that we are going to run on the Propeller Board to interface with the virtual peripherals via serial ASCII human-readable commands.

There are a number of good terminal programs out there—of course, the most obvious is HyperTerminal, but it is very buggy, locks up, and is a system-resource hog. I suggest the following terminal programs:

PuTTY: http://chiark.greenend.org.uk/~sgtatham/putty/download.html

ZOC: http://www.emtec.com/zoc

Absolute Telnet: http://www.celestialsoftware.net

You can download all of them from these links and/or from Download.com. In all cases, we are going to set them up for serial communications over COM*nn* (where *nn* is the COM port your USB driver installed the driver at) with the following settings:

- Baud: 9600
- Parity: None
- Data: 8 bits
- Stop Bits: 1
- Terminal type: VT100
- Local echo: Off
- Handshaking: Off.

For example, Fig. 10-3 shows a screenshot of PuTTY's setup for my PC, which has the USB serial port on COM27.

Tip: Use Control Panel → System → Hardware → Device Manager to locate the new USB serial port. Typically, it will be a rather high number, like COM27 or something ridiculous. You will need this to set up your serial terminal program.

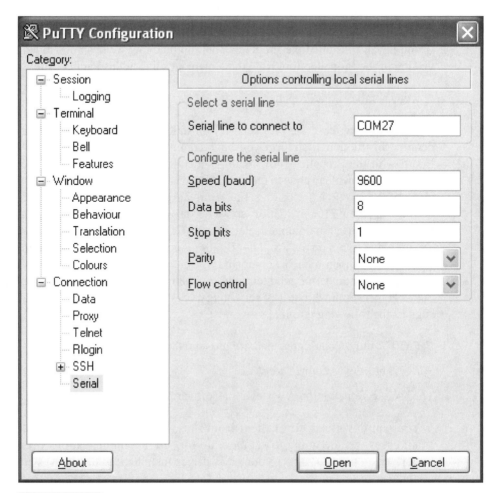

Figure 10-3 A screenshot of PuTTY setup for my PC with the USB serial port on COM27.

TEASER DEMO

Let's go ahead and try the system out and see if we can get it working before delving into how the system works and design issues. To get the program to work, you need to set up the hardware, including the Propeller Demo Board and all the peripheral connections to TV, VGA, keyboard, and so on. Then the software must be compiled and loaded into the board itself via the Propeller Tool. Finally, you need to launch a serial terminal and connect it to the USB serial port that the FTDI chip in the Propeller Demo Board is currently using. Then with the terminal program you will make a connection to the Propeller Board and start throwing commands at the CCP, which processes the commands, parses them, and then passes calls to each of the cores running the various media drivers. Basically, you need to make sure that the serial terminal is on the right COM port and then press Reset on the Propeller Demo Board so everything starts up.

Then you will see the CCP display some initialization, as well as see output on the NTSC and VGA, and hear some sounds—these are all the systems booting.

Compiling the Demo To load the demo into your Propeller Demo Board, locate the top-level source file in the following directory on the ftp.propeller-chip.com site:

PCMProp/Chapter_10/Source/prop_serial_slave_010.spin

This file, of course, relies on a number of other objects, but they are all within the same /Source directory, so the program should compile and download with no problems. Also, there is an archived .zip of the entire project within the /Source directory.

Once you have compiled and downloaded the program into the flash memory of the Propeller Demo Board (F11 is the shortcut on the Propeller Tool), the program will immediately start booting and displaying on the VGA and NTSC, making sounds, and outputting to the serial terminal. If you don't have all these devices hooked up, no worries—just hook them up and press the Reset button on the Propeller Board.

Putting the Demo through Its Paces Figure 10-4 shows photos of what you should see on your NTSC and VGA monitors as the system boots.

Your terminal should also show some initialization and startup strings, along with a final prompt, as shown in Fig. 10-5.

At this point, we are ready to go and start typing commands to the CCP. Let's try a couple of commands and see if things are working. There are two kinds of commands: *local commands* and *remote commands*. Local commands simply control the terminal program and are processed locally by the terminal core processor that is running the command-line interpreter. Local commands are used to talk to the terminal, set things, change fonts, query the time, and so forth. I have only implemented a couple of local terminal commands for illustrative purposes, but you can add more. In a moment, we will see these in action.

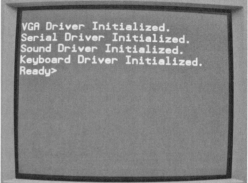

Figure 10-4 Boot process as the system starts out as output on the NTSC (left) and VGA (right) monitors.

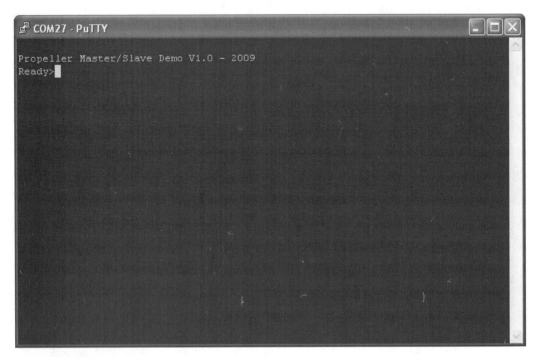

Figure 10-5   The serial terminal output on startup (after a reset).

The remote commands are more useful and do the actual heavy lifting. These commands actually send messages to the various cores (cogs) running the media peripherals and hence allow us to control the cores via the serial terminal. We can do things like print to the NTSC terminal, play a sound, read the keyboard, and so on. All of these things are being processed by the cores of the Propeller in parallel, but we can access them remotely over the serial terminal via commands.

Local Commands Sample   HELP—This command simply prints out the system HELP. Go ahead and type "HELP" into the CCP and watch the HELP scroll by, as shown in Fig. 10-6.

PROMPT {string}—This command allows you to change the actual prompt. It is more of a fun command than a useful one. Try redefining the default prompt by typing the following:

Ready>"Prompt C:\".

If successful, your CCP will look like a DOS prompt now!

C:\

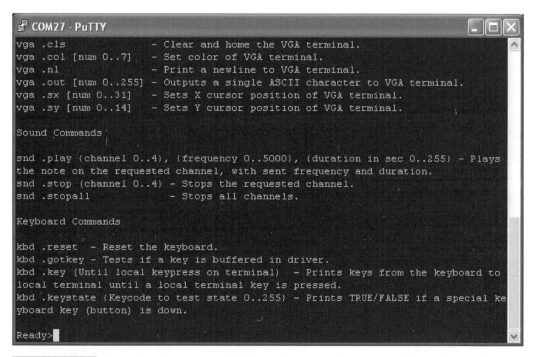

```
COM27 - PuTTY                                                          _ □ X
vga .cls            - Clear and home the VGA terminal.
vga .col [num 0..7]  - Set color of VGA terminal.
vga .nl             - Print a newline to VGA terminal.
vga .out [num 0..255] - Outputs a single ASCII character to VGA terminal.
vga .sx [num 0..31]  - Sets X cursor position of VGA terminal.
vga .sy [num 0..14]  - Sets Y cursor position of VGA terminal.

Sound Commands

snd .play {channel 0..4}, {frequency 0..5000}, {duration in sec 0..255} - Plays
the note on the requested channel, with sent frequency and duration.
snd .stop {channel 0..4} - Stops the requested channel.
snd .stopall         - Stops all channels.

Keyboard Commands

kbd .reset  - Reset the keyboard.
kbd .gotkey - Tests if a key is buffered in driver.
kbd .key {Until local keypress on terminal}  - Prints keys from the keyboard to
local terminal until a local terminal key is pressed.
kbd .keystate {Keycode to test state 0..255} - Prints TRUE/FALSE if a special ke
yboard key (button) is down.

Ready>
```

Figure 10-6   The help menu printing out to the serial terminal.

To change it back, try "Prompt Ready>" and you should see this:

```
READY>
```

Notice that all commands are case-insensitive, the parser capitalizes all inputs.

Remote Commands Sample NTSC .PRINT {string}—This command prints a single string (no spaces allowed) to the NTSC terminal. Try printing your name to the NTSC terminal by typing this:

```
Ready>NTSC .Print Andre
```

Of course, replace my name with yours! If all goes well, you will see your name on the TV connected to the NTSC terminal.

SND .PLAY {channel}, {frequency}, {duration}—The command uses the sound driver to play a pure tone on the speaker (audio output). You control the channel (0...4), frequency (0...5000 Hz), and duration (0...255 s). Here's how you would start a 1 kHz tone on channel 2 that is 5 s long:

```
Ready>SND .PLAY 0, 1000, 5
```

You should hear a nice piercing 1 kHz sound coming out of the TV (or whatever you have connected to the audio out port of the Propeller Demo Board).

KBD .KEY—This command reads the keyboard connected to the Propeller Demo Board and echoes back the keys pressed continuously. Give it a try with the following command:

```
Ready>KBD .KEY
```

Try hitting keys on the keyboard plugged into the Propeller Demo Board, and you will see them on the CCP. To stop the loop, hit a key on the PC's terminal keyboard.

TRYING NOT TO CONFUSE YOURSELF WITH THE LOCAL AND REMOTE KEYBOARDS—OH MY!

One of the traps that is easy to fall into when using remote communications along with multiple keyboards is to forget what is talking to what! I do this all the time, so remember: The keyboard connected to the Propeller Demo Board is the *remote* keyboard. The PC's keyboard is connected only to the serial terminal program, and you are typing to the CCP with it.

That completes our little demonstration of the CCP. As you can see, it's pretty cool and there are lots of possibilities with this technology. You can more or less create a product right now that would allow users to throw commands via serial to a Propeller chip and render video, audio, and more. And with a binary rather than ASCII protocol, you could speed it up 10- to100-fold and make a rather robust system. But let's not get ahead of ourselves yet. It's time now to look at the actual engineering of the whole system and all its pieces.

System Architecture and Constructing the Prototype

As noted earlier in the chapter, it was decided not to build a custom piece of hardware for this project, but to use one of the Propeller Demo Boards (Rev. C). Actually, revisions A and B will work as well, but Revision C and later versions are clean and have built in USB-to-serial. Figure 10-7 shows the reference design for the Propeller Demo Board for reference.

The software is written in such a way that it assumes the connections as shown in the Propeller Demo Board, but you can always change these, as mentioned. In any event, using the schematic as a reference, let's take a look at the overall system design, as shown in Fig. 10-8.

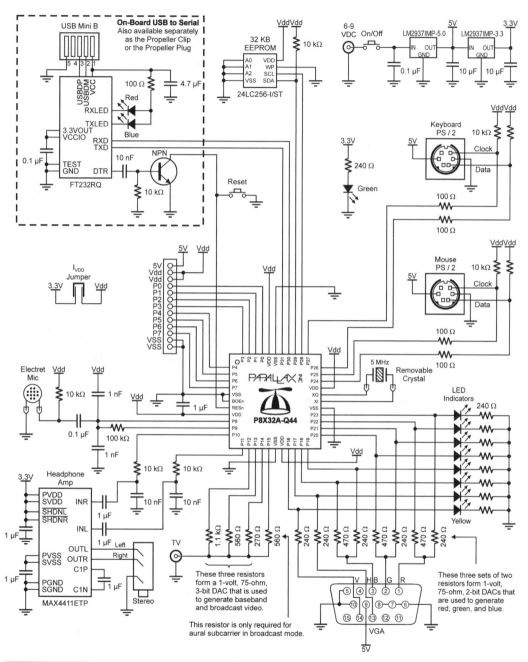

Figure 10-7 **Propeller Demo Board Rev. D schematic.** (*Courtesy of Parallax Inc.*)

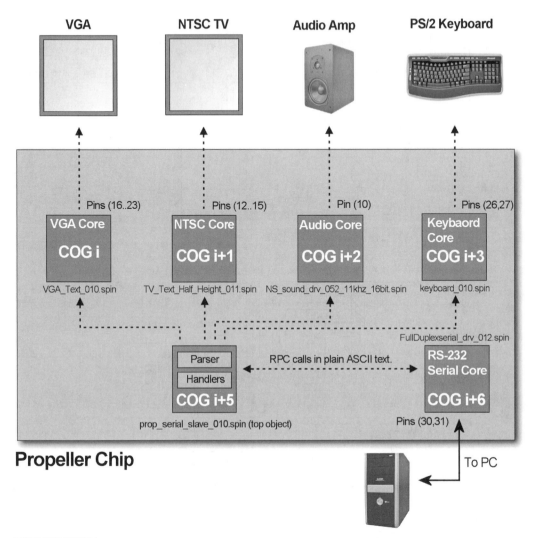

Figure 10-8 System-level modular schematic.

For this project, it was decided to use NTSC, VGA, audio, serial, and PS/2 keyboard devices. This uses up a minimum of five processing cores, leaving us three cores for other things and for future expansion. Of course, one core is used for the CCP; thus, we are left with two ultimately for expansion. As shown in the figure, the NTSC signal is on pins 12, 13, and 14; VGA is on pins 16...23; audio is on pin 10 (or 11); the PS/2 keyboard is connected to pins 26 and 27; and finally, the serial is on pins 30 and 31. This is basically a standard Propeller Board Rev. C/D pin map. Table 10-1 shows the pin map in more detail.

However, if you want to design your own system and move things around, this is no problem; you simply have to make slight changes in each of the drivers to accommodate

TABLE 10-1   THE I/O PIN MAP FOR THE CONNECTIONS TO MEDIA DEVICES

I/O DEVICE	PROPELLER I/O PINS USED
NTSC video	12, 13, 14 (video LSB to MSB)
VGA video	16(V), 7(H), 18(B1), 19(B0), 20(G1), 21(G0), 22(R1), 23(R0)
PS/2 keyboard	26 (data), 27 (clock)
Sound	10 (PWM at left channel of stereo headphones)
Serial coms	30 (TX), 31 (RX)

the pin changes. In most cases, you can simply open up the driver, and the PUB start() method can be investigated to see what the start-up pins are and how they are passed into the driver. The video drivers take a little more work, since you have to decide not only which pins to use, but a couple of other settings, like upper/lower bank, broadcast, baseband, and so on. However, inspection of the driver comments usually gives hints on how to do this.

COMMAND CONSOLE OVERVIEW

Moving on from the hardware connections of the Propeller Demo Board itself, we see that Fig. 10-8 shows some more modules, including the command console program (CCP) and the serial terminal. The CCP runs on its own core on the Propeller chip and listens to the serial line for user input. As the user types on the PC's serial terminal, the CCP buffers the text and acts like a single-line text editor. When the user hits <RETURN>, the CCP tries to tokenize the input line and make sense of it. In other words, does the input have commands in it? Are they valid? And so forth. If valid commands are found, they are passed to the various device drivers running on the multiple cores and the commands are executed. Thus, the CPP has three major components:

- User input handing (editing, line input, echoing)
- Parsing and tokenization of the user input
- Execution of the requested command; messages are passed to the drivers

We will cover the CCP in more detail in the following sections.

SELECTING THE DRIVERS FOR THE VIRTUAL PERIPHERALS

The drivers for each of the media devices were selected based on functionality and popularity. There are definitely better drivers for many of the devices—for example, more robust graphics drivers, more advanced sound drivers, and so on. However, this project isn't about using the best, but more about system integration. Thus, it needs drivers that are easy to interface to and easy to control with a simple subset of commands. In the sections that follow we will cover the exact drivers used, but the point is that they were chosen more or less for ease of use and user base.

COMPLETE DATA FLOW FROM USER TO DRIVER

Now that you have seen all the pieces of the puzzle from a hardware and software point of view, let's review exactly how these pieces all fit together. To begin with, the Propeller is running a number of drivers on multiple cores: NTSC graphics terminal, VGA graphics terminal, keyboard driver, and sound driver. Each of these drivers takes a single core. Now, by themselves, they don't do much. So the "glue" of the system is the CCP, which is pure Spin code that not only issues commands to the drivers, but also is the interface from the serial terminal running off-chip (the PC) and the Propeller chip itself.

Thus, the CCP runs on its own core, controlling the media cores, as well as handing user input from the serial line, which at its other end has a serial terminal running in VT100 mode, like PuTTY was used here. With the system in this known state, let's review exactly what happens in each phase when the system boots.

Initialization. After reset, the software simply loads each of the drivers for NTSC, VGA, keyboard, and serial. Finally, the main PUB start() method of the CCP is entered and the system begins listening to the serial line.

User input loop. As the user types over the serial terminal, the CCP listens to the input. Thus, there is a little toy editor that understands character input, backspace (editing), and the <RETURN> key, which means "process this line."

Processing and tokenization. When the user enters a line of text, the CCP has no idea what it means; thus, the text needs to be "processed" and tokenized into meaningful strings of data. The processing and tokenization phase results in an array of token strings—these might be text, numbers, or symbols—that are ready to be passed to the command processor for inspection.

Command processing and execution. The command processor is a handler that interrogates the previously tokenized input data and looks for commands in the stream. If a command is found, it continues to process the command pattern and looks for the parameters that should follow it. The parameters (if any) are extracted, and then the proper "handler" is entered. The handler is where the action happens. Each handler is connected to its respective driver and can send and receive messages from it. So when an NTSC command is parsed, for example, the handler simply calls the NTSC driver and passes the request for execution.

Remote Procedure Call Primer

Even if you don't have a degree in computer science, you have probably used remote procedure calls, or RPCs, in one form or another, or even invented them unknowingly!

The idea of a remote procedure call came about in the 1970s actually, so it's a rather old concept. The basic idea is simple: for one process/program to be able to call/use a subroutine or function in another process/program. It's more or less a form of inter-process communication.

RPCs are a little different from DLLs or libraries since they are passive entities that are loaded on demand. RPCs are more like making calls to another running program and using its resources (subroutines). Thus, there is a client-server relationship here, and you can think of the RPC call as "message passing," as shown in Fig. 10-9. There are various forms of the technology, and it's more of a concept than a specific algorithm or methodology. For example, in some RPC setups, the RPC call mimics the actual binary footprint of the function call. Assuming the C/C++ programming language, here's a solid example:

```
float DotProduct(float ux, float uy, float uz, float vx, float vy, float vz );
```

Looking at this function, depending on the compiler directives, the parameters are passed by value on the stack from right to left if we assume six floats, each four bytes, there is a return value requiring another four-byte float; thus, we need to "pack" the parameters up into a single record and pass it along and then wait for a single float (four-byte result).

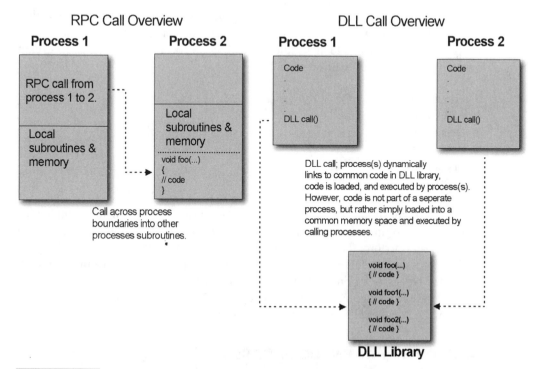

Figure 10-9   RPC call from process to process in contrast to a DLL call.

So we might do something like this to "pack" the parameters up into a contiguous memory space:

```
// This is used as transport buffer for RPC calls
CHAR RPC_input_buff[256];

// Now pack the parms into the array one at a time
memcpy( RPC_input_buff+=4, @ux, sizeof(float) );
memcpy( RPC_input_buff+=4, @uy, sizeof(float) );
memcpy( RPC_input_buff+=4, @uz, sizeof(float) );
memcpy( RPC_input_buff+=4, @vx, sizeof(float) );
memcpy( RPC_input_buff+=4, @vy, sizeof(float) );
memcpy( RPC_input_buff+=4, @vz, sizeof(float) );

// Finally, a call would be made to the "RPC interface"
RPC_Interface( "DotProduct", RPC_input_buff, RPC_output_buff );
```

Now that the parameters are packed into a single data structure, we simply call the RPC interface and pass the starting address of the structure. The RPC interface in this case takes a string as the function name and then two pointers: one to the input parameters and one to where the output results are stored. There is obviously an "agreement" and a set of conventions between the caller and receiver on this function and how it works, so the client can make RPC calls and the server can respond to them. In this case, the server or other process reads the first string, determines the RPC function, and then calls a handler with the two pointers. It's up to the handler to "know" how to unpack the parameters and generate the results via calling the local function. Thus, RPC calls necessitate a number of extra steps, including:

1 Encoding
2 Transporting to server
3 Decoding
4 Executing
5 Encoding
6 Transporting back to client

This is obviously not the fastest thing in the world; however, if the computation workload is two times or more than all the interface steps, or if the local process or machine can't perform the computation, it's worth it. Thus, RPC calls and technology allow a process or machine to use subroutines and resources running in another process or another processor, or on an entirely different machine.

In our case, we are going to use the concept of RPCs to make calls to another processor from the PC's serial interface, thus it's a machine-to-machine call.

ASCII OR BINARY ENCODED RPCs

When designing an RPC system, you can make it really complex or really simple. The main idea is that you want to be able to call functions in another process, processor, or machine. Decisions have to be made about the "RPC protocol" and how you are going to do things. There are no rules for RPC calls, unless you are using a Windows, Linux, Sun, or similar machine and want to use one of the operating system's RPC call application programming interfaces (APIs). When you design your own RPC protocol, it's up to you.

In our case, I decided that since we are going to have a human on one end making the calls via a serial terminal, the RPC protocol should be ASCII. This, of course, requires more bandwidth and is slower than binary. Next, since it is human-readable, the RPC calls take a format that look more like commands rather than strings of bytes representing data.

Thus, our models used in this project should be easy to use and remember for a *human*. The next step up would be to still use ASCII-formatted data that is human-readable, but to make it more abstract. For example, instead of having a command like this:

```
NTSC .Print Hello
```

we might encode "NTSC" as a single number and ".Print" as another number, and then the "Hello" would stay as is:

```
25 0 Hello
```

As you can see, this version of the ASCII protocol is much smaller—we have saved bytes already! It's still human-readable, but not as warm and fuzzy. The entire RPC string is 11 bytes (including NULL terminator). But can we do better? Sure, if we encode in binary, we don't send the longer ASCII text; we send the actual byte data. Therefore, the binary encoding would look like this:

```
Byte 0, Byte 1, 'H', 'e', 'l', 'l','o'
```

. . . where byte 0 and 1 would represent the NTSC and Print subfunctions. In this case, the entire RPC call costs seven bytes, but isn't human-readable anymore since byte 0,1 are in binary and could be anything, depending on what we picked them to be. The point is that if you were using another processor or process to make the RPC calls and not a human at a serial terminal, then there is no reason to use ASCII coding. However, that said, I still prefer sending things in ASCII format during the debugging phase so I can at least look at my data strings on the other end and see what's going on.

COMPRESSING RPC FOR MORE BANDWIDTH

The whole idea of RPC technology is to use it; thus, you might have a program running that makes 100 local calls with 100 RPC calls every time through the main loop. You want the RPC transport process to be quick; hence, compression of the RPC data and/or caching is in order. For example, if you are sending large chunks of text or data that has repeating symbols, it's better to compress it locally, transport it, and then decompress and execute since computers are typically thousands (if not millions) of times faster than the communications links.

In addition, advanced RPC systems might use the *same* data over and over; thus, there is no need to keep sending the data to the server. A caching system should be employed where on the first RPC call, the caller indicates that a data structure being passed is static and cacheable. Thus, the server caches it after its first use. Then on subsequent calls, the client need not send the structure until it needs refreshing. For example, say you have a 3-D database that you want an RPC to perform calculations on. The database never changes, so there is no need to keep sending it over and over; once it's in the server's memory space, you can save the bandwidth.

OUR SIMPLIFIED RPC STRATEGY

Considering all these interesting methods, the method used for our project is human-readable, ASCII encoded, noncompressed RPC calls. This way, you can connect a serial terminal up to the Propeller chip on two pins and start calling the virtual peripherals running on the Propeller and get them to do work and return results. Therefore, our RPC calls all look like ASCII strings starting with a command, maybe a subcommand, followed by parameters and then a NULL terminator. This is the basic "unit" of information the CCP running on the Propeller interprets as an RPC call.

Later, you might want to make a pure "binary" version of the protocol that is not human-readable and much faster. Also, you might want a high-speed "chip-to-chip" version based on SPI (serial peripheral interface) or I²C (inter-integrated communications). More on this idea later in the chapter.

Virtual Peripheral Driver Overview

There isn't much to say about the drivers used on the project other than I went to the Parallax Object Exchange Web site located here:

http://obex.parallax.com/objects

and hunted around for appropriate objects to use for this project based on my experience with developing objects and using them. The objects aren't the fastest, the coolest, or the best necessarily—they just work and get the job done, and in most cases, are the reference objects developed by Parallax initially with small changes by myself and other authors. The idea was to have the NTSC, VGA, audio, and keyboard all running at the same time and be able to access these devices. In the future, you might want to use other objects or improve these for more specific needs.

In any event, referring to Fig. 10-10, the objects used are shown in Table 10-2.

All of the drivers and their subobjects are included in the Source directory for this chapter, located in the /PCMProp/Chapter_10/Source/ directory.

If you look on the Parallax Object Exchange site, you should be able to find all these drivers; however, we will use the ones from my chapter and my sources since I made slight modifications to each of them to make things easier.

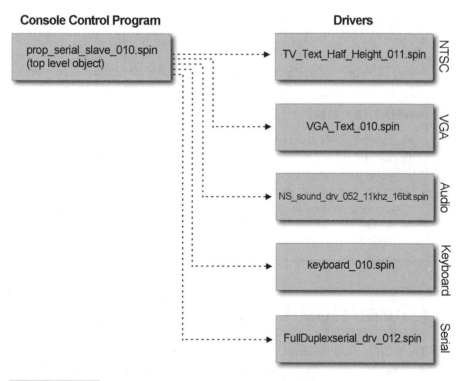

Figure 10-10   The virtual drivers used for the project.

TABLE 10-2 OBJECTS USED IN PROJECT		
FUNCTION	**VERSION**	**TOP OBJECT FILE NAME***
CCP	1.0	prop_serial_slave_010.spin
NTSC	1.1	TV_Text_Half_Height_011.spin
VGA	1.0	VGA_Text_010.spin
Audio	5.2	NS_sound_drv_052_11khz_16bit.spin
Serial	1.2	FullDuplexSerial_drv_012.spin
PS/2 Keyboard	1.0	keyboard_010.spin
* Many of the drivers include other subobjects.		

NORMALIZATION OF DRIVERS FOR COMMON RPC CALLS IN THE FUTURE

The last thing I want to discuss about the drivers is the interfaces to all of them. Since this is a pieced-together system of other people's drivers, each driver obviously has its own methodology and API. For example, the NTSC calls look entirely different from the keyboard calls, and so forth. Alas, if you were to develop a system from the ground up and design drivers for NTSC, VGA, keyboard, and so on, you would be wise to design all the APIs in a similar fashion, with conventions for function calls, inputs, and outputs so that technologies like RPC calls and others could be implemented more easily.

WHAT IS COM?

Microsoft's COM technology, which stands for *component object model,* is actually an example of *normalization* of data and methods. The idea here is that COM defines a binary pattern for data structures and interfacing the methods; thus, any language that follows the specification can use COM objects. Moreover, COM objects can be used remotely, just like RPC calls in a way, using a technology called D-COM or *distributed COM.* So the idea of interface and data structure normalization is a good one.

Client/Host Console Development

The CCP is the ringleader of the whole project; it ties together the user input (RCP commands), drivers, and initialization and monitoring into a single process running on a single core. In this section, we are going to discuss the program in brief and take a look at some of the code. Referring to Fig. 10-11, the CCP is a Spin program that begins with the usual Propeller declarations to initialize the Propeller chip for the target hardware. In this case, that's a Propeller Demo Board. The init code is shown here:

```
CON
_clkmode = xtal1 + pll16x ' Enable ext clock and pll times 16
_xinfreq = 5_000_000     ' Set frequency to 5 MHz
 _stack  = 128           ' Accommodate stack
```

Next, there are some constants to make parsing easier that define a number of ASCII symbols; here's an excerpt:

```
CLOCKS_PER_MICROSECOND = 10       ' Used for delay function

' ASCII codes for ease of parser development
ASCII_A      = 65
ASCII_B      = 66
ASCII_C      = 67
ASCII_D      = 68
ASCII_E      = 69
ASCII_F      = 70
ASCII_G      = 71
ASCII_H      = 72
ASCII_0      = 79
```

Cog 0 (Core 0)

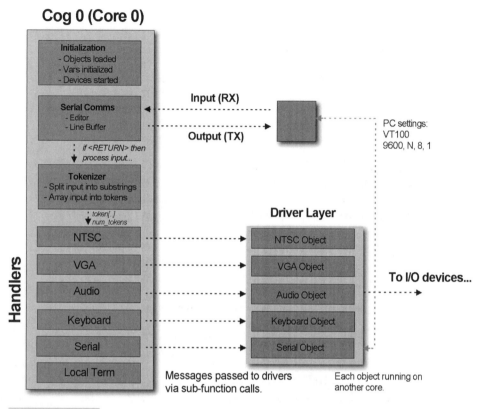

Figure 10-11   The command console program at a glance.

```
ASCII_BS     = 127 ' Backspace
ASCII_LF     = $0A ' Line feed
ASCII_CR     = $0D ' Carriage return
ASCII_ESC    = $1B ' Escape
ASCII_HEX    = $24 ' $ for hex
ASCII_BIN    = $25 ' % for binary
ASCII_LB     = $5B ' [
ASCII_SEMI   = $3A ' ;
ASCII_EQUALS = $3D ' =
ASCII_PERIOD = $2E ' .
ASCII_COMMA  = $2C ' ,
ASCII_SHARP  = $23 ' #
ASCII_NULL   = $00 ' Null character
ASCII_SPACE  = $20 ' Space

' Null pointer, null character
NULL         = 0
```

The VAR section has very few globals. The majority of them are to support the parser and keep state globally for the parser and tokenizer. Here is a listing of the VAR section:

```
VAR

  long cogon, cog        ' Ids for cogs

  byte input_buff[80]   ' Storage for input buffer
  long input_buff_index ' Index into current position
                        ' of command buffer

  byte tok_buff[80]     ' Storage for token buffer
                        ' during processing
  byte prompt[32]       ' Storage for user prompt
  long tok_buff_index   ' Index into current position
                        ' of tokenbuffer
  long token_ptr        ' Used to point to output token
                        ' from tokenizer
  long tokens[ 16 ]     ' Array of pointers to parsed tokens
                        ' ready for processing
  long num_tokens       ' Number of tokens in token array

  long cmd_token        ' A single command token

  long cmd_data_ptr     ' ptr to command token
  long cmd_parse_index  ' Index of command token in array

  ' General parameters used during parameter extraction
  long arg1, arg2, arg3, arg4

  ' State vars for strtok_r function, basically static
  ' Locals that we must define
  long strtok_string_ptr
  long strtok_string_index
  long strtok_string_length
```

Next up are the object includes for all of the drivers we need for our virtual RPC system: NTSC, VGA, keyboard, serial, and sound:

```
OBJ

  serial   : "FullDuplexserial_drv_012.spin"     ' The full duplex
                                                 ' serial driver

  term_ntsc : "TV_Text_Half_Height_011.spin"     ' The NTSC driver

  term_vga  : "VGA_Text_010.spin"                 ' The VGA driver

  kbd      : "keyboard_010.spin"           ' The PS/2 keyboard driver

  snd      : "NS_sound_drv_052_11khz_16bit.spin" ' Sound driver
```

INITIALIZATION

The initialization process for the program consists of initializing a few variables, and then it falls into starting up each of the drivers on their own processing cores (cogs). As each device is started, messages are printed out to the NTSC, VGA, and serial communication lines. Here is the initialization code:

Note: The "←" symbol is used to denote code that wrapped to the next line due to book layout constraints.

```
' Initialize variables section
' for parser
input_buff_index := -1

' Copy default prompt
bytemove( @prompt, @ready_string, strsize (@ready_string)+1)

' Let the system initialize
Delay( 1_000_000 )

' Initialize the NTSC graphics terminal
term_ntsc.start(12,0,0,40,15)
term_ntsc.ink(0)

'Print a string to NTSC
term_ntsc.newline
term_ntsc.pstring(@ntsc_startup_string)

' Initialize VGA graphics terminal
term_vga.start(%10111)

'Print a string VGA
term_vga.newline
term_vga.pstring(@vga_startup_string)

' Initialize serial driver
' (only works if nothing else is on serial port)
' receive pin, transmit pin, baud rate
serial.start(31, 30, %0000, 9600) baud rate

' Announce serial is good to go
term_ntsc.pstring(@serial_startup_string)
term_vga.pstring(@serial_startup_string)

' Start sound engine up on pin xx
snd.start(24)
```

```
' Announce sound is good to go
term_ntsc.pstring(@sound_startup_string)
term_vga.pstring(@sound_startup_string)

' Give everything a moment to initialize
Delay( 1_000_000 )

' Play a little sound to let user know sound is good to go
snd.PlaySoundFM(0, snd#SHAPE_SQUARE, snd#NOTE_C4,←
              ( Round(Float(snd#SAMPLE_RATE) * 1.0)),←
              200, $3579_ADEF )←

repeat 100_000
snd.PlaySoundFM(1, snd#SHAPE_SQUARE, snd#NOTE_C5,←
              ( Round(Float(snd#SAMPLE_RATE) * 1.0)),←
              200, $3579_ADEF )←

repeat 100_000
snd.PlaySoundFM(2, snd#SHAPE_SQUARE, snd#NOTE_C6,←
              ( Round(Float(snd#SAMPLE_RATE) * 1.0)),←
              200, $3579_ADEF )←

'Start keyboard up, 2-pin driver
kbd.start(26, 27)

' Announce keyboard is good to go
term_ntsc.pstring(@keyboard_startup_string)
term_vga.pstring(@keyboard_startup_string)

' Final strings
term_ntsc.pstring(@ready_string)
term_vga.pstring(@ready_string)

' Initial splash text to terminal
serial.tx( ASCII_CR )
serial.tx( ASCII_LF )
serial.txstring( @system_start_string)
serial.tx( ASCII_CR )
serial.tx( ASCII_LF )
serial.txstring( @prompt )
```

If you want to change where signals are on your particular development board, the initialization section is the place to start. For example, the serial driver is started on pins (31,30), which are the standard TX/RX pins for the programming of the Propeller chip. However, you might want to add another serial port and connect it to pins (1,2); thus, you would simply change the (31,30) to (1,2), and you're in business!

SERIAL COMMUNICATIONS, PARSING, AND TOKENIZATION

After initialization, the code immediately falls into the main repeat loop, which begins with the following code:

```
repeat

  ' Attempt to retrieve character from serial terminal
  ch := serial.rxcheck

  ' Character ready in receive buffer?
  if (ch <> -1)
  ' Process character, test for carriage return,
  ' or basic EDITing characters like back space

    case ch
      .
      .
      .
```

This code is more or less the input port to the program. The call to serial. rxcheck checks if a character is ready in the receive buffer; if so, returns it; otherwise, it returns −1. If a character is received, the case ch statement is entered, which has three primary cases:

Case 1: Is the character a non-edit character and non-return? If so, simply insert it into the edit buffer and echo it out on the serial terminal. This is actually the last case in the code itself, and is part of the other case (default for you C/C++ programmers). Here's the code:

```
other:
  ' Insert character into command buffer for processing
  input_buff[ ++input_buff_index ] := ch

  ' Echo character to terminal (terminal must be
  'in non ECHO mode, otherwise you will see input 2x!)
  serial.tx( ch )
```

Case 2: Is the character an "editing" character, such as <BACK_SPACE>? If so, erase the character in the buffer and echo it out as well. This gives the user a crude edit feature; he can at least back up a bit and retype something. Here's the code for that:

```
ASCII_BS:        ' Backspace edit command
  ' Insert null
  if (input_buff_index => 0)
    input_buff[ input_buff_index-- ] := ASCII_NULL

    ' Echo character
    serial.tx( ch )
```

Case 3: The user is done entering data into the command-line buffer and wants the console to process it and hopefully send an RPC to the appropriate driver. This is where all the action takes place and the tokenization of the input string into whole substrings so that the command handlers can query the string and test for properly formatted command strings. Here's the code that is executed when the user hits <RETURN> at the serial terminal:

```
ASCII_LF, ASCII_CR: ' Return

  ' Newline
  serial.tx( ASCII_CR )
  serial.tx( ASCII_LF )

  ' Print the buffer
  if (input_buff_index > -1)
    ' At this point we have the command buffer,
    ' so we can parse it
    ' copy it
    bytemove( @tok_buff, @input_buff, ++input_buff_index )

    ' Null terminate it
    tok_buff[ input_buff_index ] := 0
    tok_buff_index := input_buff_index

    ' Reset buffer
    input_buff_index := -1

    ' Tokenize input string
    num_tokens := 0

    ' Start the tokenization of the string
    ' (this function mimics the C strtok_r function more or less
    strtok_r(@tok_buff, string(","), @token_ptr)

    ' Continue tokenization process now that
    'First token has been found
    repeat while (token_ptr <> NULL)

      ' Upcase the token before insertion
      StrUpper( token_ptr )

      ' Insert token into token array
      tokens[ num_tokens++ ] := token_ptr

      ' Get next token
      strtok_r( NULL, string(","), @token_ptr)

    ' End repeat tokenization...
```

The code fragment does not perform any kind of syntactic analysis whatsoever or validation. All we are interested in is striping delimiters from the string and then breaking out the tokens in the string. For example, here's a sample command line:

```
"sound        .play        0, 100, 10"<RETURN>
```

The parser breaks this into an array of string pointers that looks like this:

```
tokens[0]  →"SOUND",0
tokens[0]  →".PLAY",0
tokens[0]  →"0",0
tokens[0]  →"100",0
tokens[0]  →"10",0
```

. . . along with computing `num_tokens` and setting it to 5. Notice that the strings are converted to uppercase and the comma and white space are removed.

THE COMMAND-LINE INTERFACE

From the user's perspective, the command-line interface is a prompt on the serial terminal running VT100 emulation software. This interface gives the user a powerful terminal editing and rendering capability, if we wish to support the entire VT100 command list.

VT100 TERMINALS

VT100 is a specification that was developed by Dec Equipment Corp. in the 1970s as a method of how a mainframe or mini-computer would interact with "dumb" terminals. Figure 10-12 shows a vintage VT100 terminal. Back then, people couldn't afford to put complete computers on everyone's desk, so "terminals" were in wide use. The VT100, like many other specifications, is a command language that allows the terminal to draw text on the screen, scroll, change colors, make sounds, and so forth. These capabilities are more than enough to support user input, crude editing, and even games! Thus, most serial terminals support the VT100 standard as we use it. Of course, we aren't using much of the standard. In fact, the only VT100 commands we are using are to clear the screen and home the cursor. The interesting thing about the VT100 commands is how they are sent to the terminal. There are a couple ways to do it, but the most commonly used method is to send the "ESCape" character 27, followed by "[" and then the command string. Thus, most commands look like "ESC[xx..x," where xx are ASCII characters. For example, the clear screen command is "2J," so to send this to the terminal, you would send "ESC[2J" to the terminal. This string will not print, but is interpreted by the VT100 emulator and clears the screen. If you are interested in learning more about VT100 commands, try the Wikipedia entry here: http://en.wikipedia.org/wiki/VT100.

Figure 10-12 A DEC VT100 terminal.

Our console application is simple; thus, it doesn't need or use many of the VT100 features, but they are there if needed. The main features we need are for the user to be able to type and edit (via BACKSPACE), and for the console application to perhaps clear the screen, scroll, or change a color (not supported yet). Thus, the interface from the user's perspective is a simple prompt (similar to DOS) that the user can type commands into.

ISSUING COMMANDS TO DRIVERS

The previous two sections described how the user input is entered, parsed into tokens, and stored into the tokens[...] array, so we have everything we need to discuss the command handlers and their communication with the drivers. To review, the idea is that once the tokens[...] array has the tokens in it, the handlers need to query the strings and test for valid commands. Once found, the associated handler is entered and the strings are extracted from the array and converted into the appropriate format (integers, values, and so on). The next step for the handler is to call the associated driver by sending messages to it with the parameters and execute the commands that the user initially requested.

This whole process is rather interesting, so let's take a look at a couple of examples to see what's going on: one example of a locally processed command and one that sends messages to the drivers.

Local Command Processing Example   Local commands are between the user and the console program, and do not send messages to the drivers. In the CCP, there are only three local commands so far: one clears the screen, one prints out the help menu, and the third redefines the prompt (I know—how useless—but, it's cool!). Let's take a look at the CLS (clear screen) command, since it's not only a local command, but also uses VT100 codes and is the first command tested by the CCP handler loop. Here's the code for the CLS handler:

```
' Clear the terminal screen command
if ( strcomp( tokens[0], string ("CLS") ) )
  ' Send the standard VT100 command for clearscreen
  serial.tx( ASCII_ESC )
  serial.txstring( string ("[2J"))    ' Clear screen
  serial.txstring( @system_start_string)
  serial.tx( ASCII_CR )
  serial.tx( ASCII_LF )
```

Let's take a moment to ponder its simplicity. We know that all the user input has been split into the `tokens[...]` array, so the first entry should be the command itself; therefore, the parser tries to match "CLS" in this case. Assuming it has a hit, the handler is entered (ignores any other strings) and simply starts sending the VT100 clear-screen command to the terminal: "ESC [2J." That's it! Of course, there is no error handling, so the user could type "CLS" as well as "CLS your momma," and both would work since the "your" and "momma" strings would be ignored. If you like, you can add more error handling yourself.

Let's take a look at one more local command: the "PROMPT {string}" command. This command allows you to redefine the prompt to something different than "Ready>." This is really for fun and serves no earth-shattering purpose. Here's the code:

```
' Prompt redefinition function
elseif ( strcomp( tokens[0], string ("PROMPT") ) )
  bytemove (@prompt, tokens[1], strsize(tokens[1] ) + 1 )
```

The parser tries to match "PROMPT" in this `case` statement and then enters the handler if a match is found. Now something interesting happens next: The handler needs more data from the `tokens[...]` array—in fact, the next element in the array after the command will be the string the user wants to change the prompt to, so the code uses `tokens[1]` as the string pointer and updates the global prompt string.

Remote Command Processing Example   This is where things get really interesting. Each remote command is tested, just as the local commands, but the difference is that instead of doing something locally to the user interface experience, the command is processed, parameters extracted and converted to proper formats, and then the handler(s) call down to the drivers on the other cores. Let's start with the "SOUND" command, since it's one of the simplest, relatively speaking. The parser is looking for "SOUND" to begin with. Once found, the parser needs to find one of the subcommands ".PLAY" or ".STOP" or ".STOPALL." Each of these subcommands might have parameters as well, so there is a lot going on, but if you write the parsing functions cleanly in each handler, it's a snap. Let's take a look at the code for this handler:

```
{{
Sound Commands

snd .play {chan 0..4}, {freq 0..5000}, {dur in sec 0..255} - Plays the note
on the requested channel, with sent frequency and duration.
snd .stop (channel 0..4) - Stops the requested channel.
snd .stopall          - Stops all channels.
}}
elseif ( strcomp( tokens[0], string ("SOUND") ) or strcomp( tokens[0], string
( "SND") ))

  ' Case .PLAY
  if ( strcomp( tokens[1], string(".PLAY") ) )

    ' SND [channel 0..3] [freq 0..4K] [duration 0...secs]
    ' Enter handler

    ' Channel 0..3
    arg1 := atoi2( tokens[ 2 ] )

    ' Frequency to play
    arg2 := atoi2( tokens[ 3 ] )

    ' Duration in seconds
    arg3 := atoi2( tokens[ 4 ] )

    ' Issue command to sound core
    if (arg2 > 0)
      snd.PlaySoundFM(arg1, snd#SHAPE_SINE, arg2,
                  snd#SAMPLE_RATE * arg3, 255, $3579_ADEF )
    else
      snd.StopSound( arg1 )
```

```
' Case .STOP
if ( strcomp( tokens[1], string (".STOP") )  )

  ' [channel 0..3]
  ' Enter handler

  ' // Channel 0..3
  arg1 := ( atoi2( tokens[ 2 ] ) // 4)

  ' Issue command to sound core
  if (arg1 > 0)
    snd.StopSound( arg1 )

' Case .STOPALL
if ( strcomp( tokens[1], string (".STOPALL") ) )
  repeat arg1 from 0 to 3
    snd.StopSound( arg1 )

' // End if sound
```

There is a lot going on here, so let's focus on one subfunction: the .PLAY code. Locate the highlighted fragment in the previous sample that looks like this:

```
' Case .PLAY
if ( strcomp( tokens[1], string (".PLAY") ) )
```

Once the match for .PLAY has been found, we are good to go and simply need to convert the next three values into numbers representing channel, frequency, and duration. This is where more error handling is needed. Currently, I assume the user has typed numbers into these spots, such as:

```
SOUND .PLAY 0, 100, 10
```

. . . which would indicate to play a 100 Hz tone on channel 0 with a duration of 10 seconds. But he could have typed:

```
SOUND .PLAY  Alpha beta zulu
```

This would get passed to the parser, and the custom atoi2 (...) conversion functions I wrote would blow up. Thus, to make this more robust, the atoi2 (...) functions that convert strings to integers need to test if a string is numeric. In any event, assuming the parameters are valid, tokens[2], tokens[3], and tokens[4] are converted to integers, and then we have all we need to make a call to the sound driver, send the message, and make the RPC call. The actual call is made with the other highlighted fragment copied here:

```
' Issue command to sound core
if (arg2 > 0)
  snd.PlaySoundFM(arg1, snd#SHAPE_SINE, arg2,←
                     snd#SAMPLE_RATE * arg3, 255, $3579_ADEF )←
```

Note: The call to `snd.PlaySound(...)` should be on a single line, but due to the restraints of this book's layout, we had to put the code on two lines. Spin does not allow this in most places, and code must be on the same line (comments can be on multiple lines with the { } or {{ }} syntax).

The call is made to the sound driver with the parameters, and presto—it generates the sound! Thus, the entire process of typing into the serial terminal, user input, tokenization, parsing, and finally command execution are complete!

Before moving on, let's look at the most complex of the commands, which are the graphics commands to the video drivers. Again, it's more of the same: test for the command string, then test for subcommands, parse out the appropriate parameters from the `tokens[...]` array, and make the calls to the driver. Here's the code:

```
{{
NTSC Commands - These commands are processed and issued to NTSC terminal
core.

ntsc .pr [string]      - This command prints a single string
                         token to the NTSC terminal.
ntsc .cls              - Clear and home the NTSC terminal.
ntsc .col [num 0..7]   - Set color of NTSC terminal
ntsc .nl               - Print a newline to NTSC terminal.
ntsc .out [num 0..255] - Outputs a single ASCII character to NTSC terminal.
ntsc .sx [num 0..39]   - Sets X cursor position of NTSC terminal.
ntsc .sy [num 0..29]   - Sets Y cursor position of NTSC terminal.
}}
elseif ( strcomp( tokens[0], string ("NTSC") ) )

  ' Case .PR "print"
  if ( strcomp( tokens[1], string(".PR") ) or
       strcomp( tokens[1], string(".P") ))
    term_ntsc.pstring( tokens[ 2 ] )
    term_ntsc.out( ASCII_CR )

  ' Case .CLS "clear screen"
  elseif ( strcomp( tokens[1], string(".CLS") ) )
    ' Clear screen
    term_ntsc.out( $00 )
    term_ntsc.newline
    term_ntsc.pstring(@prompt)
    ' Set color to normal
```

```
      term_ntsc.out ( $0C )
      term_ntsc.out ( 0 )

   '  Case .OUT "output a single character"
   elseif ( strcomp( tokens[1], string (".OUT") ) or
            strcomp( tokens[1], string (".O") ) )
     term_ntsc.out (byte[ tokens[2] ][0] )

   '  Case .NL "newline"
   elseif ( strcomp( tokens[1], string (".NL") ) )
     term_ntsc.newline

   '  Case .COL "set color"
   elseif ( strcomp( tokens[1], string (".COL") )  )
     ' Extract color 0..7
     arg1 := atoi2 ( tokens[ 2 ]) // 8
     ' Send driver set color command, [$0C, col 0..7]
     term_ntsc.out ( $0C )
     term_ntsc.out ( arg1 )

   '  Case .SX "set cursor x position"
   elseif ( strcomp( tokens[1], string (".SX") )  )
     ' Extract x
     arg1 := atoi2 ( tokens[ 2 ]) // 40
     ' Send driver set x command, [$0A, x 0..39]
     term_ntsc.out ( $0A )
     term_ntsc.out ( arg1 )

   '  Case .SY "set cursor y position"
   elseif ( strcomp( tokens[1], string (".SY") )  )
     ' Extract y
     arg1 := atoi2 ( tokens[ 2 ]) // 30
     ' Send driver set x command, [$0B, y 0..29]
     term_ntsc.out ( $0B )
     term_ntsc.out ( arg1 )
```

Look through the code carefully; if you understand this handler, you can create any additional handlers you need. Once again, you will notice there is virtually no error handling or value range testing. These things must be added if you need them. As an example, let's take a look at the ".OUT" subcommand, which is copied here and highlighted in the previous listing for reference:

```
   '  Case .OUT "output a single character"
   elseif ( strcomp( tokens[1], string (".OUT") ) or
            strcomp( tokens[1], string (".O") ) )
     term_ntsc.out (byte[ tokens[2] ][0] )
```

The ".OUT" command connects directly to the NTSC driver's PUB out(c) method, which is listed next. All we need to do is extract a character from the user input and pass it directly to the PUB out(c) method on the driver side.

```
PUB out(c) | i, k

" Output a character
"
"      $00 = clear screen
"      $01 = home
"      $08 = backspace
"      $09 = tab (8 spaces per)
"      $0A = set X position (X follows)
"      $0B = set Y position (Y follows)
"      $0C = set color (color follows)
"      $0D = return
"    others = printable characters

  case flag
    $00: case c
           $00: wordfill(@screen, $220, screensize)
                col := row := 0
           $01: col := row := 0
           $08: if col
                   col--
           $09: repeat
                   print(" ")
                while col & 7
           $0A..$0C: flag := c
                        return
           $0D: newline
           other: print(c)
    $0A: col := c // cols
    $0B: row := c // rows
    $0C: color := c & 7
  flag := 0
```

In fact, through the PUB out(c) command, the user can set the cursor, change colors, and clear the screen. However, since these functions are so common, I decided to bring them out as separate commands and add to the command list. Therefore, you can actually make the user experience from a command point of view easier, since now the user (or RPC client) doesn't have to remember some cryptic "clear screen" code, but simply needs to remember ".CLS." This is converted by the handler into the appropriate driver messages, and the results are the same.

Exploring the Command Library to the Slave/Server

In this section, we are going to briefly look at all the commands supported by the CCP in a reference fashion. Some simple conventions for the format of commands:

- Parameters are shown in braces { }.
- All numbers are in ASCII format in human-readable format.
- Commands and subcommands are case-insensitive.

LOCAL COMMANDS TO TERMINAL

These commands control the "terminal" itself and do not send messages to any of the drivers.

`Cls`	Clears the terminal screen
`help`	Prints the help menu
`prompt {string}`	Allows the user to redefine the prompt

The next set of commands is remote commands that send messages to the cores running the virtual peripherals; thus, all the work is performed with these commands.

NTSC COMMANDS

These commands are processed and issued to the NTSC terminal core.

`ntsc .pr {string}`	This command prints a single string token to the NTSC terminal
`ntsc .cls`	Clears and homes the NTSC terminal
`ntsc .col {num 0..7}`	Sets the color of the NTSC terminal
`ntsc .nl`	Prints a newline to the NTSC terminal
`ntsc .out {num 0..255}`	Outputs a single ASCII character to the NTSC terminal
`ntsc .sx {num 0..39}`	Sets the X cursor position of the NTSC terminal
`ntsc .sy {num 0..29}`	Sets the Y cursor position of the NTSC terminal

VGA COMMANDS

These commands are processed and issued to the VGA terminal core.

`vga .pr {string}`	This command prints a single string token to the VGA terminal
`vga .cls`	Clears and homes the VGA terminal
`vga .col {num 0..7}`	Sets the color of the VGA terminal
`vga .nl`	Prints a newline to the VGA terminal
`vga .out {num 0..255}`	Outputs a single ASCII character to the VGA terminal
`vga .sx {num 0..31}`	Sets the X cursor position of the VGA terminal
`vga .sy {num 0..14}`	Sets the Y cursor position of the VGA terminal

SOUND COMMANDS

These commands control the sound driver core and play simple tones on up to five channels.

`snd .play {channel 0..4}, {frequency 0..5000}, {duration in sec 0..255}`	Plays the note on the requested channel, with sent frequency and duration
`snd .stop {channel 0..4}`	Stops the requested channel
`snd .stopall`	Stops all channels

KEYBOARD COMMANDS

These commands are used to read the state of the keyboard.

`kbd .reset`	Resets the keyboard
`kbd .gotkey`	Tests if a key is buffered in the driver
`kbd .key` **(Until local keypress on terminal)**	Prints keys from the keyboard to the local terminal until a local terminal key is pressed
`kbd .keystate` **{Keycode to test state 0..255}**	Prints TRUE/FALSE if a special keyboard key (button) is down

If you have trouble getting any of the commands to work, simply review the code and make sure you are entering the parameters correctly for the parser to process them properly.

Enhancing and Adding Features to the System

This demo project is just a starting point—there are so many ways to add to it and improve it. For example, you can add more drivers and peripherals, like a mouse, servos, GPS, Ethernet, and so forth. To add a new device, you would simply include it as an object in the main program, initialize it, and then write a handler for it in the CCP. Of course, you need to have enough cores or cogs to run the new object on since there are only eight. However, another approach would be to have an "on-demand" system where you could start and stop cogs with other objects as needed to manage the number of cogs running at once. For example, say you don't have room for both a keyboard and mouse at the same time but you want to support them. Thus, you would make a function that when you tried to talk to the mouse, if it wasn't running, it would terminate the keyboard processor then start the mouse and vice versa.

ON-DEMAND DRIVERS

Another more advanced idea is *on-demand drivers*. In this design, you would have a multitude of drivers on EEPROM or secondary storage, then you would pull them into memory on-the-fly, start them up on processors, and then the console or RPC client would be able to send commands to them.

SPEEDING THINGS UP

We briefly discussed the merits of human-readable ASCII text and why the decision to go that route was used. However, we also touched upon using ASCII to encode commands not in human-readable form and finally going to 100 percent binary. In the exercises section at the end of the chapter there will be a couple challenges to take what we have now and speed it up, but start thinking about it now. Next, we'll talk about more advanced chip-to-chip protocols and take the serial terminal out of the loop and create a truly high-speed system that runs in the MHz range and is appropriate for chip-to-chip communications.

Exploring Other Communications Protocols

This project uses a simple RS-232 serial communications system since it's well understood, every computer has a serial port (well, they used to!), and the signaling is easy and relatively high-speed (up to 115,200 bps). However, even with pure binary-encoded RPC calls and 115,200 bps, that results in a throughput of RPC calls of about 360/s based on a 32-byte RPC call serial record. This is hardly enough to do much more than print terminal text, play some sound, read keyboards, and so on. No real work can be accomplished with standard serial. Hence, we need to take it to the next level and explore some of the more advanced serial communications protocols that are faster than RS-232. Two schemes come to mind immediately: SPI (*serial peripheral interface*) and I²C (*inter-integrated circuit*). Both of these protocols are electrical and packet-based, and support speeds in the MHz range (25 MHz roughly is common for SPI). Therefore, with either of these protocols, we can increase our command bandwidth for RPC calls by a factor of nearly 10- to 100-fold!

Before we talk about how we might use these protocols and rewrite the current serial code, let's take a few moments and review a crash course on both protocols for those readers who aren't familiar with the technical specifications of each protocol.

SPI BUS BASICS

SPI was originally developed by Motorola. It's one of two popular modern serial standards, including I²C by Phillips. SPI—unlike I²C, which has no separate clock—is a clocked synchronous serial protocol that supports full-duplex communication. However, I²C only takes two wires and a ground, whereas SPI needs three wires, a ground, and potentially chip-select lines to enable the slave devices. But SPI is much faster, so in many cases, speed wins and the extra clock line is warranted. The advantage of I²C is that you can potentially hook hundreds of I²C devices on the same two-bus lines since I²C devices have addresses that they respond to. The SPI bus protocol, on the other hand, requires that every SPI slave has its own chip-select line.

Figure 10-13 shows a simple diagram between a master (left) and a slave (right) SPI device and the signals between them, which are:

SCLK	Serial Clock (output from master)
MOSI/SIMO	Master Output, Slave Input (output from master)
MISO/SOMI	Master Input, Slave Output (output from slave)
SS	Slave Select (active low; output from master)

SPI is fast, since not only is it clocked, but it's also a simultaneous full-duplex protocol, which means that as you clock data out of the master into the slave, data is clocked from the slave into the master. This is facilitated by a transmit-and-receive bit buffer that constantly recirculates, as shown in Fig. 10-14.

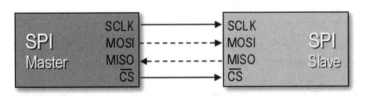

Figure 10-13   The SPI electrical interface.

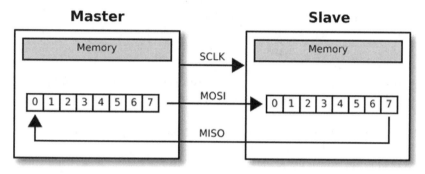

Figure 10-14   SPI circular buffers.

The use of the circular buffers means that you can send and receive a byte in only eight clocks rather than clocking out eight bits to send and then clocking in eight bits to receive. Of course, in some cases, the data clocked out or in is "dummy" data, meaning when you write data and are not expecting a result, the data you clock in is garbage and you can throw it away. Likewise, when you do an SPI read, typically, you would put a $00 or $FF in the transmit buffer as dummy data since something has to be sent and it might as well be predictable.

Sending bytes with SPI is similar to the serial RS-232 protocol: You place a bit of information on the transmit line, then strobe the clock line (of course, RS-232 has no clock). As you do this, you also need to read the receive line since data is being transmitted in both directions. This is simple enough, but the SPI protocol has some specific details regarding when signals should be read and written—that is, on the rising or falling edge of the clock, as well as the polarity of the clock signal. This way, there is no confusion about edge, level, or phase of the signals. These various modes of operation are logically called the SPI mode, and are listed in Table 10-3.

Mode Descriptions

- **Mode 0**—The clock is active when HIGH. Data is read on the rising edge of the clock and is written on the falling edge of the clock (default mode for most SPI applications).

TABLE 10-3 SPI CLOCKING MODES		
MODE #	CPOL (CLOCK POLARITY)	CPHA (CLOCK PHASE)
0	0	0
1	0	1
2	1	0
3	1	1

- ■ **Mode 1**—The clock is active when HIGH. Data is read on the falling edge of the clock and is written on the rising edge of the clock.
- ■ **Mode 2**—The clock is active when LOW. Data is read on the rising edge of the clock and is written on the falling edge of the clock.
- ■ **Mode 3**—The clock is active when LOW. Data is read on the falling edge of the clock and is written on the rising edge of the clock.

Note: Most SPI slaves default to Mode 0, so typically, this mode is what is used to initiate communications with an SPI device.

That about sums it up for SPI. Of course, the Propeller chip does not have any SPI hardware built into it; thus, if we want to talk to the Propeller chip with an SPI interface, we have to write a virtual SPI interface—just like the serial driver itself. Therefore, we have to "bit-bang" the SPI interface. Assuming we want to get each bit of information into the Propeller chip, plus a little overhead, maybe five instructions per bit—that means that a 20 MIP (core is going to be able to keep up with a 4 MHz SPI interface, so something to keep in mind). It's still better than the 115,200 serial connection! For more information about SPI interfacing and protocols, take a look at some of the documents located on the FTP site in the Chapter_10/Docs/spi/ directory.

I²C BUS BASICS

The I²C bus is a little more complex than the SPI bus interface and protocol. The reason is that the I²C bus uses only two signal lines (SDA–data, SCL–clock); thus, more protocols and conventions must be followed to avoid bus contention and other issues. Second, I²C supports up to 128 devices simultaneously connected to the bus. This feature makes I²C superior for "daisy chaining" devices together, as well as cheaper. Of course, you never get something for nothing, and the I²C bus is not without its shortcomings. First, it is nowhere near as fast as SPI. SPI can operate at 25 MHz and even up to 50 MHz. I²C, on the other hand, averages around 100 kHz+, with 400 kHz being fast, and with many new devices supporting 1 to 2 MHz. Thus, SPI is at least 25 times faster. But that's not the whole story. The added overhead that the I²C protocol attaches to communication (addressing, commands, and so on) slow down the protocol even more. Thus, I²C devices tend to find their way into "slower" peripherals where speed isn't an issue but addressing many devices is. For example, with serial memories and sensors, where the device itself is slow, the 100 to 400 kHz average speed of I²C is more than enough for our application. However, you will see SPI devices in very high-speed applications, such as video and audio.

Figure 10-15 shows an architectural diagram of how the I²C bus is laid out in relation to the master device and to the slaves on the line.

Electrically, the I²C bus consists of any number of *masters* and *slaves* on the same bus. Masters initiate communication, while slaves listen and respond. Masters can transmit and receive from a slave, but a slave cannot initiate communications. In addition, to enforce that masters are in charge, the clock line SCL can only be controlled by a master, furthermore placing the slave(s) into a passive role. Multiple devices can

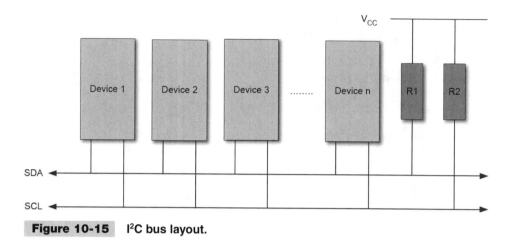

Figure 10-15   I²C bus layout.

be connected to the same two-signal bus from an electrical point of view, so both SDA and SCL are open-drain; thus, "pull-up" resistors must be connected from SDA and SCL to V_{cc} via a 5 to 10 K resistor on both lines. Note that with the SPI bus, all SPI devices share the MISO (master input), MOSI (master output), and SCLK (serial clock) lines; however, the CS, or chip-select, lines for each device control the selection of the target device (not an address, as with I²C). When a device is selected, its bus is active, while any other deselected devices go tristate. Thus, the I²C bus is always active and arbitration is achieved through an open-drain design, where SDA and SCL can only be pulled down by the master and SDA alone by the slave.

The addressing of devices is achieved by a seven-bit address sent down the I²C bus to all slave devices; only the listening device with the matching address responds, and then communication begins. The protocol is rather complex and beyond the scope of this chapter, but some good references are located on the FTP site in Chapter_10/Docs/i2c/ directory.

SPI/I²C AND LOW-LEVEL CHIP INTERFACING

Now that we have discussed some of the technical details of SPI and I²C, let's take a moment and talk about how we might design or alter our current system to take advantage of the technology. First, our current system uses a serial terminal (controlled by a primate) to send and receive commands and data to and from the Propeller chip, which is running the CCP and the virtual peripherals. Thus, we need to remove the serial terminal and human out of the loop, and create a tighter low-level interface based on either SPI or I²C. Let's assume SPI for now.

As Fig. 10-16 shows, this is what we need to implement. The CCP essentially works the same, but now, instead of an RS-232 communications driver, we need to write or obtain an SPI driver. Then, as the SPI driver receives bytes, we need to monitor it for commands, execute them, and return results. This is all going to happen 100 times faster than the serial interface; thus, assembly language might need to be employed to

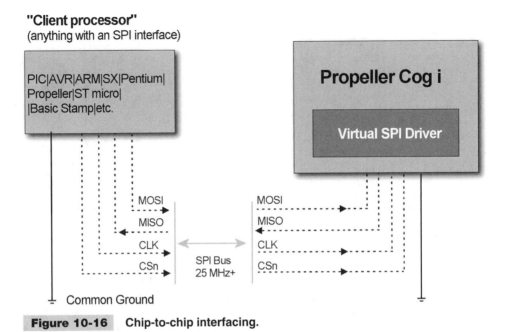

Figure 10-16 Chip-to-chip interfacing.

get maximum performance. However, with this setup, now the command packets (RPC calls) would be encoded as command bytes followed by data bytes, so the interface would be nonhuman-readable, but much faster.

Now the Propeller chip truly acts like a virtual peripheral, and as long as the host/ client chip/system has an SPI interface, we can send commands to the Propeller! In other words, we turn the Propeller into a "black box" that can be controlled with an SPI interface that has a rich set of multimedia capabilities. For example, you could hook a BASIC Stamp up to the SPI interface or another processor and then leverage the power of the Propeller chip over a single SPI interface with a simple set of commands.

USING TCP/IP TO SERIAL FOR FUN

The last topic I want to discuss is more for fun than for anything else, but it's kind of a cool hack. Currently, we are using a PC and a serial terminal program to talk to the Propeller chip, but the serial terminal is running locally on the machine, so at best, we can put a 10- to 20-ft cable out there and connect it to the Propeller Demo Board. But what if we wanted to truly be remote? Well, we could create a real Ethernet connection and connect an Ethernet module to the Propeller Demo Board, but that would require a lot of work and time. However there is a cool trick though, that we can play with TCP/IP and Ethernet—the trick is to load a "serial over TCP/IP" connection on two computers,

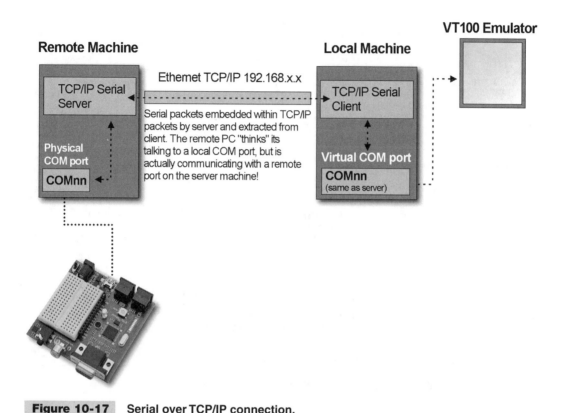

Figure 10-17   Serial over TCP/IP connection.

and then we can create a virtual serial connection over the Internet to connect our serial port to. Figure 10-17 shows this setup.

The idea is that we run a program on the PC that is connected directly to the Propeller's serial port. Then we run another program on a remote PC—these programs link up and then "tunnel" the serial port through TCP/IP transparently, so the COM port on the remote PC seems as if it were local on the remote machine! Presto—we can send commands to our little project over the Internet!

A number of programs can do this. Go to www.download.com and search for "serial over internet, serial over TCP/IP, serial port redirector" and see what comes up. One of my favorites is called the "Eltima Software Serial to Ethernet Connector," located at www.eltima.com/products/serial-over-ethernet.

It allows you to set up a "server" on one machine that serves out the serial port and then a client on another machine that accesses the serial port over TCP/IP. Remember, turn off your firewall, or make sure to poke a hole in the port(s) you use for the connection. Figure 10-18 is a screenshot of the program's graphical user interface (GUI); it's rather slick.

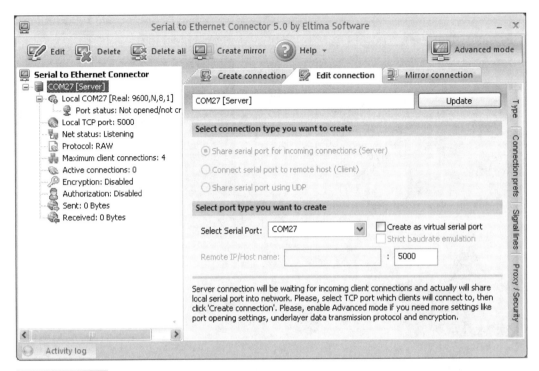

Figure 10-18 A screenshot of the Eltima program in server mode.

Summary

At this point, hopefully, you see the true power of networking Propeller chips and using them as slaves from a client/host computer. Not only can simple serial protocols like RS-232 be used, but lower-level chip-to-chip protocols, such as SPI, I²C, and CAN (control area network), can be employed to truly transform the Propeller chip into a *virtual peripheral,* where the user need not know what's inside it since the behavior is exposed via the interface. Moreover, with the platform developed here and the ideas therein, you can easily see how on-demand virtual peripherals and real-time changes to the cores can be achieved, creating a limitless set of possibilities for the client/host controller to command the Propeller chip and its multicore architecture.

Exercises

1 Add mouse support to the CPP and commands so that you can stream the state of the mouse (x, y, buttons) until a key is pressed on the local serial terminal.

2 Try writing a Visual BASIC, C/C++, Perl, PHP, or other program that opens a serial port up and then controls the CCP by throwing commands at it by building up the command strings programmatically.

3 Convert the CPP to work with an SPI interface. Then connect another Propeller chip and see if you can slave one Propeller from the other.

4 Create a "clock" counter on the Propeller and add a command in the CCP to read it out.

5 Try increasing the serial baud rate and see how fast you can get and maintain reliable communications.

6 Add another USB2SER port to the project, and run the serial terminal through this serial port while using the built-in serial port on the Propeller Demo Board only for programming and downloading code.

7 Add error handling to all the commands in the CCP so the correct number of parameters, types, and values are checked.

8 Using the NTSC commands, see if you can write a *Pong*-style game by drawing and removing characters from the screen. Control it from Visual BASIC, C/C++, Perl, PHP, or some other language running on the PC with an open serial port to the CPP.

11

THE HVAC GREEN HOUSE MODEL

Vern Graner

Introduction

In this chapter, we will explore a Propeller-powered HVAC (heating, ventilation, air conditioning) energy-saving "green" house model. This experimental platform takes advantage of the Propeller chip's low cost and high versatility, allowing us to experiment with intelligently managing a scale model of a typical residential central heating and cooling system.

Through a combination of low-cost components and some powerful software objects, we can easily explore different methods to boost HVAC system efficiency, reduce operating costs, and ultimately make the environment more comfortable for the building's inhabitants. During our exploration, we will be covering the following topics:

- Examining the makeup and drawbacks of central air HVAC systems
- Creating the HVAC green house model
- Designing the control electronics: Scalability and real-world considerations
- Software creation considerations: What gets processed where?
- Modular programming using Propeller objects
- Electronics design and implementation

We'll start out by getting familiar with some inherent problems in a typical residential HVAC system and see how using the Propeller as a system management tool can be educational, interesting and fun!

Resources: Demo code and other resources for this chapter are available for free download from ftp.propeller-chip.com/PCMProp/Chapter_11.

Exploring the Problem

On April 23, 1886, an inventor named Albert Butz patented a furnace regulator he called the "damper flapper." When a room cooled below a predetermined temperature, a switch energized a motor to open a furnace's air damper so the fire would burn hotter. When the temperature rose above the preset level, the switch automatically signaled the motor to close the flapper, damping the fire so it burned cooler. This simple yet ingenious thermostat was the seed that eventually grew into a little company named "Honeywell."

Information: To learn more about Albert Butz and Honeywell, read the article referenced here at www51.honeywell.com/honeywell/about-us/our-history.html.

Though the "damper flapper"-inspired thermostat design is now more than 100 years old, the typical residential central air system controllers of today rely on essentially the same principles. Though the modern thermostat may contain a microprocessor, a real-time clock, timers, and factory-calibrated solid-state temperature sensors, the primary operating mode remains the same; when the air temperature nears a preset point, the thermostat simply turns an HVAC system off or on. Though this method of temperature management is widely used, it has observable issues regarding comfort and efficiency.

One problem with the "binary" nature of a thermostat is that the optimal temperature point is "chased," following a predictable sinusoidal path as it oscillates between "too cold" and "too hot." The reason for this cannot entirely be laid at the feet of the thermostat module itself, but instead may be attributed in part to the design of the heating and cooling portions of a typical central-air system. These systems are usually tuned to produce a *specific* amount of heated or cooled air (usually in cfm, or cubic feet per minute) when activated.

This precludes providing a *proportional* response to temperature needs, as fan speeds and heating/cooling element temperatures are typically set at a fixed value (though some newer HVAC fans do offer high-/low-speed options in a rudimentary attempt to provide some proportionality). As long as the thermostat is calling for heat or cool, most HVAC systems will blindly produce the hottest or coldest air that they can until the thermostat tells them to stop. To make matters worse, it is rare that a residential HVAC system has more than one thermostat, so a single sample is used to provide the standard for the entire building.

Without a way to monitor the temperature in each room, managing the distribution of conditioned air so that all rooms attain the same temperature requires someone to manually visit one room at a time, measure the room temperature, and then adjust the air register in that room in an attempt to "balance" it with the others.

TRYING TO STRIKE THAT BALANCE

This "balancing" process would seem to be relatively simple to perform and not require much maintenance after it has been completed. However, there is more to this process than you might imagine. When balancing airflow in a building, you enter the first room and either increase the opening of the air register to allow more airflow or decrease it to create the opposite effect. As pointed out earlier, in most central air systems, you have a set amount (in cubic feet per minute) of conditioned air being created. So when you reduce the amount of air allowed into one room, air is then diverted to the *other* rooms in the building.

This causes unwanted (and usually unpredictable) changes in temperature elsewhere in the building. The inverse is true as well: When you open a register more fully, you may "steal" air from rooms that were otherwise already comfortable. Determining the amount that an adjustment in one room may affect other rooms is further complicated by factors including air duct size, the distance from the air production source, and the size of the return air path for the rooms.

If this room-to-room "balancing" process is completed by a competent HVAC professional, this complex interrelationship between air register settings can result in reasonably consistent air temperature between all rooms. However, the adjustments are based on the readings taken at a specific point of time and assume that the entire system of rooms is *static*. Unfortunately, in real-world HVAC environments, this is just not the case.

The reality is that a room's heating and cooling needs will change over time as many variables come into play. For example, sunlight shining on an exterior wall, through a window, or on the roof will likely increase the room temperature. The contents of a room can also have a considerable impact on the room temperature. Consumer electronic equipment such as stereo systems, televisions, and computers all create heat when operating. Other typical household devices, such as clothes dryers, refrigerators, stoves, and coffee makers, all release heat as well. Even if a room is relatively devoid of heat-producing equipment, it is extremely likely to have some type of lighting system, with low-voltage halogen lights and incandescent light bulbs that generate lots of heat as a by-product of creating light.

Of course, we can't overlook the inhabitants themselves. Humans give off a surprising amount of heat. It's generally accepted that a typical adult human being gives off about the same amount of heat as a 100-watt incandescent light bulb! Then, just when you thought it couldn't get more complex, think of what happens to room temperatures when you open doors between rooms or, worse yet, open exterior doors and windows!

With so many different variables affecting room temperature at any given time, the idea of attaining consistent, balanced temperatures among all the rooms in a building by the application of a one-time "balancing" of individual registers seems rather unlikely. Even if you could manage to get all the rooms to reach a specific temperature, the ultimate arbiter of temperature is not the thermostat or even a thermometer. It's the *inhabitants* of a given room that decide if the temperature is "right" or would need to be altered to make them more comfortable.

Personal perceptions of "warm" and "cold" vary widely, and it is rare that all of a building's inhabitants will be comfortable at the same room temperature. For example, a sedentary person may be comfortable in a room at 76 degrees, but an active person may feel the room is too warm. Even a sedentary person, when placed in a room with lots of electronic equipment, may require more cooling in order to be comfortable.

The sad reality is that, in most cases, a single thermostat in a multiple-room building just isn't up to the task of creating a balanced, comfortable temperature for everyone. If the thermostat in the building decides to make it cooler or warmer based on its single sample and that causes some rooms to be too hot or cold, that's just too bad. In some situations, it's the "worst case" room that controls the temperature for the whole building. A good example would be cranking up the AC to deal with one hot room in the house and wasting lots of energy "freezing out" other rooms.

So given such an extremely variable and dynamic environment to deal with, is it possible to design an upgrade to a standard HVAC system by adding some technology "smarts" to deal with all these situations? To find out, we built a scale-model HVAC green house on which we could experiment.

The HVAC Green House Model

Two main considerations were taken into account when building the model house for this project. First, of course, we wanted to have a test bed to experiment with mechanical, electrical, and software designs to deal with the HVAC problems outlined previously. Second, we hoped to end up with an attractive and educational display piece we could use to showcase the concept of a dynamically managed HVAC system. We wanted to visually demonstrate that a small amount of technology could make a large difference in energy efficiency as well as the comfort level of the building inhabitants.

CHOOSING THE STRUCTURAL/MECHANICAL ELEMENTS

A number of smaller decisions needed to be made before the construction of the model house began. For example, we wanted all the electronic and mechanical devices to be visible, so using quarter-inch-thick clear plastic panels for the walls, ducts, and structural components was a must. In Fig. 11-1, you can see the preliminary sketch with clear panels.

The trade-off was that this type of material doesn't do a very good job of insulating the rooms from ambient temperatures external to the model or from adjacent rooms. To compensate for this, we did away with windows and doors in the rooms to make each one a "sealed" environment. Each room would only receive air from a single air register and would exhaust air through a single return air vent. We also "oversized" some components, in effect using "brute force" to overcome heat loss/gain from the construction materials. For example, at our scale size of approximately "1 in = 1 ft," the air registers in each room are nearly 1-1/2 ft wide by 1 ft tall, and the air circulation fan would be a whopping 5 ft tall!

The HVAC Itself   The next consideration for this test system was exactly how we were going to cool or heat the air. We searched in vain for an actual Freon-based HVAC system with an associated external evaporator/condenser small enough to fit in our "dollhouse" scale building, but were unable to discover any inexpensive units this small. Luckily, we were able to find a small-scale solid-state heat pump that used a Peltier

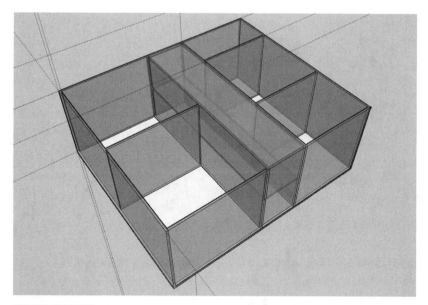

Figure 11-1   Preliminary Google sketch of HVAC green house.

device. Typically used in portable electric "iceless" coolers for RVs and cars, Peltier devices can act as both a cooling and heating device.

The Peltier device we used (Fig. 11-2) consisted of a 40-mm-square thermoelectric heat pump. The device conveniently operates on 12 volts DC and draws 5 A of current. On our test power supply running at approximately 12.3 V, we measured the hot side at 183.6°F and the cold side at 56.2°F. In order to reverse the hot and cold sides, all you need to do is reverse polarity of the applied 12 volts DC.

The Cooling Tower   We mounted the Peltier device in a clear polycarbonate "cooling tower" with 12 V high-speed fans on both the external "waste" heat sink and the internal "supply" heat sink that would serve to heat or cool the inside of our model house. This cooling tower (Fig. 11-3) was made as a separate enclosure, allowing us to remove it from the model for maintenance or cleaning and also for access to the send and return air ducts and the center wiring channel (aka "wire chase").

The Air Registers   As we now had the capability to cool and warm air, we needed a way to remotely adjust the *amount* of air delivered to each room. This was a key concept in allowing us to experiment with balancing how much air was delivered to a given room. Though commercially available air registers used in residential construction consist of a series of adjustable louvers, we decided that a simple sliding panel would allow us to adjust the size of the entry port to the room while also making it clear to an observer exactly how much of the air vent was occluded. We fabricated brackets and used servo motors with linkages to adjust the sliding port cover as shown in Fig. 11-4.

Figure 11-2 12-volt "Peltier" solid-state thermoelectric heat pump.

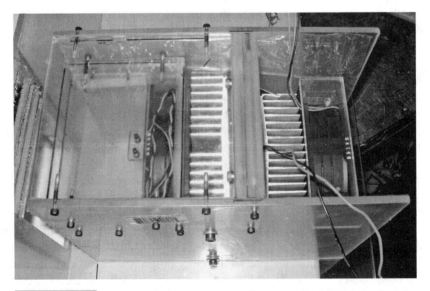

Figure 11-3 Clear polycarbonate "cooling tower" with external and internal blower fans mounted.

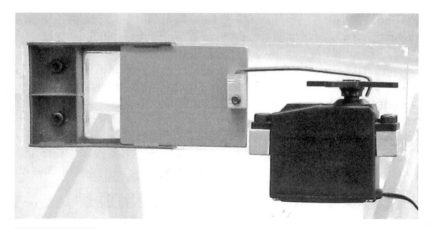

Figure 11-4   Servo-controlled air register prototype.

The Air Bypass System   One of the last things to be implemented was an air bypass system (Fig. 11-5) that would allow us to either recirculate the existing air in the house for normal operation or draw air from outside the house, pass it through all rooms, and then exhaust it out the opposite end of the house for "fresh air" mode of operation.

The idea behind this feature was twofold: First, in the event that the external ambient air temperature was sufficient to cool or heat the inside of the house, engaging this

Figure 11-5   Photo of cooling tower servo-controlled "bypass" doors.

bypass and activating only the blower fan would allow the house temperature to be brought to a comfortable level without expending energy operating the heat pump. This part was a key piece of making our HVAC system more "green."

Second, in the event that the air inside the house was found to be unhealthy or dangerous (i.e., a natural gas leak or a buildup of carbon monoxide), the system could react by sounding an alarm, moving the bypass doors into "fresh air" position, moving every room air register to 100 percent open, and activating the blower fan, thereby purging unhealthy air from the entire house.

To implement this bypass system, two servo-controlled doors were added to the cooling tower that, when open, allow air to recirculate normally through the system (see Fig. 11-6).

When closed, they block the return-air plenum path to the fan while opening side vents in the cooling tower. These side vents allow external air to be drawn in to the fan from outside the cooling tower (a source outside the house), pushed through all the rooms in the house, and then exhausted via a vent flap at the opposite end of the house as shown in Fig. 11-7.

If you look closely at the CAD drawings of the house, you may notice that when the bypass doors are placed in the "fresh air" mode (i.e., the return air duct pathway is blocked), the air delivered to each of the rooms has no place to exit. To deal with this,

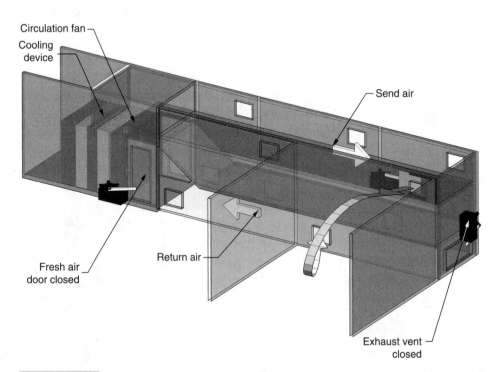

Figure 11-6   CAD "cutaway" drawing of cooling tower bypass doors in "recirculate" mode.

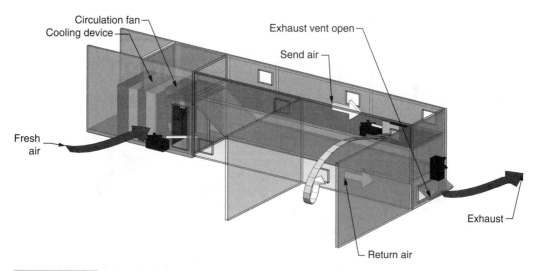

Figure 11-7   CAD "cutaway" drawing of cooling tower bypass doors in "fresh air" mode.

a servo-controlled exhaust vent "flap" (Fig. 11-8) was added to the end of the return air duct. When this flap is opened, the air exiting each room is vented out the opposite side of the house from the fresh air intakes on the cooling tower.

Besides being used for emergency house ventilation or for cooling/heating the house with external air, the cooling tower bypass could also be activated manually, creating a healthier atmosphere by bringing more fresh air into the house or even as a way to reduce odors from cooking or smoking.

Figure 11-8   Servo-controlled exhaust vent flap at the end of the return air duct (in closed position).

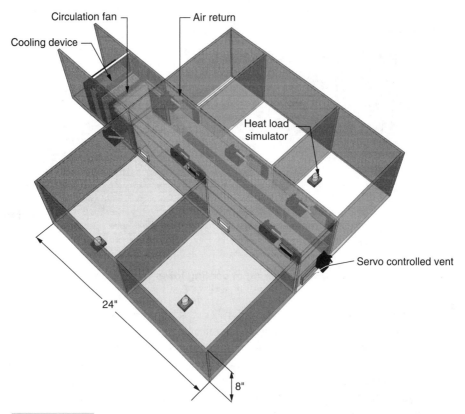

Circulation fan

Air return

Cooling device

Heat load
simulator

Servo controlled vent

24"

8"

Figure 11-9   CAD drawing of final design configuration.

The Final Layout   In the final design, we settled on five rooms, each one a different size to simulate different-sized rooms in a typical house as shown in Fig. 11-9.

A servo-controlled air register was created for each room and mounted in a central plenum that spanned the house from the cooling tower in front all the way to the back of the building. Beneath the "send" air plenum, we created a wire chase to allow us to run wire from each room back to a master control area.

The Hinged Roof Panels   In addition to being able to control the amount of air delivered to each room, we needed a way to set a target temperature per room. We created hinged roof panels for each room and mounted pushbuttons that would allow an observer to alter the target temperature up or down simply by pressing one of the buttons. Above the buttons we added 2 × 16 LCD panels that would display both the current temperature and the target temperature to the observer as shown in Fig. 11-10.

Once all these mechanical components were mounted, it was time to design the electrical systems that would bring this creation to life.

Figure 11-10   Hinged top panels with RED/BLUE temperature control buttons and LCD displays.

DESIGNING THE CONTROL ELECTRONICS: SCALABILITY AND REAL-WORLD CONSIDERATIONS

Though we are working with a model system, we don't want to preclude the possibility that this design may be implemented on a large scale, possibly even being used in an actual residential HVAC system. Since the Propeller chip has the ability to control multiple servos and to read multiple sensors, we sketched one of the first designs using a single Propeller to operate the entire system. This would require a "star" wiring design where each room would have a separate line running all the way back to a central wiring closet.

When planning an installation of technology such as this in a new home, or retrofitting an existing home for something similar, wiring requirements are a substantial issue, and the downside to a "star" design is the cost. There is both the cost of the wire itself and of the associated labor to run all the lines. Though the "star" wiring scheme was inherently expensive, it wasn't a deal killer in and of itself.

However, the second strike against this design had to do with the length of the cable runs from each room back to the wiring closet. Typically, the distance between a Propeller chip and devices such as the 5 V transistor-transistor logic (TTL)-level serial LCD unit or pulse width modulation (PWM)-controlled servo motors would be on the order of a few inches to maybe a couple of feet. At real-world distances of 50 ft or more, there was a very good chance these devices would become unstable or completely fail to operate due to voltage loss over distance and/or wire capacitance "rounding" the edges of the square wave signals.

The final nail in the coffin for this design was that a single Propeller chip simply did not have sufficient pins to support many rooms, even when using low pin-count "intelligent" devices. For example, if each room had a 2 × 16 serial LCD display, a servo motor, two pushbutton switches, a temperature sensor, and two indicator lights, the minimum pin count per room would be as follows:

1 5 V+ power
2 GND
3 Serial LCD signal
4 Servo motor control
5 Button 1-to-gnd
6 Button 2-to-gnd
7 "1-wire" temperature sensor
8 LED1
9 LED2

As the Propeller chip typically has 28 available pins, this design would only allow *three* rooms to be monitored before all the pins on the chip had been allocated! The one-Propeller design also called for more expensive components, such as serially controlled LCD displays and 1-wire temperature sensors, as well as additional expenses in cabling because readily available four-pair CAT-5 cable doesn't contain enough wires. So, for implementation, either *two* CAT-5 cable runs would need to be run from each room back to the wiring closet or special-order wire containing at least nine leads would need to be purchased.

If One Propeller Is Good...   As the single-Propeller design was clearly not feasible, we started to look for other design ideas. The next logical step was to see if we could place the sensors and controls closer to the processor. We started looking at placing one Propeller in *each room,* as this idea solved multiple problems and provided additional benefits. First, the wiring issue would become much less complex, as we need only supply power, ground, and TX/RX lines for digital communication. Second, we would be able to reduce costs by moving away from the more expensive "intelligent" peripherals and substituting a standard "parallel" LCD display for the serially controlled unit and replacing the 1-wire temperature sensor with an inexpensive DS-1620 chip.

Having the Propeller in the room placed it in close proximity to the RC servo motor, eliminating worries of signal degradation affecting performance. Also, driving the LCD from the Propeller eliminated concerns of long wire runs corrupting serial TTL signals to the LCD. As we mentioned earlier, because the Propeller has 28 available pins, we have the option of adding items such as the inexpensive HS-1101 humidity sensor, more indicator LEDs, and even a speaker for acoustic feedback and/or alarms.

The "one Propeller per room" design also lent itself to using a network protocol and having multiple units share a single cable. This eliminated the need for "star"-type wiring, allowing all the units to be "daisy chained" on a single cable from room to room. This reduces wiring cost in new construction and greatly simplifies installation in existing buildings. So, using our "one Propeller per room design," our new cable pair consumption would simply be:

Pair 1: +5 V/GND

Pair 2: TX/RX

This new reduced wiring consumption also means less expensive two-pair wire (i.e., telephone cable) could be used to connect the rooms to the wiring closet. In our prototype, we decided to use four-pair CAT-5 cable to support the possible current draw that we may see from the servo motors. In addition, we decided to put +12 V on one pair to drive 12 V indicator lights in each pushbutton switch.

Developing Control Board Prototypes   At this point, we decided to build a test "room control" board (Fig. 11-11) to see how well it would function. The first one was

Figure 11-11   The first prototype "room control" board with LCD display, temperature sensor, humidity sensor, servo driver, and pushbutton switches.

Figure 11-12   The second prototype "room control" board.

built using the low-cost Parallax Propeller Proto Board USB, and some rudimentary code was written to display current temperature and humidity, the "target" temperature for the room, and the position of the servo motor that would control the register.

Based on the success of the first board, a second board (Fig. 11-12) was carefully constructed by hand based on the layout of the first prototype.

As we expected these boards to communicate via CAT-5 cable, we needed to add RJ-45 jacks. The only PC-board-mount RJ-45–style connectors we could find did not have .100" pin spacing, making them rather difficult to install on the Proto Boards. Hand routing wires also led to a couple of mistakes, resulting in time being spent troubleshooting the second Proto Board. At this point, it was becoming clear that building each board by hand was going to be more time-consuming than we originally thought and more error-prone than we would like, especially considering how many boards would need to be built.

There would be five rooms in the model house requiring a room control board, and we wanted to have at least one board prebuilt and programmed as a spare to make it easy to repair the model if there were failures. In addition, we wanted to have some boards to use for testing and programming without having to be in possession of the entire model (which, at 3 ft × 4 ft in size, was decidedly nontrivial to move!). We hit upon the idea of creating "daughterboards" that would attach to the Propeller USB Proto Boards via female pins soldered around the surface-mounted Propeller chip, as shown in Fig. 11-13.

Figure 11-13   Female .100" headers soldered onto the Propeller USB Proto Board.

We then laid out a schematic for the room control board (Fig. 11-14) where all the components would be populated on the daughterboard and would connect to the underlying Propeller Proto Board through standard .100" spaced pins. Not only would this design allow us to create multiple room boards, it would also allow us to use the PC-board-mount RJ-45 jacks. This design had the added benefit of being "repairable." In the event of a failure of the surface-mount Propeller chip, we could simply "swap" the Propeller Proto Board with a new one, as none of the room board components were soldered to the Propeller Proto Board.

Using the daughterboard design had enough advantages that we wanted to forge ahead. However, we didn't want the associated "lag time" and expense of having a commercial board house etch the boards for us, so we went ahead and auto-routed the schematic to create a printed circuit board but we then converted to a "trace isolation" style board (Fig. 11-15) so it could be cut and drilled using the three-axis CNC system available in-house.

A Three-Board Solution   As this method of creating boards worked rather well, we decided to use the same process to create daughterboards for the additional functions that would be required. For example, each room board would need to be able to send data back to a "master board." The master board collects room temperatures and calculates servo positions to control the air registers in each room. A new schematic was created for the master board, and it was then cut on the CNC system using the same board size and pin footprint as the "room board."

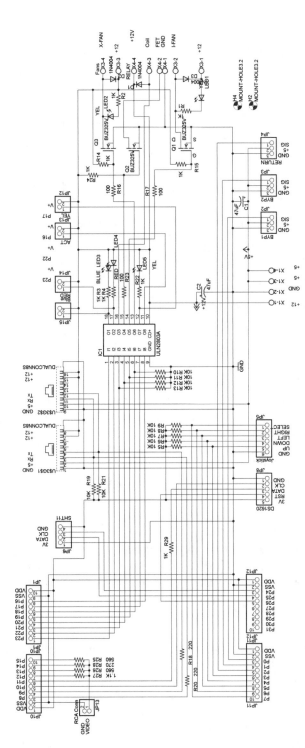

Figure 11-14 Preliminary schematic of the "room board" daughterboard.

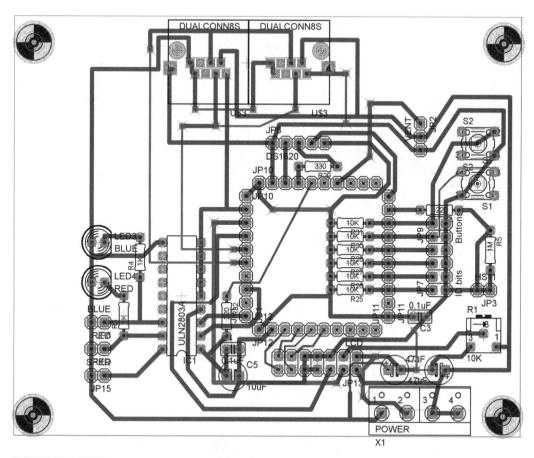

Figure 11-15   Room board daughterboard design.

At first, this master board had both an NTSC output to be used for a GUI and the components to control the blower fans and heat pump. After thinking about "real-world" requirements, we decided that it would be more realistic to have the "control" components located outside the house or in the attic near the HVAC equipment. Subsequently, we decided that a third board would need to be created that would host all the high-amperage control components that would operate the heat pump and the blower fans. This board would also host a sensor that would read the *outdoor* ambient temperature and humidity to be used for determining if the outside air were suitable to use for indoor heating or cooling. As this board would be nearer the blower fans (in the cooling tower), we placed headers for servo motors to control the cooling tower bypass doors, as well as the return-air bypass vent flap. With its feature set complete, we dubbed this new board the "control board."

Another round of autorouting, and CNC cutting/drilling now left us with three board styles: "room," "master," and "control," all with the same physical size and pin configuration but with different component loads and roles. Figure 11-16 shows the results.

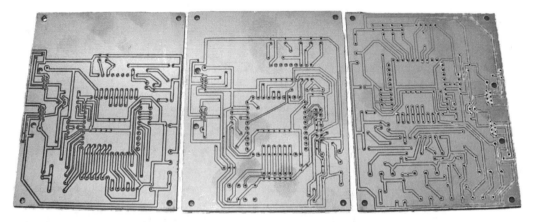

Figure 11-16 The "room," "master," and "control" daughterboards before being stuffed and soldered.

After stuffing, soldering, and carefully testing the daughterboards and the master board, we attached them to the Propeller Proto Boards and used aluminum standoffs on the four corners to stabilize the connections. The resulting board sets, shown in Figs. 11-17 and 11-18, were robust, compact, repairable, and reproducible.

So, now that we had the hardware designed for the house, it was time for some software design.

Figure 11-17 The "control" daughterboard mounted on the Propeller Proto Board.

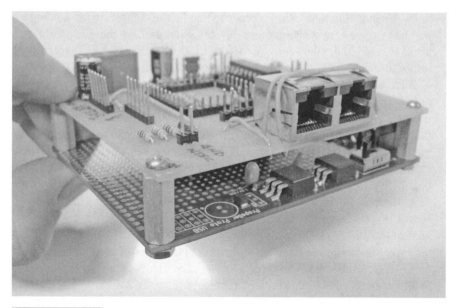

Figure 11-18   The "master" daughterboard mounted on the Propeller Proto Board.

SOFTWARE DESIGN CONSIDERATIONS: WHAT GETS PROCESSED WHERE?

Having processing power in each room made some functions intuitively obvious. Here is a listing of the functions that could easily be handled by the single Propeller chip in each room:

- Servo PWM generation
- Parallel LCD control
- Pushbutton detection/debouncing
- Temperature sensor polling
- Humidity sensor polling
- Status light control
- Target temperature display
- Communications (TX/RX)

Something to remember is that the design of the Propeller chip lends itself well to providing *parallel* functions and services. For example, servo motors require a steady stream of pulses in order to maintain their position. Therefore, it is possible to use an object to constantly supply these pulses. You need only tell the servo control object at what position the servo should be, and the object takes care of the rest. The same can be said for the parallel LCD. There are Propeller objects that make interacting and updating a parallel LCD as simple as interacting with the more expensive serial-controlled LCD units. A serial communication object can handle the complexity of sending and receiving data over a network with other devices. So, based on these capabilities, the room board would handle the servo motor, the LCD, the temperature sensors, the humidity sensor, user interface status lights, and network communications.

Since we knew there would be more than one room board, and we were hoping to reduce the wiring costs by using a daisy-chain wiring configuration, we decided to use an "open collector" communication approach and allow all the boards to share the transmit and receive pins. To make sure that only one board at a time would be transmitting, a polling method of communication was chosen where the master board would call each room board and only then would the room board send an answer to the master board. To differentiate the room boards, we placed a bank of four jumpers connected to Propeller pins P0-P3 to allow us to set an ID number, allowing a maximum of 16 room boards on our "network" at one time. The room boards would simply ignore any poll request that did not begin with their unique address.

Since the master board needs only two pins to perform its duties as the network master (i.e., just transmit and receive lines), it has plenty of pins (and processing power!) to allow for an advanced intuitive graphical user interface. By using the NTSC video output object, it is a straightforward process to display statistical information about system performance and present an advanced graphical user control interface through an embedded NTSC video monitor. For our experimental platform, we are currently implementing the NTSC video output to display current temperature, target temperature, and vent position for each room on a Parallax Mini LCD A/V color display (Fig. 11-19).

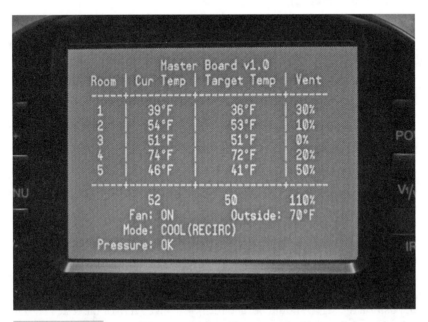

Figure 11-19 The NTSC monitor with the preliminary system test data displayed.

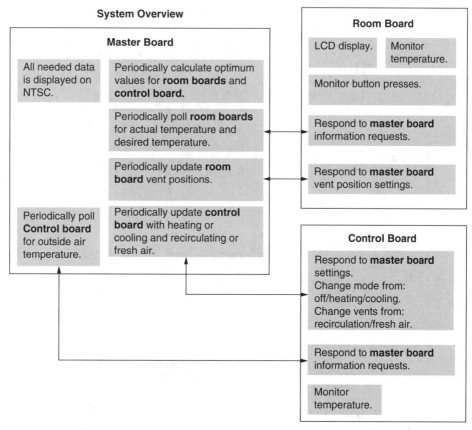

Figure 11-20   System overview flow chart.

The main job of the master board is to poll each of the room boards, retrieve their current temperature, retrieve the desired temperature, and then perform calculations that are used to direct the operations of the rest of the system. A quick recap of all the boards, their functions, and their interactions is shown in Fig. 11-20.

The point of building the HVAC green house was to create a test platform that would allow experimentation with different approaches to air system management and to observe how these approaches perform in this small-scale environment. Though full-scale environments may react differently, experimentation on a small scale can be quite useful in determining what approaches to use in a full-scale environment.

The flow chart in Fig. 11-21 is the first attempt to use an algorithm to balance and maintain the air temperature in the building.

As you can see in Fig. 11-21, we first calculate the average internal air temperature and then compare that with the outdoor air temperature to determine if the system should be in "heating" or "cooling" mode. The Spin code to accomplish that is listed here:

Note: Lines marked with the "←" symbol are intended to be on a single line and wrap here due to space constraints.

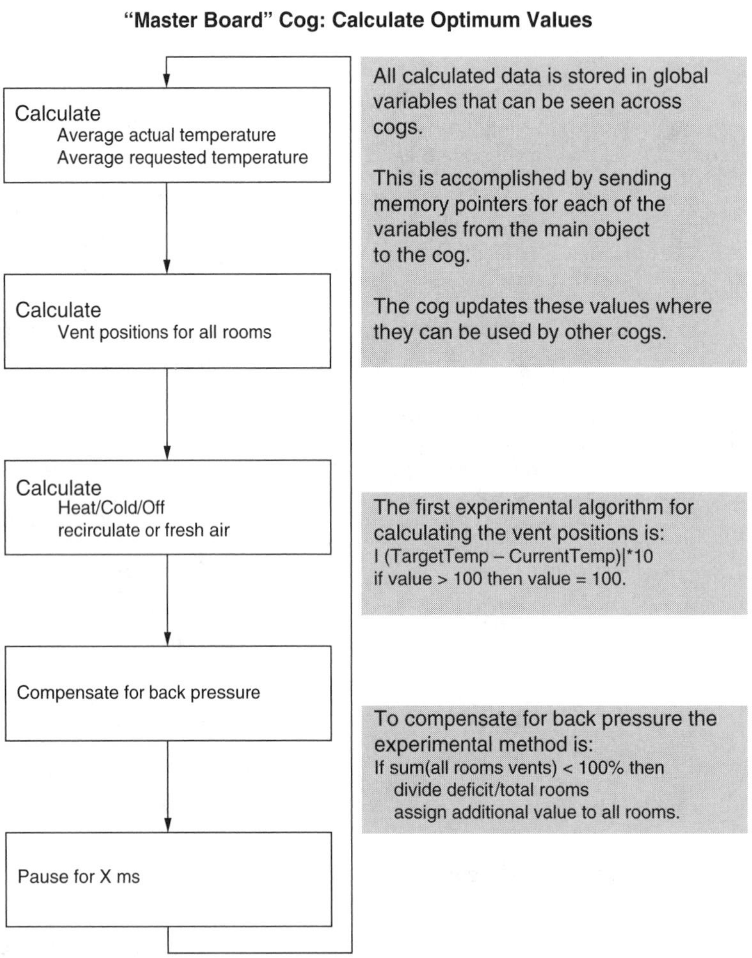

"Master Board" Cog: Calculate Optimum Values

Figure 11-21   Master board cog flow chart.

```
' Calculate Room averages
OverallAverageTemp10X:=0
OverallAverageDesiredTemp10X:=0
repeat count from 0 to 4
  OverallAverageTemp10X:=OverallAverageTemp10X+ ←
    LONG[RoomCurrentTemp10X+(count*4)] ←
  OverallAverageDesiredTemp10X:=OverallAverageDesiredTemp10X+ ←
    LONG[RoomTargetTemp10X+(count*4)] ←

OverallAverageTemp10X:= OverallAverageTemp10X/5
OverallAverageDesiredTemp10X:= OverallAverageDesiredTemp10X/5
```

Vent positions are then calculated for each room based on its "need" according to its temperature reading. Again, the Spin code to accomplish this is straightforward:

```
' Calculate Vent positions
 totalvent:=0
 repeat count from 0 to 4
   temp:= || (LONG[RoomTargetTemp10X+(count*4)]- ←
     LONG[RoomCurrentTemp10X+(count*4)]) ←
   if temp > 10
     temp:=10
   temp:=temp*10
   LONG[RoomCurrentVentPosPercent+(count*4)]:=temp
   totalvent:= totalvent + temp
```

Next, using the calculated average internal air temperature compared to the external air temperature allows us to automatically determine if the Peltier solid-state heat pump should be operating in a "cooling" or "heating" mode:

```
'''''''''''''' HEATING ''''''''''''''
 if OverallAverageDesiredTemp10X > OverallAverageTemp10X ' Heating
   if OverallAverageDesiredTemp10X < outsideTemp
     'Fresh
     LONG[BlowerFanPower] := POWER_ON
     LONG[PeltierPower] := POWER_OFF
     LONG[PeltierDirection] := PELTIER_HOT_INSIDE
     LONG[AirSupply] := FRESH_AIR
   if OverallAverageDesiredTemp10X > outsideTemp
     'Recirc
     LONG[BlowerFanPower] := POWER_ON
     LONG[PeltierPower] := POWER_ON
     LONG[PeltierDirection] := PELTIER_HOT_INSIDE
     LONG[AirSupply] := RECIRCULATE

'''''''''''''' COOLING ''''''''''''''
 if OverallAverageDesiredTemp10X < OverallAverageTemp10X ' Cooling
   if OverallAverageDesiredTemp10X > outsideTemp
     'Fresh
     LONG[BlowerFanPower] := POWER_ON
     LONG[PeltierPower] := POWER_OFF
     LONG[PeltierDirection] := PELTIER_COOL_INSIDE
     LONG[AirSupply] := FRESH_AIR
   if OverallAverageDesiredTemp10X < outsideTemp
     'Recirc
     LONG[BlowerFanPower] := POWER_ON
     LONG[PeltierPower] := POWER_ON
     LONG[PeltierDirection] := PELTIER_COOL_INSIDE
     LONG[AirSupply] := RECIRCULATE
```

Once the mode of operation is determined, a comparison between the average desired temperature and the current outdoor temperature could be made. If bringing the outside air into the rooms would allow them to reach their target temperature without activating the heat pump, the control board may be instructed to open the cooling tower bypass doors, allowing outside air to be sent into the rooms.

This "fresh air" mode of operation could also be used in conjunction with various gas sensors to react to an unhealthy air situation. For example, a buildup of carbon monoxide due to leaving a fire burning, a propane leak from a stove/heater, or even a broken sewer line allowing methane into the building all could be dangerous, even life-threatening situations. It would be possible, using sensors currently available from Parallax, to detect these situations and not only alert the building occupants by sounding an alarm, but also activate the "fresh air" mode.

This would bring fresh air indoors and vent the dangerous gases out of the building. In the event that a fire is detected, every room vent could be instructed to close in order to reduce the spread of smoke through the building and keep from feeding "fresh air" to the fire.

Another real-world requirement has to do with *backpressure*. When the blower fan turns on, it generates a specific number of cubic feet per minute of air. In order to operate properly, a certain amount of air must be allowed to exit the ducting system. If the room air registers are closed, backpressure may develop, possibly causing the condenser coil to freeze up, the blower fan to overheat, and the compressor to be damaged. With such serious consequences, avoiding backpressure in the system is a major concern.

Though a Peltier solid-state heat pump does not suffer from backpressure issues the way a typical HVAC compressor-based system would, we decided to implement the backpressure detection correction system in anticipation of dealing with real-world issues.

To make the concept both easy to implement and to understand, we decided to use the equivalent value of "100 percent open" of one air register as the minimum outlet amount that must be available whenever the blower fan and heat pump were running. This makes it simple, as it can be expressed as four of the room's registers set to 0 percent and one of the room's registers set to 100 percent, or as each of the five rooms' air registers set to a position of 20 percent open. A simple backpressure routine was added to the master board software to manage distributing the "error" so backpressure would automatically be regulated by the system.

The Spin code to implement our backpressure detection is shown here:

```
' Check for Pressure Buildup
  pressure:= 0 ' Reset amount of Vent Openings
   ' We assume here that at least 1 vent must be at 100% or
   ' the total sum of all open vents should add to 100%
  repeat count from 0 to 4 ' Calculate Total Vent Openings
    pressure := pressure + RoomCurrentVentPosPercent[count]

  if pressure < 100   ' If this is true then there is not enough venting
     ' need to allocate the remainder
    totalvent:=0
    overpressure := 100-pressure
    splitpressure:= overpressure/5
```

```
  pressure:= overpressure
  repeat count from 0 to 4
    RoomCurrentVentPosPercent[count]:= ←
      RoomCurrentVentPosPercent[count]+ splitpressure ←
    totalvent:= totalvent + RoomCurrentVentPosPercent[count]
else
  pressure:= 0
```

Summary

The Propeller-powered HVAC green house, shown in Fig. 11-22, allowed us to explore parallel processing and distributed computing. We delved into sensor sampling, network communications, servo motor control, video display, and data recording.

The simplicity of the Spin language, combined with the power of objects, makes it straightforward to convert theories into test cases. It also allows us to avoid *labor inertia*—that reluctance we all have to abandon a design after discovering a fatal flaw or a simpler/better way to accomplish the task. The Propeller chip, with its multiple processors and large library of preexisting software objects, allows us to change direction easily and reduce the amount of time required to see results.

The "Exercises" section lists a small sample of the things we have considered exploring with our HVAC green house research platform. As real-world design criteria were taken into account when building the system, it may well be possible to implement the system

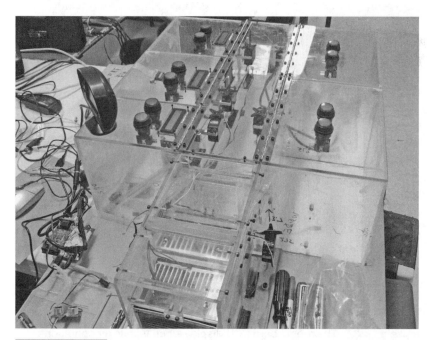

Figure 11-22   The Propeller-powered HVAC green house hardware.

in a new home or to retrofit an existing home. I would imagine it will only be a matter of time until someone implements a system such as this in an attempt to see if they can cut their heating and cooling costs, as well as make their home more comfortable.

IN CONCLUSION

At the time of this writing, we are continuing to develop software and test hardware for the model. We have recently obtained the liquid propane, methane, and carbon monoxide detectors, and plan to add the emergency venting routines to the model, as shown in Fig. 11-23.

> **Note:** If you would like to download the complete source code sets for all the boards in the HVAC green house model, as well as view high-resolution color pictures of the unit's construction and videos of the unit in action, please visit: ftp. propeller-chip.com/PCMProp/Chapter_11.

I would also like to take this opportunity to thank the team of amazing folks responsible for the creation of the HVAC green house model:

Rick Abbot—Machining and mechanical fabrication

Paul Atkinson—Schematic design, PCB fabrication and assembly

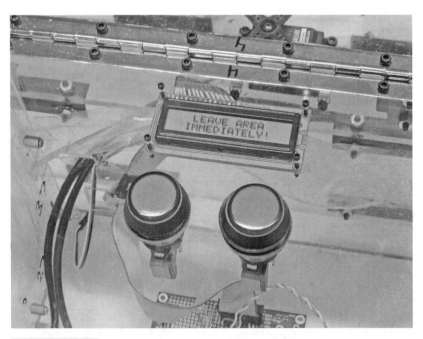

Figure 11-23   The room board LCD display when unsafe air conditions are detected.

James Delaney—Software design, 3-D CAD design, and system assembly/testing

Jake Ivey—Carpentry

Gray Mack—Hardware prototyping, software design and implementation

Parallax Inc.—Component supplier

I would also like to thank André LaMothe and Ken Gracey for inviting me to participate in this book and Chip Gracey for making the Basic Stamp and the Propeller chip a reality!

Exercises

During the creation of the house and the various discussions that ensued, some ideas were tossed around but not implemented that I thought might be a good "jumping off" point for anyone wanting to explore a system similar to this one. Some of the idea would require little or no modification to the basic system as built.

1 Adding an IR (infrared) sensor to the master board. By adding an inexpensive IR receiver module and using the SIRC object to decode Sony IR signals, you could essentially use a handheld IR transmitter to control the system. In addition, the NTSC video output could be routed to a conveniently located TV monitor. This would allow the operator to review and change the system settings from their home entertainment system or to monitor the HVAC conditions using "picture in picture" technology available on many TV sets.

2 Adding an IR sensor to the room boards. The same IR sensor and SIRC object, when added to the room board, would allow the room occupant to control the temperature using a simple IR remote control (for example, use "channel up" and "channel down" signals to adjust the temperature set point).

3 Add speech synthesis to provide voice alerts. During an alarm condition (i.e., detected unhealthy air or fire), using verbal alerts would allow the room occupant to know to what the alarm issue relates without having to be close enough to the board to read the LCD screen. In addition, if an IR remote was being used to adjust the target temperature for the room, voice feedback could be used to "tell" the user what temperature they have selected

4 Select some rooms for "fresh" air and others for recirculated air. If some occupants preferred "fresh" air in their rooms, while others preferred recirculated air, it would be possible to meet both preferences by alternating the air-handling method between the rooms. To accomplish this, the air register in a selected room (or rooms) would be set to 100 percent open, while the air registers in the remaining rooms would be placed in the 100 percent closed position. Next, position the "bypass" doors to the "fresh air" positions, and start the blower fan. After a few moments, reverse the process for the rooms that have selected recirculated air.

5 Mix recirculated and fresh air. As the bypass doors are controlled by servomotors, they may be commanded to hold the doors in any position from 100 percent open to 0 percent open. So, it would be possible to set a "mixture" amount if you want to assure a certain amount of fresh air is moved through the building on a scheduled or regular basis to reduce odors and to remove "stale" air from the building.

6 Error detection and alerts. If a room fails to reach the desired temperature in a reasonable amount of time, the master board could show an alert or trouble condition. Another possible error-detection function could be if the room board has not been polled by the master board in a specific period. When this "timeout" has been reached, the board could show a "trouble" message on the LCD to alert the room occupant that the board needs diagnostics and repair.

7 Alert required preventive maintenance. The master board is able to track the "run time" for the blower fans so it would be possible to alert the user when filter replacement intake cover cleaning and other maintenance may be required.

8 Go wireless! As was shown earlier in this book, it is possible to forgo the use of CAT-5 cables for communication altogether. If the room boards, master board and control board were all equipped with ZigBee modules and powered by a small "wall-wart"-type power supply, it may be possible (depending on the distances involved) to retrofit an existing home HVAC system without the additional cost and disruption of running cable through the building.

12

SYNTHESIZING SPEECH WITH THE PROPELLER

Chip Gracey

Introduction

Have you ever wondered how to make a human voice in software? I've always been tantalized by the idea, since speech is perhaps the *ultimate* analog communication protocol. It certainly predates computers by a long time! In this chapter, we are going to "look" at actual speech and then synthesize it with the Propeller using small amounts of data. Here's what's in store:

- Using spectrographs to "see" speech
- Synthesizing formants, vowels, and making speech sound like speech
- Exploring the VocalTract object

Resources: Demo code and other resources for this chapter are available for free download from ftp.propeller-chip.com/PCMProp/Chapter_12.

Using Spectrographs to "See" Speech

In order to synthesize speech, we need to know what's in it. We could start by looking at some waveforms of recorded speech. Figure 12-1 shows a partial waveform from the utterance "to."

I found this type of analysis to be a recipe for madness, as you easily fall into the morass of trying to replicate details that are not important and then missing what matters.

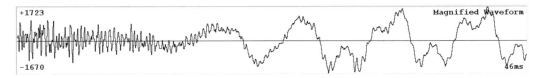

Figure 12-1   A partial waveform from the utterance "to."

Instead of worrying about waveforms, we need to look at the spectrum of speech over time, since that is where our real information is. Figure 12-2 shows a spectrograph of me counting from 1 to 10.

What you're looking at is called a *spectrograph.* This is a 3-D representation of a time-varying signal. From left to right, you have time. From bottom to top, you have frequency. The intensity at each point is the energy of that frequency at that time. Looking at speech from this perspective gives us the insight we need to go about synthesizing it. In fact, this is what your ear "sees," rather than a sputtering waveform.

Note the horizontal bands of darkness in the spectrograph. Those are harmonics of my voice's glottal pitch. I tend to talk at about 110 Hz, so there's nominally about 110 Hz of separation between each band. You can see the bands spread out and compress as my voice deviates up and down in tone.

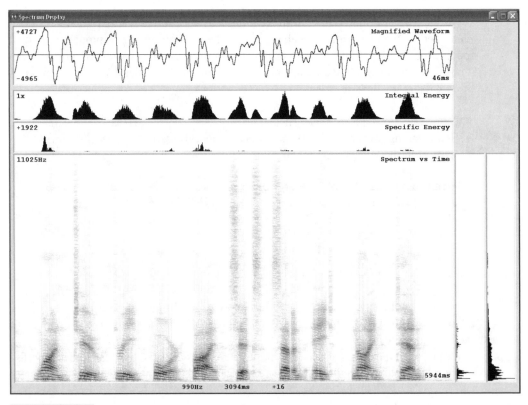

Figure 12-2   A spectrograph of me counting from 1 to 10.

Now notice the concentrations of dark bands distributed vertically. Those are the result of resonances within my vocal tract. As I talk and change the shape of my vocal tract, they move around—some up, some down. Each is a *formant*. The key to speech synthesis is synthesizing these formants.

Information: These spectrographs are made using the Spectrum Display software written by the author. It is included in the Chapter_12/Tools folder, and the sample programs are in the Chapter_12/Source folder.

SYNTHESIZING FORMANTS AND VOWELS AND MAKING SPEECH SOUND LIKE SPEECH!

Synthesizing formants was very challenging for me, and I spent a lot of time trying to do it. The first and most obvious approach was to synthesize individual sine waves for each harmonic of the glottal pulse. This was way too much work and computationally impractical. I did, much later, come up with one approach that was really simple, but falls into the "hack" category, as it generated discontinuities in the output signal: Just generate sine waves at the center frequency of each formant and reset their phases at each glottal pitch interval—like magic, the sine waves diffuse and the pitch harmonics distribute nicely around the formants' centers, giving you instant vowel sounds! Good enough for a talking watch, but not a musical instrument.

Let's look at some vowel sounds. In Fig. 12-3, I will make my glottal pitch very low (like a frog) so that the pitch lines will be highly compressed, making it easier to see

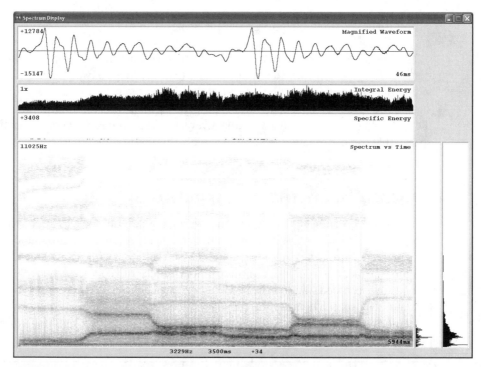

Figure 12-3   A spectrograph of me making some vowel sounds.

TABLE 12-1 FORMANT FREQUENCIES OF SOME SPOKEN VOWELS					
	BEET	HAT	HOT	SOAP	BORROW	BALL
F4	3700	3400	3200	3200	4700	3600
F3	3100	2500	2400	2400	1500	2600
F2	2000	1700	1050	950	1200	850
F1	310	730	750	530	580	560

the formants. In sequence, I will pronounce the vowels "b__ee__t," "h__a__t," "h__o__t," "s__oa__p," "bo__rr__ow," "ba__ll__." Yep, "r" and "l" are vowels in synthesis.

Looking at the spectrograph, we can glean the formant frequencies for each vowel. Table 12-1 lists the first four formant frequencies (F1..F4) observed for each vowel.

If we can synthesize these formants, we can re-create those vowel sounds.

Now, I'll introduce the VocalTract object. This is the neatest piece of software I've ever written, and it's the kind of program that the Propeller chip was *designed* for. All the demos for this chapter were developed on the Propeller Demo Board, but they will work with a minimal Propeller chip setup, as long as you have audio output connections to pins 10 and a simple R/C filter. The featured spectrograph program and all code examples can be downloaded from this FTP address:

ftp://ftp.propeller-chip.com/PCMProp/Chapter_12

The VocalTract object synthesizes a human vocal tract in real time using a simple frame of 13 byte-size parameters. All you have to do is change any parameters of interest within the frame, then tell it how fast to transition to the new frame. It will internally snapshot and queue up each frame you give it, then linearly interpolate each of the 13 parameters from frame to frame, over specified periods. Out comes continuous speech that can be made very expressive, with amazingly little input!

A WORD ON SYNTHESIS ALGORITHMS

It used to be that synthesis algorithms were king several decades ago. With the advent of big, inexpensive memory, however, rote recording and regurgitation have become the norm for many areas in which algorithmic synthesis would be much more flexible, not to mention more memory-efficient. Many speech synthesizers today use recorded snippets of speech, which are concatenated at runtime to create continuous speech. These programs and data often require many megabytes of memory. In contrast, VocalTract is only 0.0013 MB in size (334 longs of code + 44 longs of variables per instance).

Exploring the VocalTract Object

Figure 12-4 shows some documentation from the VocalTract object. Note the schematic diagram and the parameters associated with each section.

The ASPIRATION section produces turbulence in the vocal tract. This is for breathy sounds, but is always needed for natural-sounding speech. The GLOTTAL section produces continuous glottal pulses for voiced speech. The VIBRATO section modulates the glottal pitch for singing and helps regular speech sound natural, as only robots are known to speak in pure monotone. The pitch parameters are all registered to the musical scale, with 48 steps per octave, making every fourth value a chromatic tone. Note that the ASPIRATION and GLOTTAL sounds are summed before feeding into the formant resonators F1..F4 that make vowel sounds. The NASAL section is an anti-resonator used for M, N, and NG sounds. Lastly, the FRICATION section is for white-noise sounds, like T, TH, SH, S, and F that occur at the tongue, teeth, and lips, or after the resonators.

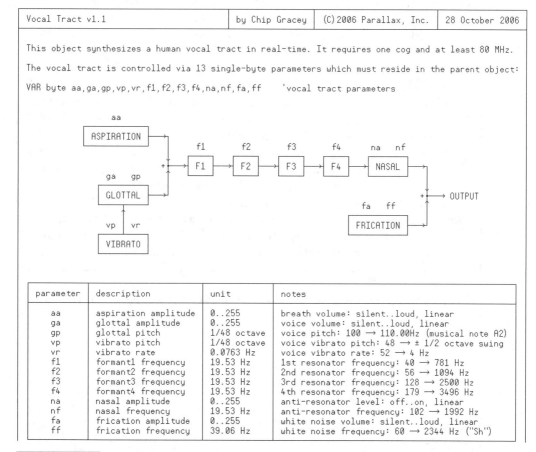

Figure 12-4   Documentation from the VocalTract object.

The FRICATION output is summed with the NASAL output to make the final product: a 32-bit, 20-kHz-updated audio signal, available as both a single-ended or differential PCM signal, and as a continuously updated LONG in hub RAM.

Though all parameters are 8-bit within a frame, they are interpolated with 24 sub-bits, so intraframe transitions are very smooth. Also, frames connect seamlessly with the same precision, so there are never any discontinuities in the output signal. One thing a user must be aware of is the likelihood of numerical overflow when volume levels get high. You'll know if this happens, believe me.

THE QUEST TO GENERATE FORMANTS

The resonators in VocalTract that make the formants are very simple, but it took a lot of work and luck to figure out how to make them. Most resonator algorithms require transcendental math operations, which need extreme precision around the nearly asymptotic +/−1 sine and cosine points. This is fine for floating-point hardware systems, but a disaster in fixed-point math architectures. Also, you could easily spend more time computing ever-changing resonator coefficients than you would actually computing the resonator state for each output sample.

I needed a simpler way to build a resonator. Fortunately, the CORDIC algorithm came to my rescue! *CORDIC* is an ingenious technique developed in the 1950s by Jack E. Volder that uses binary shifts and adds to perform the transcendental functions. In a nutshell, it can rotate (X,Y) coordinates efficiently, without the edgy numerical sensitivities of commonly used resonator algorithms. In each of VocalTract's resonators, the incoming signal is summed into Y of an (X,Y) point, which defines the state of the resonator. Then, that point is CORDIC-rotated by an angle directly proportional to the formant's center frequency. The resulting Y of the (X,Y) point is passed on to the next resonator. When in-band excitation is received, the (X,Y) point grows further from (0,0) as it rotates around. Out-of-band excitation causes the amplitude to decay to the point where the resonator pretty much just passes its input through to its output. If this wasn't simple enough, since the step-angle fed to each formant's CORDIC rotator is directly proportional to the formant's center frequency, simple linear interpolation is used to slide each formant from one frame's setting to the next. Goodbye, math headaches!

Let's now re-create those six vowel sounds that I recorded. Here's all you need to do (see Example1.spin):

```
CON

  _clkmode = xtal1 + pll16x
  _xinfreq = 5_000_000

OBJ

  v     : "VocalTract"
```

```
VAR

  'vocal tract parameters
  byte   aa,ga,gp,vp,vr,f1,f2,f3,f4,na,nf,fa,ff

PUB start

  v.start(@aa, 10, 11, -1)      'Start tract, output to pins 10,11

  repeat                        'Keep repeating talk
    gp := rnd(50, 100)          'Set random glottal pitch
    vp := rnd(4, 24)            'Set random vibrato pitch
    vr := rnd(4, 10)            'Set random vibrato rate
    talk                        'Talk

PRI talk

  setformants(310, 2000, 3100, 3700)    'Set formants for bEEt
  v.go(1)

  aa := 3                               'Set breathiness
  ga := 30                              'Set voiced volume
  v.go(100)                             'Ramp up
  v.go(500)                             'Sustain

  setformants(730, 1700, 2500, 3400)    'Set formants for hAt
  v.go(100)                             'Transition
  v.go(500)                             'Sustain

  setformants(750, 1050, 2400, 3200)    'Set formants for hOt
  ga := 12                              'Volume down a little
  v.go(100)                             'Transition
  v.go(500)                             'Sustain

  setformants(530, 950, 2400, 3200)     'Set formants for sOAp
  v.go(100)                             'Transition
  v.go(500)                             'Sustain

  setformants(580, 1200, 1500, 4700)    'Set formants for boRRow
  v.go(100)                             'Transition
  v.go(500)                             'Sustain

  setformants(560, 850, 2600, 3600)     'Set formants for baLL
  v.go(100)                             'Transition
  v.go(500)                             'Sustain
```

```
       aa := 0                              'Taper down to silence
       ga := 0
       v.go(100)

       v.go(500)                            'Pause


PRI setformants(s1, s2, s3, s4)

    f1 := s1 * 100 / 1953
    f2 := s2 * 100 / 1953
    f3 := s3 * 100 / 1953
    f4 := s4 * 100 / 1953


PRI rnd(low, high)

    return low + ||(?cnt // (high - low + 1))
```

Recording the audio output of this program into our spectrograph, we get the data shown in Fig. 12-5.

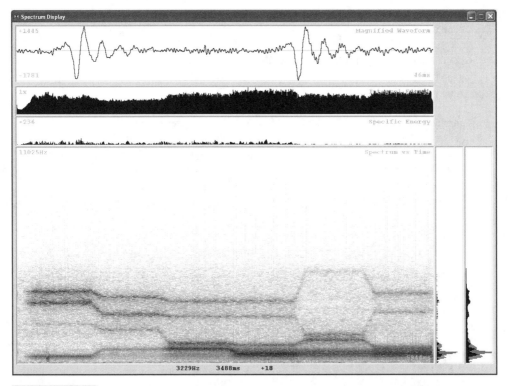

Figure 12-5   A spectrograph of the vowels synthesized by Example1.spin.

You can see the formants transition as they did in the original voice recording. Compare the waveforms in the top panes as well. They look a little different, but they sound similar. Speech is a rather inexact phenomenon, thankfully.

By making ga := 0 in the previous program, you will hear the vowels whispered, since there is some breath noise via aa. You can tell from listening to the synthesis that the transition from one vowel to another makes nice diphthongs. For example, going from "bEEt" to "hAt" makes "yeah." Vowel speech is pretty straightforward in this way.

How about some breathing sounds? Try modifying the PRI talk method in the previous program as follows (see Example2.spin):

```
PRI talk

    setformants (880,1075,2350,3050)        'Set "inhale" formants
    v.go (1)

    aa := 30                                'Slowly ramp up
    v.go (1000)

    v.go (500)                              'Sustain

    aa := 0                                 'Slowly ramp down
    v.go (1500)


    setformants (500,800,1100,2325)         'Set "exhale" formants
    v.go (1)

    aa := 10                                'Ramp up
    v.go (500)

    aa := 0                                 'Slowly ramp down
    v.go (2500)

    v.go (500)                              'Pause between breaths
```

This should clear up any confusion about the Propeller being able to sleep.

What about consonant sounds? I've made a lot of little discoveries about synthesizing consonants by observing spectrographs and then experimenting with VocalTract. You can easily do the same. Sometimes it's necessary to speak very slowly and distinctly into the spectrograph in order to see otherwise-subtle phenomena. Looking at the actual waveform can give important hints, too—particularly around *plosives* (i.e., "t," "p," and "k") where there is silence.

It turns out there is just a handful of basic recipes you need to know in order to make VocalTract pronounce any consonant. The biggest divider among consonants is whether they are voiced or unvoiced. That is, is the glottal pulse going or is there just breathiness? This distinction halves the number of basic consonant recipes.

TABLE 12-2 RELATED CONSONANTS	
VOICED	**UNVOICED**
<u>b</u>url	<u>p</u>earl
<u>d</u>o	<u>t</u>o
<u>v</u>an	<u>f</u>an
<u>g</u>irl	<u>c</u>url
<u>z</u>oo	<u>s</u>ue
<u>j</u>ock	<u>ch</u>alk
<u>th</u>is	<u>th</u>in
mea<u>s</u>ure	pre<u>ss</u>ure

Take a look at Table 12-2. Say each of these word pairs aloud, and note the voiced/ unvoiced difference in the underlined consonant.

The only difference within these sets of consonants is *when* your glottal pulses are going. You'll note that the rest of your vocal tract is in the same exact state when making either sound. This means that controlling the ga parameter differently will be the key to making one sound versus the other.

Let's do "girl" and "curl." To make these g/c sounds, you can initially set f1 to zero and set f2 and f3 to the average of the following vowels' f2 and f3. In other words, F1 will sweep upwards rapidly and F2 and F3 will begin in the center and then head down and up, respectively, to their final vowel positions. This is the base recipe for "g" and "c" sounds. The difference between "g" and "c" is when the glottal pulses start: early for "g" and late for "c."

Substitute the following method for PRI talk in our program (see Example3.spin):

```
PRI talk

  setformants(0, 1350, 1350, 4700)   'Set formants for G/C before R
  v.go(1)

  aa := 20                           'Set breathiness
  v.go(25)                           'make brief G/C turbulence

  setformants(580, 1200, 1500, 4700) 'Set formants for R
  ga := 30                           'Voice up for G / skip for C
  v.go(150)                          'Transition to R

  ga := 30                           'Sustain / voice up for C
  v.go(100)                          'Sustain / voice up for C
  v.go(200)                          'Sustain
```

```
setformants(560, 850, 2600, 3600)  'Set formants for L
v.go(250)                          'Transition
v.go(100)                          'Sustain

aa := 0                            'Taper down to silence
ga := 0
v.go(100)

v.go(500)                          'Pause
```

You should hear "girl." If you comment out the first `ga := 30`, you will hear "curl," as the glottal pulses will start later. Commenting out both `ga := 30` lines will result in a whisper that sounds like it could be "girl" or "curl."

We can make this say "burl" and "pearl" as well. For "b" and "p" sounds, `f1` and `f2` must be set very low and swept upwards quite rapidly, while `f3` and `f4` can be the same as the following vowel's. Two changes must be made to our `PRI talk` method to make "b" and "p." First, replace `setformants(0, 1350, 1350, 4700)` with `setformants(0, 200, 1500, 4700)`. Second, replace `v.go(150)` with `v.go(50)` to speed up the transition. Now you should hear "burl." By commenting out the first `ga := 30`, you should hear "pearl."

While I won't go over every consonant recipe here, as I'd have to rediscover them, myself (thereby depriving you of the pleasure), we should cover the nasals "m" and "n." Let's try to say the word "monster." That way, we can slip in the "s" and "t" sounds to utilize the FRICATION section, as well as the NASAL, completing our tour of VocalTract's parameters. First, I'll say "monster" very slowly in a froggy voice (see Fig. 12-6).

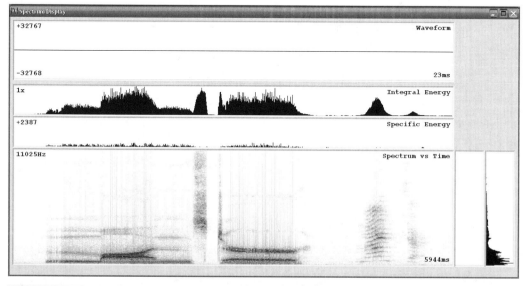

Figure 12-6   A spectrograph of me saying "monster," followed by a baby "monster."

After I said "monster," our two-year-old daughter repeated "monster" unexpectedly. You can see from those pitch lines how shrill her voice is.

Looking at the spectrograph, you can see the initial "m" followed by the open-mouth "o" and then the "n." After those nasal-vowel-nasal sounds, you can see the tall noise burst of the "s" followed by silence, then the brief "t" burst going into the "r."

The "m" and "n" sounds are enhanced by using the nasal anti-resonator. It sucks up in-band energy rather than accumulating it, like the formant resonators do. This creates a limited-spectrum vacuum, which can make things sound nasally.

By studying the spectrograph and incrementally building the word, I made a new `PRI` `talk` method to say "monster" (see Example4.spin):

```
PRI talk

  setformants(170, 1100, 2600, 3200)     'Set formants for M
  nf := 2900 * 100 / 1953                'Set nasal for M
  na := 255
  v.go(1)


  aa := 10                               'Ramp up M
  ga := 30
  v.go(150)


  setformants(700, 900, 2800, 4500)      'Set formants for O
  na := 0                                'Nasal off
  v.go(50)                               'Transition to O
  v.go(150)                              'Sustain


  setformants(200, 1550, 2500, 4500)     'Set formants for N
  nf := 1550 * 100 / 1953                'Set nasal for N
  na := 255
  v.go(100)                              'Transition to N
  v.go(100)                              'Sustain


  aa := 0                                'Ramp down N
  ga := 0
  v.go(100)


  na := 0                                'Nasal off
  ff := 120                              'Set frication for S
  v.go(1)


  fa := 10                               'Ramp up S
  v.go(150)


  fa := 0                                'Ramp down S
  v.go(75)
```

```
ff := 80                              'Set frication for T
v.go(50)                              'Sustain silence briefly

fa := 10                              'Ramp up T
v.go(20)

setformants(580, 1200, 1500, 4700)    'Set formants for R
fa := 0                               'Ramp down T
v.go(20)

aa := 10                              'Ramp up R
ga := 30
v.go(50)
v.go(150)                             'Sustain

aa := 0                               'Taper down to silence
ga := 0
v.go(100)

v.go(500)                             'Pause
```

The timing of the "s" and "t" sounds took me a lot of trial and error to get tuned right. Even so, you can see that the amount of data involved in synthesizing words is quite low compared to making audio recordings of them (less than 1 percent).

To utilize VocalTract, it will take some experimentation on your part. Once mastered, though, you will have gained untold insight into the mechanics of speech that you'll be eager to share with others who will be less interested. As a side benefit, you will be able to add speech to your Propeller applications using minimal resources.

In parting, here is an odd application that uses the StereoSpatializer object to mix and place four separate VocalTracts into a stereo image. This demonstrates how the Propeller's multiple processors can be easily put to use in a cooperative fashion to perform complex tasks of trivial value.

Top object called "SpatialBlah" (see SpatialBlah.spin):

```
CON

  _clkmode     = xtal1 + pll16x
  _xinfreq     = 5_000_000

  voices       = 4               'Voices can be 1..4
  buffer_size  = $1000           'Spatializer buffer size

OBJ

  v   [voices]  : "Blah"
  s             : "StereoSpatializer"
```

```
VAR

  'Spatializer parameters
  word  input[4], angle[4], depth[4], knobs

  'Spatializer delay buffer
  long buffer[buffer_size]

PUB start | i, r

  repeat i from 0 to voices - 1     'Start voices and connect
    input[i] := v[i].start          '..to spatializer inputs

  knobs := %000_011_100_101         'Start spatializer
  s.start(@input, @buffer, buffer_size, 11, -1, 10, -1)

  repeat                            'Scan for done voices
    repeat i from 0 to voices - 1
      if v[i].done                  'If voice done
        angle[i] := ?r              '..Set new random angle
        depth[i] := ?r & $FFF       '..Set new random depth
        v[i].go                     '..Start new "blah"
```

Lower object called "Blah" (see Blah.spin):

```
OBJ

  v    : "VocalTract"


VAR

  'Vocal tract parameters
  byte  aa,ga,gp,vp,vr,f1,f2,f3,f4,na,nf,fa,ff

PUB start

  v.start(@aa, -1, -1, -1)    'Start tract, no pin outputs
  return v.sample_ptr         'Return tract sample pointer


PUB go
                              'Say "blah" randomly
  gp := rnd(60, 120)
  vp := rnd(4, 48)
  vr := rnd(4, 30)

  setformants(100, 200, 2800, 3750)
  v.go(rnd(100, 1000))
```

```
    setformants (400, 850, 2800, 3750)
    aa := 10
    ga := 20
    v.go (20)


    v.go (80)


    setformants (730, 1050, 2500, 3480)
    aa := 20
    ga := 30
    v.go (50)


    v.go (rnd (200, 1000))

    aa := 0
    ga := 0
    v.go (100)


PUB done

    return v.empty

PRI setformants (s1, s2, s3, s4)

    f1 := s1 * 100 / 1953
    f2 := s2 * 100 / 1953
    f3 := s3 * 100 / 1953
    f4 := s4 * 100 / 1953


PRI rnd (low, high)

    return low + || (?cnt // (high - low + 1))
```

Summary

I hope you had fun reading this chapter and running the example programs. Speech synthesis is a uniquely challenging and rewarding endeavor. It's fascinated me for a long time, and after writing this chapter, I just want to jump back in again and spend a few more months writing more code to speak numbers, units, and who knows what else. I know that if I got into it again, I'd probably gain all kinds of new knowledge that would further my understanding of how speech works and how it can be done even more efficiently. You can do this yourself, if you have the interest and patience. It would be a fun time!

Exercises

1 Synthesize the word "he." Hint: You'll only use one set of formants.

2 Synthesize the word "she." Hint: Add to your "he" program.

3 Synthesize "la, la, la…" to sing "Row, Row, Row Your Boat."

4 Synthesize numbers 1 through 10.

5 Make a talking clock.

PROPELLER LANGUAGE REFERENCE

Categorical Listing of Propeller Spin Language Elements

Elements marked with a superscript "a" are also available in Propeller Assembly.

BLOCK DESIGNATORS

CON	Declare constant block
VAR	Declare variable block
OBJ	Declare object reference block
PUB	Declare public method block
PRI	Declare private method block
DAT	Declare data block

CONFIGURATION

CHIPVER	Propeller chip version number
CLKMODE	Current clock mode setting
_CLKMODE	Application-defined clock mode (read-only)
CLKFREQ	Current clock frequency
_CLKFREQ	Application-defined clock frequency (read-only)
CLKSET[a]	Set clock mode and clock frequency

_XINFREQ	Application-defined external clock frequency (read-only)
_STACK	Application-defined stack space to reserve (read-only)
_FREE	Application-defined free space to reserve (read-only)
RCFAST	Constant for _CLKMODE: internal fast oscillator
RCSLOW	Constant for _CLKMODE: internal slow oscillator
XINPUT	Constant for _CLKMODE: external clock/osc (XI pin)
XTAL1	Constant for _CLKMODE: external low-speed crystal
XTAL2	Constant for _CLKMODE: external med-speed crystal
XTAL3	Constant for _CLKMODE: external high-speed crystal
PLL1X	Constant for _CLKMODE: external frequency times 1
PLL2X	Constant for _CLKMODE: external frequency times 2
PLL4X	Constant for _CLKMODE: external frequency times 4
PLL8X	Constant for _CLKMODE: external frequency times 8
PLL16X	Constant for _CLKMODE: external frequency times 16

COG CONTROL

COGID[a]	Current cog's ID (0–7)
COGNEW	Start the next available cog
COGINIT[a]	Start, or restart, a cog by ID
COGSTOP[a]	Stop a cog by ID
REBOOT	Reset the Propeller chip

PROCESS CONTROL

LOCKNEW[a]	Check out a new lock
LOCKRET[a]	Release a lock
LOCKCLR[a]	Clear a lock by ID
LOCKSET[a]	Set a lock by ID
WAITCNT[a]	Wait for system counter to reach a value
WAITPEQ[a]	Wait for pin(s) to be equal to value
WAITPNE[a]	Wait for pin(s) to be not equal to value
WAITVID[a]	Wait for video sync and deliver next color/pixel group

FLOW CONTROL

`IF`
`...ELSEIF`
`...ELSEIFNOT`
`...ELSE`
	Conditionally execute one or more blocks of code

`IFNOT`
`...ELSEIF`
`...ELSEIFNOT`
`...ELSE`
	Conditionally execute one or more blocks of code

`CASE`
 	
`...OTHER`
	Evaluate expression and execute block of code that satisfies a condition

`REPEAT`
 	
`...FROM`
`...TO`
`...STEP`
`...UNTIL`
`...WHILE`
	Execute block of code repetitively an infinite or finite number of times with optional loop counter, intervals, exit, and continue conditions

`NEXT`	Skip rest of `REPEAT` block and jump to next loop iteration
`QUIT`	Exit from `REPEAT` loop
`RETURN`	Exit `PUB`/`PRI` with normal status and optional return value
`ABORT`	Exit `PUB`/`PRI` with abort status and optional return value

MEMORY

`BYTE`	Declare byte-sized symbol or access byte of main memory
`WORD`	Declare word-sized symbol or access word of main memory
`LONG`	Declare long-sized symbol or access long of main memory
`BYTEFILL`	Fill bytes of main memory with a value
`WORDFILL`	Fill words of main memory with a value
`LONGFILL`	Fill longs of main memory with a value
`BYTEMOVE`	Copy bytes from one region to another in main memory
`WORDMOVE`	Copy words from one region to another in main memory
`LONGMOVE`	Copy longs from one region to another in main memory

LOOKUP	Get value at index (1..N) from a list
LOOKUPZ	Get value at zero-based index (0..N−1) from a list
LOOKDOWN	Get index (1..N) of a matching value from a list
LOOKDOWNZ	Get zero-based index (0..N−1) of a matching value from a list
STRSIZE	Get size of string in bytes
STRCOMP	Compare a string of bytes against another string of bytes

DIRECTIVES

STRING	Declare in-line string expression; resolved at compile time
CONSTANT	Declare in-line constant expression; resolved at compile time
FLOAT	Declare floating-point expression; resolved at compile time
ROUND	Round compile-time floating-point expression to integer
TRUNC	Truncate compile-time floating-point expression at decimal
FILE	Import data from an external file

REGISTERS

DIRA[a]	Direction register for 32-bit port A
DIRB[a]	Direction register for 32-bit port B (future use)
INA[a]	Input register for 32-bit port A (read-only)
INB[a]	Input register for 32-bit port B (read-only) (future use)
OUTA[a]	Output register for 32-bit port A
OUTB[a]	Output register for 32-bit port B (future use)
CNT[a]	32-bit system counter register (read-only)
CTRA[a]	Counter A control register
CTRB[a]	Counter B control register
FRQA[a]	Counter A frequency register
FRQB[a]	Counter B frequency register
PHSA[a]	Counter A phase-locked loop (PLL) register
PHSB[a]	Counter B phase-locked loop (PLL) register
VCFG[a]	Video configuration register

VSCL[a]   Video scale register

PAR[a]   Cog boot parameter register (read-only)

SPR   Special-purpose register array; indirect cog register access

CONSTANTS

TRUE[a]   Logical true: −1 ($FFFFFFFF)

FALSE[a]   Logical false: 0 ($00000000)

POSX[a]   Maximum positive integer: 2,147,483,647 ($7FFFFFFF)

NEGX[a]   Maximum negative integer: −2,147,483,648 ($80000000)

PI[a]   Floating-point value for PI: ~3.141593 ($40490FDB)

VARIABLE

RESULT   Default result variable for PUB/PRI methods

UNARY OPERATORS

+   Positive (+X); unary form of Add

−   Negate (−X); unary form of Subtract

− −   Predecrement (− −X) or postdecrement (X− −) and assign

+ +   Preincrement (++X) or postincrement (X++) and assign

^^   Square root

|| Absolute value

~   Sign-extend from bit 7 (~X) or postclear to 0 (X~)

~~   Sign-extend from bit 15 (~~X) or postset to −1(X~~)

?   Random number forward (?X) or reverse (X?)

|<   Decode value (modulus of 32; 0–31) into single-high-bit long

>| Encode long into magnitude (0–32) as high-bit priority

!   Bitwise: NOT

NOT   Boolean: NOT (promotes non-0 to −1)

@   Symbol address

@@   Object address plus symbol value

BINARY OPERATORS

Note: All operators in the right column are assignment operators.

=	--and--	=	Constant assignment (CON blocks)
:=	--and--	:=	Variable assignment (PUB/PRI blocks)
+	--or--	+=	Add
−	--or--	−=	Subtract
*	--or--	*=	Multiply and return lower 32 bits (signed)
**	--or--	**=	Multiply and return upper 32 bits (signed)
/	--or--	/=	Divide (signed)
//	--or--	//=	Modulus (signed)
#>	--or--	#>=	Limit minimum (signed)
<#	--or--	<#=	Limit maximum (signed)
~>	--or--	~>=	Shift arithmetic right
<<	--or--	<<=	Bitwise: Shift left
>>	--or--	>>=	Bitwise: Shift right
<-	--or--	<-=	Bitwise: Rotate left
->	--or--	->=	Bitwise: Rotate right
><	--or--	><=	Bitwise: Reverse
&	--or--	&=	Bitwise: AND
|	--or--	|=	Bitwise: OR
^	--or--	^=	Bitwise: XOR
AND	--or--	AND=	Boolean: AND (promotes non-0 to −1)
OR	--or--	OR=	Boolean: OR (promotes non-0 to −1)
==	--or--	===	Boolean: Is equal
<>	--or--	<>=	Boolean: Is not equal
<	--or--	<=	Boolean: Is less than (signed)
>	--or--	>=	Boolean: Is greater than (signed)
=<	--or--	=<=	Boolean: Is equal or less (signed)
=>	--or--	=>=	Boolean: Is equal or greater (signed)

SYNTAX SYMBOLS

%	Binary number indicator, as in %1010
%%	Quaternary number indicator, as in %%2130
$	Hexadecimal indicator, as in $1AF or assembly 'here' indicator
"	String designator, as in "Hello"
_	Group delimiter in constant values, or underscore in symbols
#	Object-Constant reference: obj#constant
.	Object-Method reference: obj.method(param) or decimal point
..	Range indicator, as in 0..7
:	Return separator: PUB method : sym, or object assignment, etc.
|	Local variable separator: PUB method | temp, str
\	Abort trap, as in \method(parameters)
,	List delimiter, as in method(param1, param2, param3)
( )	Parameter list designators, as in method(parameters)
[ ]	Array index designators, as in INA[2]
{ }	In-line/multi-line code comment designators
{{ }}	In-line/multi-line document comment designators
'	Code comment designator
"	Document comment designator

Categorical Listing of Propeller Assembly Language

Elements marked with a superscript "s" are also available in Spin.

DIRECTIVES

ORG	Adjust compile-time cog address pointer
FIT	Validate that previous instructions/data fit entirely in cog
RES	Reserve next long(s) for symbol

CONFIGURATION

CLKSET[s]   Set clock mode at run time

COG CONTROL

COGID[s]   Get current cog's ID

COGINIT[s]   Start, or restart, a cog by ID

COGSTOP[s]   Stop a cog by ID

PROCESS CONTROL

LOCKNEW[s]   Check out a new lock

LOCKRET[s]   Return a lock

LOCKCLR[s]   Clear a lock by ID

LOCKSET[s]   Set a lock by ID

WAITCNT[s]   Pause execution temporarily

WAITPEQ[s]   Pause execution until pin(s) match designated state(s)

WAITPNE[s]   Pause execution until pin(s) do not match designated state(s)

WAITVID[s]   Pause execution until Video Generator is available for pixel data

CONDITIONS

IF_ALWAYS   Always

IF_NEVER   Never

IF_E   If equal ($Z = 1$)

IF_NE   If not equal ($Z = 0$)

IF_A   If above ($!C$ & $!Z = 1$)

IF_B   If below ($C = 1$)

IF_AE   If above or equal ($C = 0$)

IF_BE   If below or equal ($C \mid Z = 1$)

IF_C   If C set

IF_NC   If C clear

IF_Z   If Z set

IF_NZ	If Z clear
IF_C_EQ_Z	If C equal to Z
IF_C_NE_Z	If C not equal to Z
IF_C_AND_Z	If C set and Z set
IF_C_AND_NZ	If C set and Z clear
IF_NC_AND_Z	If C clear and Z set
IF_NC_AND_NZ	If C clear and Z clear
IF_C_OR_Z	If C set or Z set
IF_C_OR_NZ	If C set or Z clear
IF_NC_OR_Z	If C clear or Z set
IF_NC_OR_NZ	If C clear or Z clear
IF_Z_EQ_C	If Z equal to C
IF_Z_NE_C	If Z not equal to C
IF_Z_AND_C	If Z set and C set
IF_Z_AND_NC	If Z set and C clear
IF_NZ_AND_C	If Z clear and C set
IF_NZ_AND_NC	If Z clear and C clear
IF_Z_OR_C	If Z set or C set
IF_Z_OR_NC	If Z set or C clear
IF_NZ_OR_C	If Z clear or C set
IF_NZ_OR_NC	If Z clear or C clear

FLOW CONTROL

CALL	Jump to address with intention to return to next instruction
DJNZ	Decrement value and jump to address if not zero
JMP	Jump to address unconditionally
JMPRET	Jump to address with intention to "return" to another address
TJNZ	Test value and jump to address if not zero
TJZ	Test value and jump to address if zero
RET	Return to stored address

EFFECTS

NR   No result (don't write result)

WR   Write result

WC   Write C status

WZ   Write Z status

MAIN MEMORY ACCESS

RDBYTE   Read byte of main memory

RDWORD   Read word of main memory

RDLONG   Read long of main memory

WRBYTE   Write a byte to main memory

WRWORD   Write a word to main memory

WRLONG   Write a long to main memory

COMMON OPERATIONS

ABS   Get absolute value of a number

ABSNEG   Get negative of number's absolute value

NEG   Get negative of a number

NEGC   Get a value, or its additive inverse, based on C

NEGNC   Get a value or its additive inverse, based on !C

NEGZ   Get a value, or its additive inverse, based on Z

NEGNZ   Get a value, or its additive inverse, based on !Z

MIN   Limit minimum of unsigned value to another unsigned value

MINS   Limit minimum of signed value to another signed value

MAX   Limit maximum of unsigned value to another unsigned value

MAXS   Limit maximum of signed value to another signed value

ADD   Add two unsigned values

ADDABS   Add absolute value to another value

ADDS   Add two signed values

ADDX   Add two unsigned values plus C

ADDSX	Add two signed values plus C
SUB	Subtract two unsigned values
SUBABS	Subtract an absolute value from another value
SUBS	Subtract two signed values
SUBX	Subtract unsigned value plus C from another unsigned value
SUBSX	Subtract signed value plus C from another signed value
SUMC	Sum signed value with another of C-affected sign
SUMNC	Sum signed value with another of !C-affected sign
SUMZ	Sum signed value with another Z-affected sign
SUMNZ	Sum signed value with another of !Z-affected sign
MUL	<reserved for future use>
MULS	<reserved for future use>
AND	Bitwise AND two values
ANDN	Bitwise AND value with NOT of another
OR	Bitwise OR two values
XOR	Bitwise XOR two values
ONES	<reserved for future use>
ENC	<reserved for future use>
RCL	Rotate C left into value by specified number of bits
RCR	Rotate C right into value by specified number of bits
REV	Reverse LSBs of value and zero-extend
ROL	Rotate value left by specified number of bits
ROR	Rotate value right by specified number of bits
SHL	Shift value left by specified number of bits
SHR	Shift value right by specified number of bits
SAR	Shift value arithmetically right by specified number of bits
CMP	Compare two unsigned values
CMPS	Compare two signed values
CMPX	Compare two unsigned values plus C
CMPSX	Compare two signed values plus C

`CMPSUB`	Compare unsigned values, subtract second if lesser or equal
`TEST`	Bitwise AND two values to affect flags only
`TESTN`	Bitwise AND a value with NOT of another to affect flags only
`MOV`	Set a register to a value
`MOVS`	Set a register's source field to a value
`MOVD`	Set a register's destination field to a value
`MOVI`	Set a register's instruction field to a value
`MUXC`	Set discrete bits of a value to the state of C
`MUXNC`	Set discrete bits of a value to the state of !C
`MUXZ`	Set discrete bits of a value to the state of Z
`MUXNZ`	Set discrete bits of a value to the state of !Z
`HUBOP`	Perform a hub operation
`NOP`	No operation, just elapse four cycles

CONSTANTS

NOTE: Refer to Constants in the Spin Language Reference section above.

`TRUE`[s]	Logical true: −1 ($FFFFFFFF)
`FALSE`[s]	Logical false: 0 ($00000000)
`POSX`[s]	Maximum positive integer: 2,147,483,647 ($7FFFFFFF)
`NEGX`[s]	Maximum negative integer: −2,147,483,648 ($80000000)
`PI`[s]	Floating-point value for PI: ~3.141593 ($40490FDB)

REGISTERS

`DIRA`[s]	Direction register for 32-bit port A
`DIRB`[s]	Direction register for 32-bit port B (future use)
`INA`[s]	Input register for 32-bit port A (read-only)
`INB`[s]	Input register for 32-bit port B (read-only) (future use)
`OUTA`[s]	Output register for 32-bit port A
`OUTB`[s]	Output register for 32-bit port B (future use)
`CNT`[s]	32-bit system counter register (read-only)
`CTRA`[s]	Counter A control register

CTRB[s]	Counter B control register
FRQA[s]	Counter A frequency register
FRQB[s]	Counter B frequency register
PHSA[s]	Counter A phase-locked loop (PLL) register
PHSB[s]	Counter B phase-locked loop (PLL) register
VCFG[s]	Video configuration register
VSCL[s]	Video scale register
PAR[s]	Cog boot parameter register (read-only)

UNARY OPERATORS

Note: All operators shown are constant-expression operators.

+	Positive (+X) unary form of Add
–	Negate (–X); unary form of Subtract
^^	Square root
||	Absolute value
|<	Decode value (0–31) into single-high-bit long
>|	Encode long into value (0–32) as high-bit priority
!	Bitwise: NOT
@	Address of symbol

BINARY OPERATORS

Note: All operators shown are constant-expression operators.

+	Add
–	Subtract
*	Multiply and return lower 32 bits (signed)
**	Multiply and return upper 32 bits (signed)
/	Divide and return quotient (signed)
//	Divide and return remainder (signed)
#>	Limit minimum (signed)
<#	Limit maximum (signed)
~>	Shift arithmetic right

<<	Bitwise: Shift left
>>	Bitwise: Shift right
<-	Bitwise: Rotate left
->	Bitwise: Rotate right
><	Bitwise: Reverse
&	Bitwise: AND
|	Bitwise: OR
^	Bitwise: XOR
AND	Boolean: AND (promotes non-0 to −1)
OR	Boolean: OR (promotes non-0 to −1)
= =	Boolean: Is equal
<>	Boolean: Is not equal
<	Boolean: Is less than (signed)
>	Boolean: Is greater than (signed)
=<	Boolean: Is equal or less (signed)
=>	Boolean: Is equal or greater (signed)

SYNTAX SYMBOLS

%	Binary number indicator, as in %1010
%%	Quaternary number indicator, as in %%2130
$	Assembly here indicator
"	String designator, as in "Hello"
_	Group delimiter in constant values, or underscore in symbols
#	Assembly literal indicator
:	Assembly local label indicator
,	List delimiter in declared data
'	Code comment designator
"	Document comment designator
{ }	In-line/multi-line code comment designators
{{ }}	In-line/multi-line document comment designators

Reserved Word List

These words are always reserved, whether programming in Spin or Propeller Assembly.

_CLKFREQ[s]	CONSTANT[s]	IF_NC_AND_NZ[a]	MIN[a]	PLL4X[s]	SUBSX[a]
_CLKMODE[s]	CTRA[d]	IF_NC_AND_Z[a]	MINS[a]	PLL8X[s]	SUBX[a]
_FREE[s]	CTRB[d]	IF_NC_OR_NZ[a]	MOV[a]	PLL16X[s]	SUMC[a]
_STACK[s]	DAT[s]	IF_NC_OR_Z[a]	MOVD[a]	POSX[d]	SUMNC[a]
_XINFREQ[s]	DIRA[d]	IF_NE[a]	MOVI[a]	PRI[s]	SUMNZ[a]
ABORT[s]	DIRB[d#]	IF_NEVER[a]	MOVS[a]	PUB[s]	SUMZ[a]
ABS[a]	DJNZ[a]	IF_NZ[a]	MUL[a#]	QUIT[s]	TEST[a]
ABSNEG[a]	ELSE[s]	IF_NZ_AND_C[a]	MULS[a#]	RCFAST[s]	TESTN[a]
ADD[a]	ELSEIF[s]	IF_NZ_AND_NC[a]	MUXC[a]	RCL[a]	TJNZ[a]
ADDABS[a]	ELSEIFNOT[s]	IF_NZ_OR_C[a]	MUXNC[a]	RCR[a]	TJZ[a]
ADDS[a]	ENC[a]	IF_NZ_OR_NC[a]	MUXNZ[a]	RCSLOW[s]	TO[s]
ADDSX[a]	FALSE[d]	IF_Z[a]	MUXZ[a]	RDBYTE[a]	TRUE[d]
ADDX[a]	FILE[s]	IF_Z_AND_C[a]	NEG[a]	RDLONG[a]	TRUNC[s]
AND[d]	FIT[a]	IF_Z_AND_NC[a]	NEGC[a]	RDWORD[a]	UNTIL[s]
ANDN[a]	FLOAT[s]	IF_Z_EQ_C[a]	NEGNC[a]	REBOOT[s]	VAR[s]
BYTE[s]	FROM[s]	IF_Z_NE_C[a]	NEGNZ[a]	REPEAT[s]	VCFG[d]
BYTEFILL[s]	FRQA[d]	IF_Z_OR_C[a]	NEGX[d]	RES[a]	VSCL[d]
BYTEMOVE[s]	FRQB[d]	IF_Z_OR_NC[a]	NEGZ[a]	RESULT[s]	WAITCNT[d]
CALL[a]	HUBOP[a]	INA[d]	NEXT[s]	RET[a]	WAITPEQ[d]
CASE[s]	IF[s]	INB[d#]	NOP[a]	RETURN[s]	WAITPNE[d]
CHIPVER[s]	IFNOT[s]	JMP[a]	NOT[s]	REV[a]	WAITVID[d]
CLKFREQ[s]	IF_A[a]	JMPRET[a]	NR[a]	ROL[a]	WC[a]
CLKMODE[s]	IF_AE[a]	LOCKCLR[d]	OBJ[s]	ROR[a]	WHILE[s]
CLKSET[d]	IF_ALWAYS[a]	LOCKNEW[d]	ONES[a#]	ROUND[s]	WORD[s]
CMP[a]	IF_B[a]	LOCKRET[d]	OR[d]	SAR[a]	WORDFILL[s]
CMPS[a]	IF_BE[a]	LOCKSET[d]	ORG[a]	SHL[a]	WORDMOVE[s]
CMPSUB[a]	IF_C[a]	LONG[s]	OTHER[s]	SHR[a]	WR[a]
CMPSX[a]	IF_C_AND_NZ[a]	LONGFILL[s]	OUTA[d]	SPR[s]	WRBYTE[a]
CMPX[a]	IF_C_AND_Z[a]	LONGMOVE[s]	OUTB[d#]	STEP[s]	WRLONG[a]

CNT[d]	IF_C_EQ_Z[a]	LOOKDOWN[s]	PAR[d]	STRCOMP[s]	WRWORD[a]
COGID[d]	IF_C_NE_Z[a]	LOOKDOWNZ[s]	PHSA[d]	STRING[s]	WZ[a]
COGINIT[d]	IF_C_OR_NZ[a]	LOOKUP[s]	PHSB[d]	STRSIZE[s]	XINPUT[s]
COGNEW[s]	IF_C_OR_Z[a]	LOOKUPZ[s]	PI[d]	SUB[a]	XOR[a]
COGSTOP[d]	IF_E[a]	MAX[a]	PLL1X[s]	SUBABS[a]	XTAL1[s]
CON[s]	IF_NC[a]	MAXS[a]	PLL2X[s]	SUBS[a]	XTAL2[s]
					XTAL3[s]

a = Assembly element; s = Spin element; d = dual (available in both languages); # = reserved for future use

UNIT ABBREVIATIONS

Computer Memory

byte	8 bits
word	2 bytes, 16 bits
long	2 words, 4 bytes, 32 bits
Kb	kilobit, 1024 bits
KB	kilobyte, 1024 bytes
K bytes	kilobyte, 1024 bytes
K words	1024 words
K longs	1024 longs
MB	megabyte, 1024 KB
GB	gigabyte, 1024 MB

Rates

bps	bits per second
kbps	kilobits (1,000 bits) per second
Mbps	megabits (1,000,000 bits) per second
cps	cycles per second
fps	frames per second

Hz	hertz
kHz	kilohertz
MHz	megahertz
GHz	gigahertz
rpm	rotations per minute
MIPS	million instructions per second
ppm	parts per million

Time

min	minutes
s	seconds
ms	milliseconds
µs	microseconds
ns	nanoseconds

Power-related

V	volts
VDC	volts direct current
VAC	volts alternating current
mV	millivolts
A	ampere
mA	milliamperes
µA	microamperes
W	watt
mW	milliwatt
MW	megawatt
Ω	ohm

kΩ	kilo-ohm
MΩ	mega-ohm
µF	microfarads
nF	nanofarads
pF	picofarads

Distance

mm	millimeters
cm	centimeters
m	meters
km	kilometers
in	inches
ft	feet
yd	yards
mi	miles

Pressure

mbar	millibars
Pa	pascals
kPa	kilopascals

Temperature

K	Kelvin
°F	Fahrenheit
°C	Celsius

Other

mol	moles
g	acceleration
dB	decibels
dBm	decibels referenced to milliwatts

INDEX

Note: Page numbers referencing figures are followed by an "*f*"; page numbers referencing tables are followed by a "*t*".